PCBs

Human and Environmental Disposition and Toxicology

Edited by

Larry G. Hansen

and

Larry W. Robertson

University of Illinois Press
Urbana and Chicago

Library of Congress Cataloging-in-Publication Data

PCBs : human and environmental disposition and toxicology / edited by Larry G. Hansen and Larry W. Robertson.
    p. ; cm.
    Includes bibliographical references and index.
    ISBN-13: 978-0-252-03254-7 (cloth : alk. paper)
    ISBN-10: 0-252-03254-3 (cloth : alk. paper)
1. Polychlorinated biphenyls—Toxicology.   2. Polychlorinated biphenyls—Health aspects.   3. Polychlorinated biphenyls—Environmental aspects.   I. Hansen, L. G.   II. Robertson, Larry W.
[DNLM: 1. Polychlorinated Biphenyls—toxicity.   2. Environmental Exposure—adverse effects.   3. Hazardous Substances—adverse effects.   4. Polychlorinated Biphenyls—adverse effects. QV 633 P3477 2007]
RA1242.P7P277   2008
615.9'512—dc22   2007037240

Contents

Acknowledgments

As with the previous book, *PCBs: Recent Advances in Environmental Toxicology and Health Effects* (First PCB Workshop, Lexington, Kentucky), this volume was developed from presentations at the Third PCB Workshop, Champaign, Illinois, June 2004. Again, this is not a "proceedings" but a collection of contributed articles that have been vigorously peer reviewed. We appreciate the diligence and hard work of the contributing authors and of all presenters at the Third PCB Workshop (see the appendix for the workshop's program). The workshop as a whole contributed greatly to the rewriting of the presentations into this volume's excellent chapters. We particularly thank Linda Birnbaum for her outstanding summary of the workshop presentations.

We thank the many financial and other contributions from the University of Illinois, the University of Iowa, and the University of Kentucky. The PCB Workshop and this book would not have been possible without the financial support of the Superfund Hazardous Substances Basic Research Program of the National Institute of Environmental Health Sciences in Research Triangle Park, North Carolina.

1

The Third Biannual International PCB Workshop

Linda S. Birnbaum, *U.S. Environmental Protection Agency*

Concern with polychlorinated biphenyls (PCBs) is not new. The first reports of chick edema disease as well as adverse skin conditions in occupationally exposed workers were first attributed to PCBs in the 1950s. Scientific as well as societal concerns have continued because of the ubiquitous exposure to the general population and the plethora of effects reported in both wildlife and experimental animals. In summarizing the PCB Workshop, held in Urbana, Illinois, in June 2004, I will not attempt to review all of the information presented, but I will try to focus on new and novel findings, identify data gaps, and point out issues for further research and discussion.

The origins of PCBs and characterization of exposures to them are ongoing concerns. There is growing evidence that chiral forms of PCBs are environmentally relevant (Wong 2004). They can be used to trace biological processes since changes in the ratios of chiral stereoisomers do not occur due to physical factors but only biological ones. While the production of PCBs was banned in the United States in the late 1970s, approximately 70 percent of the PCBs ever made are still present, and it will take many more years for them to disappear from the environment (Hites 2004). In the industrialized world major declines in both environmental and biotic samples have occurred from peak levels in the 1960s and 1970s, but this decline may have slowed down in this century. Air and sediment are both major sources of PCBs in the food chain (Hornbuckle 2004). Lower chlorinated congeners are more volatile and therefore the atmosphere may actually serve as an ongoing sink for them, while the lower volatility of the higher chlorinated congeners makes the atmosphere a source to the food web (Totten 2004). However, vascular plants accumulate PCBs directly from the vapor phase

so that plants may also serve as a source of the more volatile PCBs that normally dominate this phase. Traditionally, sediment cores have been used to trace historical patterns of environmental contamination. Recent studies indicate that analysis of PCBs in tree bark may provide information about their historical presence (Hermanson and DeCaprio 2004; Hermanson and Johnson 2007). It is clear that congener patterns are useful tools in determining sources and pathways of exposure. Use of a limited number of indicator PCBs may fail to provide information that would allow determination of historical patterns of exposure. There are also many new analytical methods under development and in recent use that have enhanced sensitivity and specificity for PCB analysis.

Studies on two populations, one in Anniston, Alabama, and the other in eastern Slovakia, with more than background levels, focused on human exposure, health effects, and characteristic congener profiles. The situation in Anniston involves multiple contaminants and many issues of environmental justice (Sherrer 2004) (Environmental justice can be defined as the fair treatment of people of all races, cultures, and incomes with respect to the development, adoption, implementation, and enforcement of environmental laws, regulations, and policies.) Many chemicals other than PCBs were produced, stored, and dumped in Anniston beginning in the 1920s. There is extensive community involvement in the PCB problem and many issues are complicated by litigation. It is becoming clearer that during the height of PCB production, between 1935 and 1971, PCB mixtures were dumped in a landfill at high temperature. Highly chlorinated congeners, such as PCB 203 and 206, were produced by these processes and are today only seen in tree bark and people who lived in Anniston during that time period. The

Agency for Toxic Substances and Disease Registry conducted a small health study in Anniston and observed that the PCB levels in the population are highly correlated with age and residence (Orloff et al. 2004). The agency noted that the fully chlorinated congener, PCB 209, is only present in older people. One of the major objectives in a study of health effects is to determine the appropriate dose metric of exposure. Should it be the peak concentration? The area under the curve? The average concentration? We understand that the daily dose is not appropriate for a persistent bioaccumulative chemical, but different measures of body burden, or of tissue dose, may be needed for different effects. Simple back extrapolation using a first-order exponential decay model may not adequately describe the historical exposures, and more complex models may be needed (Martín-Jiménez and Hansen 2004). Preliminary data from a congressionally funded study suggest that adverse health effects may be occurring in this impacted population.

The second region in which epidemiological studies were reported is eastern Slovakia, especially the area surrounding a factory where PCBs were produced. A major integrated European Union–funded project, known as PCBRISK, has produced a large and complex data set involving both adults and children (Trnovec et al. 2004). Robust associations have been reported between PCBs and several health outcomes. There is a clear association of a variety of health risks with higher blood levels of PCBs. Thyroid problems, diabetes, and autoimmune disorders have been detected in more highly exposed adults, and problems with hearing, dentition, and neurobehavior have been observed in children who were exposed both prenatally and lactationally. Studies of eight- to nine-year-old children in this population have demonstrated neurobehavioral alterations—including sensory motor effects, memory defects, and hyperactivity—associated with higher levels of exposure (Sovčíková et al. 2004). In the adults an increase in diabetes and impaired fasting glucose has been observed (Radikova et al. 2004). In fact, normal glucose regulation appears to be impaired as PCB levels increase. It has also been suggested that endocrine disruption has occurred due to the fact that there are lower estradiol levels in serum from men in the more contaminated regions (Machala et al. 2004). Thyroid homeostasis has also been altered, with increases in thyroxine levels seen concomitant with decreases in TSH (Langer et al. 2004; Radikova et al. 2004). This is accompanied by an increase in thyroid volume in association with PCB levels. The investigators also suggest that there is autoimmune impairment due to the immunomodulatory effects of PCBs. It would be desirable for a measurement of total dioxin equivalency (TEQ) in this population. If such measurements are not possible in the entire cohort, it would be reasonable to measure the TEQ in a subset of the population and see if that can be correlated in this population to the total PCBs.

The levels of some major metabolites, OH-PCBs, are as high in this cohort as in the Faroe Islands, and they are considerably elevated compared to populations in the Netherlands (Bergman et al. 2004). The levels of certain of these metabolites are similar to those of the parent PCBs. However, information on the toxicity of these OH-PCBs is still very limited. One of the important findings of this study is that PCB contamination is not restricted to those living near the factory but is in fact widespread in eastern Slovakia. This reinforces the need to have internal measures of PCB exposure in order to conduct epidemiological studies appropriately. The population originally thought to be the "control" for those living near the factory turned out to be anything but unexposed.

A second, independent study has also been started in eastern Slovakia and involves a cohort of twelve hundred children ranging in age from newborn to eighteen months (Hertz-Picciotto et al. 2004). Blood levels of PCBs, polychlorinated dibenzo-p-dioxins (PCDDs), and polychlorinated dibenzofurans (PCDFs) and some PCB metabolites are being measured in the mothers as well as the children. Preliminary data have shown that the PCB levels are related to maternal age and parity. The levels of general population exposure in these people are among the highest reported in the world. Multiple health endpoints will be investigated in the children. Preliminary data so far indicate that there may be an excess of female offspring, a switch in the normal sex ratio that has been reported in several populations exposed to elevated levels of dioxins. Unfortunately, dioxin-like PCBs and total TEQ are not being measured in this population.

Many of the effects of PCBs involve perturbation of endocrine systems. Hormone action can be affected by synthesis, transport, and degradation of the hormone, as well as binding to high affinity receptors. For the steroid hormone nuclear superfamily, as well as others, there are factors that when bound to the hormone and dimerized act as ligand-activated transcription factors. However, there is a great deal of complexity in the actual transmission of the transcriptional signal involving as many as twenty other protein transcription factors, leading to multiple opportunities for plasticity as well as species diversity (Katzenellenbogen 2004). Cross-talk also exists with other regulatory signal transduction pathways. Hormone action is not solely a matter of affinity for the receptor but also involves the concentration of ligand derived from the exposure. All of these concepts apply not only to the steroid receptors but also to the dioxin, or aryl hydrocarbon (Ah), receptor (Cooke, Mukai, and Buchanan 2004). The Ah receptor is not a steroid receptor family member but belongs to the basic helix loop helix family of nuclear transcription factors (Bradfield 2004).

It is essential to remember that different PCBs have different effects. Some are estrogenic, while others are antiestrogenic (Cooke, Mukai, and Buchanan 2004). However, these hormonal actions are context dependent, varying with tissue, species, gender, and developmental stage. Coplanar PCBs act via the Ah receptor and have no direct interaction with the estrogen receptor. Noncoplanar PCBs are a diverse group of PCBs with varying structures. Some of the PCBs with lower molecular weights, such as 28, 52, and 74, are estrogenic, while the more highly chlorinated noncoplanars, such as 203, 153, and 187, may be anti-estrogenic.

Certain PCBs have interactions with other nuclear receptors (Hurst and Waxman 2004; 2005). For example, PCB 153, one of the most common PCBs in people, binds to the pregnane X receptor (PXR). The response to binding to this receptor appears to be species-specific, as PCB 153 is a PXR agonist in the rat but may be an antagonist in people. The peroxisome proliferator activator receptor (PPAR) may also be affected by PCBs. Activation of PPAR gamma blocks the inflammatory response caused by the Ah receptor ligand, PCB 77. In fact, PCB 77 down regulates binding of the activated PPAR receptor to DNA, as do PCBs 114 and 153, suggesting that different mechanisms may lead to modulation of the activity of this nuclear receptor.

Among the major endocrine-disrupting effects of PCBs are the effects on thyroid hormone homeostasis. Thyroid perturbations have been observed in multiple species, including humans. Reduction in circulating levels of thyroxine (T4) has been observed in many animal studies. Several different mechanisms have been shown to be involved, including enzyme induction and binding to serum transport proteins, especially transthyretin. Several isoforms of uridine diphosphoglucuronyl transferases (UGTs) are transcriptionally increased, leading to enhanced conjugation and elimination of T4. There is also some evidence of induction of sulfotransferase activity leading to increased elimination of T4. In addition, some studies have reported induction of deiodinase activity, which converts T4 to T3, the active thyroid hormone, or reverse T3, an inactive form, or from T3 to T2, also inactive. There is no clear evidence for direct binding of PCBs or any of their studied metabolites to the nuclear thyroid receptors (Gauger et al. 2004). It is important to note that different PCBs may have different mechanisms of action that alter thyroid hormone signaling.

Although dysregulation of thyroid function during adulthood can have serious effects on metabolism, and in rodents can lead to thyroid cancer, there is greater concern about thyroid perturbation during development. Subclinical effects on T4 levels in the mother are associated with deficits in cognitive development in the children. Decreases in T4 are associated with hearing deficits in rodent offspring, and this result can be caused by PCBs. Many of the developmental effects associated with exposure to PCBs in experimental as well as epidemiological studies could be associated with an antithyroid mechanism. However, recent studies have indicated that the situation may be more complex in that commercial PCB mixtures may also lead to thyro-mimetic responses, including increased expression of thyroid hormone responsive genes. What is clear is that studies need to be conducted using individual congeners during development and that levels of T4, T3, and thyroid hormone activation need to be measured in the fetus and target tissues, not only in the serum.

Recent studies have demonstrated that PCBs target tissues other than the liver, thyroid, and developing brain. The cardiovascular system is clearly affected by dioxin-like PCBs via an Ah receptor–mediated mechanism (Heideman et al. 2004). There is no evidence to date of cardiovascular effects from noncoplanar PCBs. TCDD and dioxin-like PCBs alter vascular development and lead to heart defects (Walker and Ivnitski-Steele 2004). In addition, coplanar PCBs are associated with cardiomyopathy and chronic arteritis following long-term exposure in adult rodents (Walker et al. 2004). Little attention has been focused on the cardiovascular system in human studies involving PCB exposure, but the fact that PCBs can cause oxidative stress and inflammation suggest a potential role for PCBs in atherosclerosis (Hennig, Toborek, and Robertson 2004).

A major issue in environmental health is that all exposures are not alike, nor are all people (Schantz 2004). The response to chemicals can be modulated by diet, route of exposure, timing, and so on (Seegal et al. 2004). Clearly, the genetic background influences the responses, and there are critical windows of susceptibility. Although experimental studies conducted with one dose via one route of exposure at a single point in time provide valuable information for hazard identification, dose/response studies are critical to understanding the potential for adverse effects in people. The same can be said for repeated dose exposures or treatment during key time periods of development and by the appropriate route. For mechanistic understanding, use of single congeners is essential, but PCBs are always complex mixtures, meaning that people are never exposed to one PCB in isolation of others, never mind coexposure to totally unrelated additional chemicals. We need experimental work looking at the interactions of PCBs with co-contaminants such as polybrominated diphenyl ethers (PBDEs) (Eriksson 2004) and methyl mercury (Widholm 2004), not to mention lead, pesticides, and so on. Preliminary studies suggest that there may be greater than additive interactions with certain PBDE congeners and with methyl mercury. Again, dose/response studies will be essential to determine if these interactions are additive, synergistic, or antagonistic. Some data looking at interactions of TCDD and/or dioxin-like PCBs with non-dioxin-like PCBs have shown that synergism can occur when relatively low doses of these chemicals are used, but antagonism can occur at high doses. Human studies need to control for these other contaminants. Analysis of what chemicals are in different populations may provide an opportunity to discriminate different exposure scenarios, providing an opportunity to determine sources of contamination.

Nondioxin-like PCBs have often been lumped into one class of PCBs because they do not bind the Ah receptor. This is a gross oversimplification, as evidenced by the studies examining the developmental neurotoxicity of the non-coplanar PCBs. There are multiple mechanisms of action involving multiple congeners. In fact, some congeners may themselves have multiple mechanisms. Certain PCBs bind to the ryanodine receptor and block calcium channels (Pessah and Kim 2004). These PCBs can affect the immune system (not via the Ah receptor). Other PCBs alter seizure potential in genetically susceptible mice, likely via the GABA receptor. Certain noncoplanar PCBs have been shown to have effects on the dopaminergic system; others affect the cholinergic systems. Some of these may also be impacted by dioxin-like PCBs via the Ah receptor. Depression of circulating thyroid hormone levels during a critical developmental window can be caused by both noncoplanar and dioxin-like PCBs and leads to auditory defects. The multiplicity of effects on the developing nervous system raises the question of whether PCBs could be playing a role in the increase in autism, attention deficit hyperactivity disorder, learning disability, and even Alzheimer's disease. Epidemiology studies are relatively consistent in demonstrating an association between behavioral and cognitive effects following early life-stage exposure to children. Some of the differences among studies may be explained by differences in levels and congeners of PCBs involved. It is also important to realize that exposure early in life may alter later susceptibility, not only to PCBs but also to other chemicals.

It is clear that PCBs are complex, multipotential toxicants, making the risk assessment for this class of compounds particularly challenging (Rice 2004). Since there are multiple congeners with multiple mechanisms, congener-specific toxicity and exposure data are indispensable. It is important to stress that human samples do not look like most environmental samples, nor do they resemble any commercial mixtures. Information on individual congeners and interactions between cogeners is not sufficient to conduct a highly quantitative, in-depth risk assessment. There are few clear patterns regarding what is the most sensitive response, although the

developmental neurotoxicity would be a good candidate. It is clear that PCB 126, which is a dioxin-like congener, is the most toxic on the basis of acute toxicity as well as carcinogenicity (Walker et al. 2004). Little attention has been paid to the issue of the potential risk from PCB metabolites (Nakano 2004), some of which have been shown to be present in blood at levels similar to those congeners from which they were derived. Overall, preliminary benchmark-dose analysis of epidemiology studies has not shown any threshold for the developmental neurotoxicity, a situation reminiscent of lead.

There are several key lessons that have emerged from this workshop (Birnbaum 2004). The first is that if congener profiles are different, the effects will be different. It is inappropriate to generalize from one population, or experimental study, to another unless the congener distributions are very similar. The second message is that if the effects are due to coplanar PCBs and the hypothesis is that these are Ah receptor mediated, the hypothesis should be confirmed with TCDD. If studying Ah-mediated effects, don't use PCB 77, which can be metabolized via reactive intermediates to other molecules that have biological activity. In fact, since PCB 126 is the major contributor to the dioxin-like activity of PCBs and is the dioxin-like PCB with greatest mass in people, this is the congener that should be used if the question concerns dioxin-like effects of PCBs. This is related to the third lesson, especially for human studies: It is essential to have measurements of the total TEQ, since PCBs not only always have dioxin-like PCBs present but are always contaminated to some extent by PCDFs, as well as some measure of the total PCB mass. In many cases, use of the 3–14 "indicator" PCBs may account for most of the PCB mass, but more detailed congener analysis may be important for determination of sources or time-trend information. Within a given population, the total TEQ may be highly correlated with a specific PCB congener composition, but this association varies across populations based upon the exposure sources and history. The fourth key take-home point is that small shifts in the distribution of effects in a population may not have clinical relevance to an individual but can lead to large increases in the incidence of clinical effects in the population. For example, a five-point decrease in IQ may not be significant to an individual, but across the population it puts many more people in an at-risk category. And the fifth major point is that risk assessment should be conducted for each effect of concern, using multiple metrics, and the most sensitive response should be used as the basis for regulatory action. This approach is the most protective of public health.

Although it is clear that PCBs have the potential to adversely affect health, it is important to remember that PCB levels, both in the environment and in people, are decreasing from their peak in the twentieth century. This is clear in the industrialized world, and it is due to the regulatory actions taken to ban the production and use of these substances. However, much of the PCBs produced are still in existence, and it will take decades for them to totally disappear. The reality of PCBs is that they are always mixtures, they always have dioxin-like activities, and they are always present with other chemicals. Mixtures are the real world. We need to address the issues of additivity, synergism, and antagonism. And we need to remember that while the dose makes the poison, timing may be the key.

This document has been reviewed in accordance with the U.S. Environmental Protection Agency policy and approved for publication. Mention of trade names or commercial products does not constitute endorsement or recommendation for use.

REFERENCES

Bergman, Åke, Lotta Hovander, Linda Linderholm, Maria Athanasiadou, and Ioannis Athanassiadis. 2004. "PCB Metabolites in Blood from Humans Living in a PCB Contaminated Area and Two Background Districts in the Slovak Republic." Paper presented at the PCB Workshop, Champaign-Urbana, IL. June 13. All subsequent references to the PCB Workshop are to this workshop.

Birnbaum, L. S. 2004. "Overview of 3rd Biannual International PCB Workshop." Paper presented at the PCB Workshop.

Bradfield, Chris. 2004. "Genetic and Genomic Approaches to Evaluate Models of Ah Receptor Mediated Toxicity." Paper presented at the PCB Workshop.

Cooke, Paul S., Motoko Mukai, and David L. Buchanan. 2004. "Cross-talk Between the Aryl Hydrocarbon Receptor and Estrogen Receptor Signaling Pathways." Paper presented at the PCB Workshop.

Eriksson, Per. 2004. "Neurotoxicity of Developmental Exposure to PCBs and PBDEs: Evidence for Interaction." Paper presented at the PCB Workshop.

Gauger, Kelly J., Yashihisa Kato, Koichi Haraguchi, Hans-Joachim Lehmler, Larry W. Robertson, Ruby Bansal, and R. Thomas Zoeller. 2004. "Polychlorinated Biphenyls (PCBs) Exert Thyroid Hormone-Like Effects in the Fetal Rat Brain but Do Not Bind to Thyroid Hormone Receptors." Paper presented at the PCB Workshop.

Heideman, Warren, Dagmara S. Antkiewicz, Amy L. Prasch, Sara A. Carney, and Richard E. Peterson. 2004. "AHR Agonist-Induced Cardiovascular Toxicity in Developing Zebrafish." Paper presented at the PCB Workshop.

Hennig, Bernhard, Michal Toborek, and Larry W. Robertson. 2004. "Nutrition Modulates PCB Toxicity: Implications in Atherosclerosis." Paper presented at the PCB Workshop.

Hermanson, Mark H., and Anthony P. DeCaprio. 2004. "Polychlorinated Biphenyls in Tree Bark from Anniston, Alabama." Paper presented at the PCB Workshop.

——— and Glenn Johnson. 2007. "Polychlorinated Biphenyls in Tree Bark Near a Former Manufacturing Plant in Anniston, Alabama." *Chemosphere* 68:191–98.

Hertz-Picciotto, I., Z. Yu, H-Y. Park, A. Kočan, J. Petrík, and T. Trnovec. 2004. "PCB Exposures in Eastern Slovakia in a Birth Cohort." Paper presented at the PCB Workshop.

Hites, Ron. 2004. "Trends in PCBs and PBDEs in Air over the Last Decade." PCB Workshop.

Hornbuckle, Keri C. 2004. "Use of PCB Congener Profiles to Identify Sources of PCBs." Paper presented at the PCB Workshop.

Hurst, Christopher H., and David. J. Waxman. 2004. "Interaction of Environmental Chemicals with the Nuclear Receptor PXR." Paper presented at the PCB Workshop.

———. 2005. "Interactions of Endocrine Active Environmental Chemicals with the Nuclear Receptor PXP." *Toxicol. Environ. Chem.* 87:299–311.

Katzenellenbogen, John A. 2004. "Estrogen Receptor Ligand Binding, Promiscuous or Eclectic: Opportunities for Pharmaceutical Design, but Perils for Endocrine Disruption." Paper presented at the PCB Workshop.

Langer, P., M. Tajtáková, A. Kočan, J. Petrík, J. Koška, L. Kšinantová, Ž. Rádiková, R. Imrich, M. Hučková, J. Chovancová, B. Drobná, S. Jursa, S. Wimmerová, Y. Shishiba, T. Trnovec, E. Sovčíková, and I. Klimeš. 2004. "Fundamental Views on the Association of High PCB Levels with Changes of Thyroid Volume and Thyroid Hormones and Antibody Levels in an Exposed Population." Paper presented at the PCB Workshop.

Machala, Miroslav, Martina Plisková, Rocio Fernandez Canton, Jiri Neca, Jan Vondrácek, Anton Kočan, Ján Petrík, Tomáš Trnovec, Thomas Sanderson, and Martin van den Berg. 2004. "Detection of Dioxin-like, Estrogenic and Antiestrogenic Activities in Human Serum Samples from Eastern Slovakia." Paper presented at the PCB Workshop.

Martín-Jiménez, Tomás, and Larry G. Hansen. 2004. "Toxicokinetic Extrapolation to Former Residents of Anniston, AL." Paper presented at the PCB Workshop.

Nakano, Takeshi. 2004. "Trends of PCB Isomeric Patterns in Japanese Environmental Media, Food, Breast Milk, and Blood in View of Risk Assessment." Paper presented at the PCB Workshop.

Orloff, Kenneth G., Steve Dearwent, Susan Metcalf, Wayman Turner. 2004. "Human and Environmental Contamination with Polychlorinated Biphenyls in Anniston, Alabama." Paper presented at the PCB Workshop.

Pessah, Isaac N., and Kyung Ho Kim. 2004. "Block of BABAergic Receptors in Rat Hippocampus Significantly Potentiates Non-coplanar PCB Excitotoxicity *in vitro* and *in vivo*." Paper presented at the PCB Workshop.

Radikova, S., J. Koška, L. Kšinantová, R. Imrich, A. Kočan, J. Petrík, M. Hučková, J. Chovancová, B. Drobná, S. Jursa, S. Wimmerová, L. Wsóláva, P. Langer, T. Trnovec, E. Šeböková, and I. Klimeš. 2004. "Increased Prevalence of Diabetes Mellitus and Other Dysglycemias in a Population Chronically

Exposed to Polychlorinated Biphenyls and Other Persistent Organochlorine Pollutants." Paper presented at the PCB Workshop.

Rice, Deborah C. 2004. "Relative Potency of Individual PCB Congeners: What Can the Experimental Literature Tell Us?" Paper presented at the PCB Workshop.

Schantz, S. L. 2004. "Developmental Exposure to PCBs in Humans: Questions and Controversies." Paper presented at the PCB Workshop.

Seegal, R. F., K. O. Brosch, R. J. Okoniewski, and M. Shan. 2004. "Age Matters: Developmental Effects of PCBs and Methylmercury on Striatal Synaptosomal Dopamine." Paper presented at the PCB Workshop.

Sherrer, Charles. 2004. "Anniston Alabama PCB Community Consortium." Paper presented at the PCB Workshop.

Sovčíková, Eva, Tomáš Trnovec, Anton Kočan, and Ladislava Wsólová. 2004. "Neurobehavioral Output of 8–9 Year-old Children from the Michalovce Region." Paper presented at the PCB Workshop.

Totten, Lisa A. 2004. "Importance of Atmospheric Interactions to PCB Cycling in the Hudson and Delaware River Estuaries." Paper presented at the PCB Workshop.

Trnovec, Tomáš, Anton Kočan, Ján Petrík, Jana Chovancová, Beata Drobná, Stanislav Jursa, and Kamil Conka. 2004. "PCBs in East Slovakia and the Structure and Function of the PCBRISK Project." Paper presented at the PCB Workshop.

Walker, Mary K., and Irena Ivnitski-Steele. 2004. "Inhibition of Coronary Angiogenesis by AhR Agonist." Paper presented at the PCB Workshop.

Walker, Nigel J., Michael P. Jokinen, Amy Brix, Donald M. Sells, and Abraham Nyska. 2004. "Cardiotoxicity in Rats Following Chronic Exposure to Dioxins and PCBs." Paper presented at the PCB Workshop.

Widholm, J. J., B. E. Powers, C. S. Roegge, R. E. Lasky, D. M. Gooler, and S. L. Schantz. 2004. "Auditory and Motor Impairments in Rats Exposed to PCBs and Methylmercury During Early Development." Paper presented at the PCB Workshop.

Wong, Charles S. 2004. "Environmental Chemistry of Chiral PCBs." Paper presented at the PCB Workshop.

2

Trends of PCB Congener Patterns in Japanese Environmental Media, Food, Breast Milk, and Blood in View of Risk Assessment

Takeshi Nakano, *Hyogo Prefectural Institute of Public Health and Environmental Sciences*
Yoshimasa Konishi, *Osaka Prefectural Institute of Public Health*
Rie Masho, *Center for Environmental Information Science, Tokyo*

Despite the fact that PCB levels in the environment and breast milk have shown a clear declining trend since banning of PCB production in 1972 in Japan, significant levels of PCBs still remain in human tissues, that is, 0.94 ng/g in whole blood, 200 ng/g-fat in breast milk. In Japan, 12 coplanar PCBs have been monitored regularly under the "dioxin law," while no such monitoring was enforced over the remaining 197 congeners. It is extremely time-consuming and costly to perform congener specific analysis of these PCBs, but the analysis would provide valuable information, on the origin of contamination, accuracy of the analysis, and health risk assessment, which cannot be obtained by analyzing only 12 coplanar PCBs.

Here we present the results of congeneric analysis available on Japanese samples. In general, PCB congeneric patterns in environmental media, such as air, water, and sediment, are similar to the pattern of a particular PCB product used around the area. On the other hand, among biological samples, such as fish, clams, birds, and humans, the congeneric patterns differ drastically, reflecting different half-lives of congeners in different biological systems. Although the analysis of a limited number of PCB congeners would be sufficient for a particular monitoring purpose, such data sets fall short for intermonitoring comparisons. Thus we would propose that the following list of PCB congeners be included for monitoring activities, which will greatly facilitate the global data sharing and detailed comparison: 1, 3, 4, 8, 9, 10, 11, 12, 15, 18, 19, 28, 31, 33, 35, 37, 38, 44, 49, 52, 54, 57, 66, 70, 74, 77, 78, 79, 81, 87, 95, 99, 101, 104, 105, 110, 111, 114, 118, 123, 126, 138, 149, 153, 155, 156, 157, 162, 167, 169, 170, 174, 178, 180, 187, 188, 189, 194, 195, 199, 202, 203, 205, 206, 208, 209.

BACKGROUND

PCB Production in Japan

Japanese PCB production was started by Kaneka Chemical in 1954. The infamous incident of Yusho, rice-oil poisoning, took place in western Japan in 1968. In the 1970s, regulations over PCB production and use became enforced and PCB disposal measures have been in operation since the 1980s. PCB waste destruction is still going on today.

As causal chemicals of Yusho, we now know the major toxicity was from polychlorinated dibenzo furans (PCDFs), not from PCBs. However, evidence has been accumulating that PCBs and their metabolites exert neurological or behavioral effects at low doses, and needs therefore have grown ever larger for PCB risk assessment and management through elucidating the state of environmental levels and human exposures to PCBs.

PCB Use in Japan

Figure 2.1 shows PCB use in Japan. The PCB amount used was calculated as the amount of domestic production plus import minus export for each year. Domestic use amounted to 54,000 tons through these years, and 90 percent of the entire PCBs were used between 1960 and 1972. PCBs were most extensively used in the few years before 1972, the year PCB production was banned by law in Japan. The principal uses were in electric appliances, 69 percent; heat exchange fluids, 16 percent; and carbonless paper, 10 percent.

Table 2.1. Classification of Chlorobiphenyls Based on Presence in Various Media and Compared to Estimated Production Figures.

BZ No.	Structure	Structure	CAS No.	USA[b]	Japan[c]	Persistent[b]	Episodic[b]	¹³C-Labeled	Proposed	Other
1	2	2	2051-60-7	0.5	0.011				#	Window
2	3	3	2051-61-8	0.04	0.001					
3	4	4	2051-62-9	0.2	0.004			Available	#	Window
4	2,2'	2-2	13029-08-8	1.3	0.999		Minor		#	CICAD-1999
5	2,3	23	16605-91-7	0.06	0.012					
6	2,3'	2-3	25569-80-6	0.6	0.342					
7	2,4	24	33284-50-3	0.1	0.017					TO
8	2,4'	2-4	34883-43-7	3	2.600		Minor	Available	#	CICAD-1999
9	2,5	25	34883-39-1	0.2	0.075			Available	#	
10	2,6	26	33146-45-1	0.08	0.016				#	Window
11[a]	3,3'	3-3	2050-67-1	0	0.000				#	PIG
12	3,4	34	2974-92-7	0.03	0.197					
13	3,4'	3-4	2974-90-5	0.1	0.074					
14[a]	3,5	35	34883-41-5	0	0.000					
15	4,4'	4-4	2050-68-2	0.9	0.702				#	Window
16	2,2',3	2-23	38444-78-9	1.4	2.596		Minor			
17	2,2',4	2-24	37680-66-3	1.4	2.093					
18	2,2',5	2-25	37680-65-2	4	6.303		Major		#	CICAD-1999
19	2,2',6	2-26	38444-73-4	0.4	0.560			Available	#	Window
20	2,3,3'	3-23	38444-84-7	0.3	0.388					
21[a]	2,3,4	234	55702-46-0		0.000					
22	2,3,4'	4-23	38444-85-8	1.4	2.203					
23[a]	2,3,5	235	55720-44-0		0.000					
24	2,3,6	236	55702-45-9	0.05	0.066					
25	2,3',4	3-24	55712-37-3	0.2	0.369					
26	2,3',5	3-25	38444-81-4	0.6	0.956					DCP
27	2,3',6	3-26	38444-76-7	0.2	0.302					DCP
28	2,4,4'	4-24	7012-37-5	3.5	4.840	Major		Available	#	DCP/ICES7/CICAD-1999
29	2,4,5	245	15862-07-4	0.03	0.038					
30	2,4,6	246	35693-92-6		0.002					
31	2,4',5	4-25	16606-02-3	3.8	5.447		Major	Available	#	CICAD-1999
32	2,4',6	4-26	38444-77-8	0.9	1.538		Minor			
33	2,3',4'	2-34	38444-86-9	2.4	3.506		Major		#	
34	2',3,5	2-35	37680-68-5	0.01	0.020					
35	3,3',4	3-34	37680-69-6	0.03	0.043				#	PIG
36[a]	3,3',5	3-35	38444-87-0		0.000					
37	3,4,4'	4-34	38444-90-5	1	1.372				#	Window

No.	Structure	Code	CAS							Notes
38[a]	3,4,5	345	53555-66-1		0.000				#	CICAD-1999
39[a]	3,4',5	4-35	38444-88-1		0.000					
40	2,2',3,3'	23-23	38444-93-8	0.5	0.734	Minor				
41	2,2',3,4	234-2		0.4	0.656	Minor				
42	2,2',3,4'	23-24	36559-22-5	0.8	1.136					
43[a]	2,2',3,5	235-2	70362-46-8	0.1	0.000					
44	2,2',3,5'	23-25	41464-39-5	2.7	3.881		Major		#	ATROP-I
45	2,2',3,6	236-2	70362-45-7	0.5	0.869					
46	2,2',3,6'	23-26	41464-47-5	0.2	0.460					
47	2,2',4,4'	24-24	2437-79-8	0.7	1.032					TO
48	2,2',4,5	245-2	70362-47-9	0.7	1.074					
49	2,2',4,5'	24-25	41464-40-8	1.8	2.625		Major		#	
50	2,2',4,6	246-2	62796-65-0	0.1	0.013					
51	2,2',4,6'	24-26	68194-04-7		0.206		Major			TO
52	2,2',5,5'	25-25	35693-99-3	3.5	4.715		Major	Available	#	ICES7/CICAD-1999
53	2,2',5,6'	25-26	41464-41-9	0.4	0.710					CICAD-1999
54	2,2',6,6'	26-26	15968-05-5		0.013				#	Window
55	2,3,3',4	234-3	74338-24-2	0.04	0.068					
56	2,3,3',4'	23-34	41464-43-1	1.3	1.479	Major				
57	2,3,3',5	235-3	70424-67-8	0.001	0.020				#	
58	2,3,3',5'	23-35	41464-49-7		0.001					
59	2,3,3',6	236-3	74472-33-6	0.15	0.219					
60	2,3,4,4'	234-4	33025-41-1	1.4	0.883	Major				
61[a]	2,3,4,5	2345	33284-53-6		0.000					
62	2,3,4,6	2346	54230-22-7		0.002					
63	2,3,4',5	235-4	74472-34-7	0.08	0.140					
64	2,3,4',6	236-4	52663-58-8	1.3	1.969					
65	2,3,5,6	2356	33284-54-7		0.003					
66	2,3',4,4'	24-34	32598-10-0	2.5	2.826	Major			#	
67	2,3',4,5	245-3	73557-53-8	0.08	0.127					
68	2,3',4,5'	24-35	73575-52-7	0	0.007					TO
69	2,3',4,6	246-3	60233-24-1	0	0.009					
70	2,3',4',5	25-34	32598-11-1	3.3	3.264		Major	Available	#	CICAD-1999
71	2,3',4',6	26-34	41464-46-4	0.7	0.909					
72[a]	2,3',5,5'	25-35	41464-42-0		0.000					
73	2,3',5',6	26-35	74338-23-1		0.007					
74	2,4,4',5	245-4	32690-93-0	1.5	1.747	Major			#	MDCP
75[a]	2,4,4',6	246-4	32598-12-2		0.000					
76	2',3,4,5	2-345	70362-48-0		0.002					
77	3,3',4,4'	34-34	32598-13-3	0.2	0.211		Major	Available	#	TEF/Window

(continued)

Table 2.1. Continued

BZ No.	Structure	Structure	CAS No.	USA[b]	Japan[c]	Persistent[b]	Episodic[b]	^{13}C-Labeled	Proposed	Other
78	3,3',4,5	3-345	70362-49-1		0.047				#	CHEM
79[a]	3,3',4,5'	34-35	41464-48-6		0.000			Available	#	CHEM
80[a]	3,3',5,5'	35-35	33284-52-5		0.000					
81	3,4,4',5	345-4	70362-50-4	0.01	0.012			Available	#	TEF
82	2,2',3,3',4	234-23	52663-62-4	0.4	0.408					
83	2,2',3,3',5	235-23	60145-20-2	0.2	0.157					
84	2,2',3,3',6	236-23	52663-60-2	0.8	0.771		Minor			ATROP-I
85	2,2',3,4,4'	234-24	65510-45-4	0.4	0.471					
86	2,2',3,4,5	2-2345	55312-69-1	0.03	0.023					
87	2,2',3,4,5'	234-25	38380-02-8	1.3	1.286	Major			#	ATROP-II
88	2,2',3,4,6	2-2346	55215-17-3	0	0.009					
89	2,2',3,4,6'	234-26	73575-57-2	0.08	0.025					
90	2,2',3,4',5	235-24	68194-07-0	0	0.254					ATROP-I/CICAD-1999
91	2,2',3,4',6	236-24	68194-05-8	0.4	0.400					
92	2,2',3,5,5'	235-25	52663-61-3	0.4	0.493					
93[a]	2,2',3,5,6	2356-2	73575-56-1		0.000					
94	2,2',3,5,6'	235-26	73575-55-0		0.016					
95	2,2',3,5',6	236-25	38379-99-6	2.3	1.157		Major		#	ATROP-I
96	2,2',3,6,6'	236-26	73575-54-9	0.02	0.017					
97	2,2',3,4',5'	245-23	41464-51-1	0.8	0.987	Minor				
98[a]	2,2',3,4,6	246-23	60233-25-2	0	0.000					
99	2,2',4,4',5	245-24	38380-01-7	1.1	1.032	Major			#	CICAD-1999
100	2,2',4,4',6	246-24	39485-83-1	0	0.005					DCP
101	2,2',4,5,5'	245-25	37680-73-2	2.9	2.849		Major	Available	#	ICES7/CICAD-1999
102	2,2',4,5,6'	245-26	68194-06-9	0.08	0.096					
103	2,2',4,5',6	246-25	60145-21-3		0.015					
104[a]	2,2',4,6,6'	246-26	56558-16-8		0.000				#	Window
105	2,3,3',4,4'	234-34	32598-14-4	1.1	0.863	Major		Available	#	TEF
106[a]	2,3,3',4,5	2345-3	70424-69-0		0.000					
107	2,3,3',4',5	235-34	70424-68-9		0.171					
108	2,3,3',4,5'	234-35	70362-41-3		0.039					
109	2,3,3',4,6	3-2346	74472-35-8	0.1	0.002					
110	2,3,3',4',6	236-34	38380-03-9	3	2.971		Major		#	CICAD-1999
111[a]	2,3,3',5,5'	235-35	39635-32-0		0.000			Available	#	
112	2,3,3',5,6	2356-3	74472-36-9		0.005					
113	2,3,3',5',6	236-35	68194-10-5		0.001					
114	2,3,4,4',5	4-2345	74472-37-0	0.07	0.058			Available	#	TEF

No.	Structure	Code	CAS							Notes
115	2,3,4,4',6	4-2346	74472-38-1	0.07	0.071					
116[a]	2,3,4,5,6	23456	18259-05-7		0.000					
117	2,3,4',5,6	2356-4	68194-11-6	0.07	0.068					
118	2,3',4,4',5	245-34	31508-00-6	2.2	2.281	Major		Available	#	TEF/ICES7/CICAD-1999
119	2,3',4,4',6	246-34	56558-17-9	0.02	0.038					
120[a]	2,3',4,5,5'	245-35	68194-12-7		0.000					
121[a]	2,3',4,5',6	246-35	56558-18-0		0.000					
122	2',3,3',4,5	345-23	76842-07-4	0.03	0.026					
123	2',3,4,4',5	345-24	65510-44-3	0.05	0.034			Available	#	TEF
124	2',3,4,5,5'	345-25	70424-70-3	0.09	0.111					
125	2',3,4,5,6'	345-26	74472-39-2		0.004					
126	3,3',4,4',5	345-34	57465-28-8	0.001	0.004	Major		Available	#	TEF/Window
127[a]	3,3',4,5,5'	345-35	39635-33-1	0	0.000					CHEM
128	2,2',3,3',4,4'	234-234	38380-07-3	0.4	0.405	Major				DCP
129	2,2',3,3',4,5	2345-23	55215-18-4	0.1	0.157	Minor				DCP/ATROP-II
130	2,2',3,3',4,5'	234-235	52663-66-8	0.15	0.140					ATROP-I
131	2,2',3,3',4,6	2346-23	61798-70-7	0.05	0.042					
132	2,2',3,3',4,6'	234-236	38380-05-1	1.2	0.835		Major			ATROP-I
133	2,2',3,3',5,5'	235-235	35694-04-3	0.04	0.024					ATROP-I
134	2,2',3,3',5,6	2356-23	52704-70-8	0.15	0.137					
135	2,2',3,3',5,6'	235-236	52744-13-5	0.4	0.267		Minor			ATROP-I
136	2,2',3,3',6,6'	236-236	38411-22-2	0.5	0.348					
137	2,2',3,4,4',5	2345-24	35694-06-5	0.1	0.126	Minor				ICES7/CICAD-1999
138	2,2',3,4,4',5'	234-245	35065-28-2	2.8	1.873	Major		Available	#	ATROP-II
139	2,2',3,4,4',6	2346-24	56030-56-9		0.043					
140	2,2',3,4,4',6'	234-246	59291-64-4		0.007					
141	2,2',3,4,5,5'	2345-25	52712-04-6	0.8	0.384		Minor			
142[a]	2,2',3,4,5,6	23456-2	41411-61-4		0.000					
143	2,2',3,4,5,6'	2345-26	68194-15-0		0.006					
144	2,2',3,4,5',6	2346-25	68194-14-9	0.2	0.075					ATROP-II
145	2,2',3,4,6,6'	2346-26	74472-40-5		0.0002					
146	2,2',3,4',5,5'	235-245	51908-16-8	0.4	0.267	Major				ATROP-I/CICAD-1999
147	2,2',3,4',5,6	2356-24	68194-13-8	0.02	0.052					
148	2,2',3,4',5,6'	235-246	74472-41-6		0.001					
149	2,2',3,4',5',6	236-245	38380-04-0	2.8	1.914	Major	Major		#	ATROP-I/CICAD-1999
150	2,2',3,4',6,6'	236-246	68194-08-1		0.003		Major			
151	2,2',3,5,5',6	2356-25	52663-63-5	0.8	0.331		Major			
152	2,2',3,5,6,6'	2356-26	68194-09-2		0.002					
153	2,2',4,4',5,5'	245-245	35065-27-1	3	1.655	Major		Available	#	ICES7/CICAD-1999
154	2,2',4,4',5,6'	245-246	60145-22-4		0.012					

(continued)

Table 2.1. *Continued*

BZ No.	Structure	Structure	CAS No.	USA[b]	Japan[c]	Persistent[b]	Episodic[b]	¹³C-Labeled	Proposed	Other
155[a]	2,2',4,4',6,6'	246-246	33979-03-2		0.000				#	Window
156	2,3,3',4,4',5	2345-34	38380-08-4	0.3	0.261	Major		Available	#	TEF
157	2,3,3',4,4',5'	234-345	69782-90-7	0.04	0.053	Minor		Available	#	TEF
158	2,3,3',4,4',6	2346-34	74472-42-7	0.3	0.269					
159[a]	2,3,3',4,5,5'	2345-35	39635-35-3		0.000					
160[a]	2,3,3',4,5,6	23456-3	41411-62-5		0.000					
161[a]	2,3,3',4,5',6	2346-35	74474-43-8		0.000					
162	2,3,3',4',5,5'	235-345	39635-34-2		0.006				#	
163	2,3,3',4',5,6	2356-34	74472-44-9	0.8	0.424	Major				
164	2,3,3',4',5',6	236-345	74472-45-0	0.2	0.143					
165[a]	2,3,3',5,5',6	2356-35	74472-46-1		0.000					
166	2,3,4,4',5,6	23456-4	41411-63-6		0.011					
167	2,3',4,4',5,5'	245-345	52663-72-6	0.1	0.089			Available	#	TEF
168[a]	2,3',4,4',5',6	246-345	59291-65-5		0.000					
169	3,3',4,4',5,5'	345-345	32774-16-6	<0.001	0.0004			Available	#	TEF/Window
170	2,2',3,3',4,4',5	2345-234	35065-30-6	1	0.299	Major		Available	#	
171	2,2'3,3',4,4',6	2346-234	52663-71-5	0.2	0.090	Minor		Available	#	ATROP-II
172	2,2',3,3',4,5,5'	2345-235	52663-74-8	0.15	0.044	Minor				
173	2,2',3,3',4,5,6	23456-23	68194-16-1	0.02	0.010					
174	2,2',3,3',4,5,6'	2345-236	38411-25-5	1.2	0.286		Minor		#	ATROP-I
175	2,2',3,3',4,5',6	2346-235	40186-70-7	0.04	0.012					ATROP-II
176	2,2',3,3',4,6,6'	2346-236	52663-65-7	0.15	0.042					ATROP-I, II
177	2,2',3,3',4',5,6	2356-234	52663-70-4	0.6	0.157	Minor				
178	2,2',3,3',5,5',6	2356-235	52663-67-9	0.2	0.042			Available	#	
179	2,2',3,3',5,6,6'	2356-236	52663-64-6	0.5	0.109					
180	2,2',3,4,4',5,5'	2345-245	35065-29-3	2.8	0.568	Major		Available	#	ICES7
181	2,2',3,4,4',5,6	23456-24	74472-47-2		0.003					
182[a]	2,2',3,4,4',5,6'	2345-246	60145-23-5		0.000					
183	2,2',3,4,4',5',6	2346-245	52663-69-1	0.6	0.164	Major				ATROP-II

				c	b				
184[a]	2,2',3,4,4',6,6'	2346-246	74472-48-3		0.000				
185	2,2',3,4,5,5',6	23456-25	52712-05-7	0.1	0.031				CICAD-1999
186[a]	2,2',3,4,5,6,6'	23456-26	74472-49-4		0.000				
187	2,2',3,4',5,5',6	2356-245	52663-68-0	1.4	0.267	Major	#		Window
188[a]	2,2',3,4',5,6,6'	2356-246	74487-85-7		0.000		#		TEF/Window
189	2,3,3',4,4',5,5'	2345-345	39635-31-9	0.02	0.010	Minor	#	Available	
190	2,3,3',4,4',5,6	23456-34	41411-64-7	0.2	0.066	Minor			
191	2,3,3',4,4',5',6	2346-345	74472-50-7	0.03	0.011				
192[a]	2,3,3',4,5,5',6	23456-35	74472-51-8		0.000				
193	2,3,3',4',5,5',6	2356-345	69782-91-8	0.1	0.022				
194	2,2',3,3',4,4',5,5'	2345-2345	35694-08-7	0.5	0.096	Major	#	Available	CICAD-1999
195	2,2',3,3',4,4',5,6	23456-234	52663-78-2	0.2	0.044	Minor	#		
196	2,2',3,3',4,4',5,6'	2345-2346	42740-50-1	0.3	0.038	Major			ATROP-II
197	2,2',3,3',4,4',6,6'	2346-2346	33091-17-7	0.01	0.003				ATROP-II
198	2,2',3,3',4,5,5',6	23456-235	68194-17-2	0.02	0.005				
199	2,2',3,3',4,5,6,6'	23456-236	52663-75-9	0.4	0.014	Major	#		
200	2,2',3,3',4,5',6,6'	2346-2356	52663-73-7	0.06	0.012	Minor			
201	2,2',3,3',4,5,5',6'	2345-2356	40186-71-8	0.06	0.093				
202	2,2',3,3',5,5',6,6'	2356-2356	2136-99-4	0.1	0.018	Major	#		Window
203	2,2',3,4,4',5,5',6	23456-245	52663-76-0	0.4	0.070	Major	#		
204[a]	2,2',3,4,4',5,6,6'	23456-246	74472-52-9		0.000				
205	2,3,3',4,4',5,5',6	23456-345	74472-53-0	0.02	0.004		#	Available	Window
206	2,2',3,3',4,4',5,5',6	23456-2345	40186-72-9	0.15	0.018	Minor	#	Available	Window
207	2,2',3,3',4,4',5,6,6'	23456-2346	52663-79-3		0.001				
208	2,2',3,3',4,5,5',6,6'	23456-2356	52663-77-1	0.03	0.003	Minor	#	Available	Window
209	2,2',3,3',4,4',5,5',6,6'	23456-23456	2051-24-3		0.0002	Minor	#		

[a] Nor contained in PCB product.

[b] Estimated relative amount produced. From Hansen 1999.

[c] Estimated relative amount used.

Note: DCP = dechlorination product, TEF = WHO 12, CHEM = chemical process, PIG = pigment, TO = Thermal Oxidation, ATROP I = atropisomer (236-), ATROP II = atropisomer (2346-), Window = first and last isomer, ICES7 = 7 congeners for Marine Mammals (International Council for the Exploration of the Sea), CICAD-1999 = Comisión Interamericana para el Control del Abuso de Drogas.

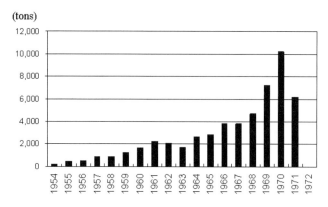

(tons)

Figure 2.1.   Trend of Japanese PCB consumption (production + imports – exports).

Congener Profile of Japanese PCB Products

PCBs consist of 209 differently chlorinated congeners. Kaneka Chemical produced PCBs under the product name of Kanechlor (KC-300, KC-400, KC-500, and KC-600). KC-300, KC-400, KC-500, and KC-600 were of different congeneric proportions and purposes: KC-300 was used in carbonless paper, KC-400 and 500 in electric appliances, and KC-600 in ship-bottom paint. Percentages of 209 PCB isomers produced in Japan were estimated based on the amount of each product used in each prefecture during 1969 and 1971, multiplying the total amount of KC-300, 400, 500, and 600 used in the whole country calculated from the data times the isomer proportions of each product (table 2.1). Although

this estimation is based on the statistical data only from 1969 and 1971, these are the years when the largest proportion of PCBs was actually used. We believe it is reasonable to consider that this estimation well reflects the isomeric proportions of the loads of PCBs to the Japanese environment. It should be cautioned, however, that physical processes (especially combustion) and biological processes such as dechlorination and other metabolism may introduce congeners and metabolites not present in the original PCB products.

Production and Isomer Characteristics

Of 209 PCB isomers, 35 were scarcely contained in PCB products (less than 0.001%); 70 congeners, one-third of the 209, were less than 0.01 percent; and 120 congeners were less than 0.1 percent. Thus only 90 congeners were present at greater than 0.1 percent of total PCB. Generally, the 19 most prominent congeners account for 60 percent, adding 6 more for 25 congeners accounts for 70 percent, 35 account for 80 percent, and 51 account for 90 percent of all PCBs. As shown in table 2.2, principal isomers differ among PCB products. For example, isomer #180 (2345-245) is characteristic to KC-600, and the ratio of #180 and #118 in samples may give information for presuming the source of PCBs.

Compared with the similar estimate for the USA by Hansen (Hansen 1999), the four most prominent congeners (#18, #31, #28 and #52) are common in both countries. Of 21 Japanese prominent isomers, 20, except #64, are among the top 22 isomers of the USA. While proportions

Table 2.2. Prominent Congeners and Abundance Ratio (%) in Each PCB Product.

KC-300		KC-400		KC-500		KC-600	
BZ No.	%	BZ No.	%	BZ No.	%	BZ No.	%
18	10.5	52	7.7	110	8.7	180	11.0
31	8.6	44	6.8	101	8.6	149	10.0
28	7.9	70	6.7	118	6.8	153	8.5
33	5.9	66	6.2	138	6.5	187	6.0
8	4.7	49	4.8	149	6.1	174	5.5
16	4.5	31	4.3	153	5.3	138	4.4
44	4.1	64	4.1	52	4.7	170	3.6
52	3.9	74	3.9	87	3.6	151	3.0
22	3.7	18	3.3	95	3.3	194	3.0
17	3.6	28	3.0	132	2.9	183	2.9
49	2.9	110	3.0	97	2.5	201	2.9
66	2.9	56	2.9	99	2.4	177	2.7
70	2.7	101	2.4	70	2.3	179	2.4
32	2.6	118	2.3	105	2.1	203	2.2
37	2.3	60	2.0	84	1.8	101	2.1
64	2.1	47	2.0	44	1.6	141	2.0

of the products used were different in the two countries, prominent isomers were almost common. PCB congener #180, which ranks as the 12th dominant isomer in the United States, is only the 44th dominant in Japan. This may reflect the wide use of Aroclor 1260 (KC-600 equivalent), which contained a high proportion of #180, in the United States.

PCB Destruction in Japan

Of 58,000 tons of PCBs produced during 1954 and 1971, 5,500 tons of liquid PCBs, less than 10 percent of production, were recovered. Although the remaining 90 percent has been put under certain supervision and keeping, there are cases of PCBs used in closed systems which now remain within stocks of large trash at waste disposal sites or those used in open systems and released into the environment (Tsuji, Nakano, and Okuno 1987; Oki et al. 1990). In various areas, PCBs in various forms such as in disintegrated electric apparatus wastes, carbonless paper, capacitors of the fluorescent lights or in paper condensers, and in ship bottom paints are transported into the environment and distributed to air, water or biological systems.

It may be surprising to know that not much PCB is needed to cause a considerable level of contamination. A concentration of 100 ng/m^3 PCBs in the indoor air of a building with 1,600m^2 floor area and 18 m high (28,800m^3 cubic content) can be reached with less than 3 mg of PCBs. On the other hand, an ordinary ceiling fluorescent light capacitor contained 50 to 100g of PCBs. The situation is the same for sediment and soil contaminations. Small amounts of PCBs are enough to cause pollution, though a large cost is required to remediate the polluted environment.

In Japan, all the PCB containing transformers and capacitors are now ordered by law to be destroyed before 2016, and the chemical destruction of liquid PCBs in these apparatuses has just started. The significance of proper PCB storage and management has ever been mounting at this time.

ENVIRONMENTAL OCCURRENCE

PCB Levels in Environmental Media

PCB migration in soil or sediment is slow due to the hydrophobicity and low water solubility of the compounds and PCBs are thus confined to local hot spots. The soil/sediment concentrations of PCBs are inclined to be high near the sources and may differ largely from point to point. On the other hand, the PCBs are distributed to air depending on the vaporization pressure of each PCB isomer and, thus, are strongly influenced by temperatures.

Ambient air. Ambient air concentrations in Japan at present are between 10^{-1} and several ng/m^3, and there is not much difference among regions. The high end of air concentration was around 100 ng/m^3 in 1974. The level declined gradually to the average of 2.8 ng/m^3 in Osaka and Kobe (industrialized cities in western Japan) in 1985 and 1986. Continuous measurements in Kobe gave the following results: averages of 0.45 ng/m^3 during June 1988 and July 1988; 0.54 ng/m^3 during December 1988 and January 1989; and 0.53 ng/m^3 during September 1996 and December 1996 (Nakano, Tsuji, and Okuno 1987, 1990b).

Sediment. While horizontal distribution of PCBs in sediment suggests diffusion from the source, vertical distribution gives information on the level of contamination at the time of disposition. In the sea-bed sediment of Osaka Bay and adjoining Harimanada Sea, PCBs are found distributed according to the vectors of tidal residual flows (constant flows).

When yearly PCB product forwarding volume was compared with the vertical distribution of PCB concentrations near the producers, the time trend of PCB concentrations in the sediment was in good correlation with the trend of PCB production. It is shown that the local hot spots in sediment tend to remain at the site without much diffusion and concentrations may differ widely between different sampling points. It is also shown that the PCB compositions in sediment reflect the compositions of the products that caused the pollution (Nakano et al. 1979; Nakano, Tsuji, and Okuno 1990a).

A persistent organic pollutant (POP) pollution survey in 2002 by the Ministry of the Environment gave the mean levels of PCBs in different media: 0.1 ng/m^3 in air (102 samples from 34 spots); 0.1 ng/L in water (114 samples from 38 spots); 10 ng/g-dry in sediment (189 samples from 63 spots); and 10 ng/g-wet in wild life (118 fish, mussel, or bird samples from 24 spots) (Nakano et al. 2004).

Human Levels and Exposure

Human milk and blood. Regarding internal exposure to PCBs, there is not much isomer-specific data on human specimens such as blood, adipose tissue, or other tissues except for breast milk. While the data on non-dioxin-like PCBs are particularly limited, Konishi and colleagues have reported the time trend for each PCB isomer in pooled milk of mothers from Osaka prefecture between 25 and 29 years old and with their first baby based on their long-term breast milk monitoring study. According to the results, the prominent isomers were #138, #153, #118, #74, #66, #99, #180, #105, #187, #60, #170, and #52 (in order of predominance) (Konishi, Kuwabara, and Hori 2001, 2003).

Hirai et al. (2005) measured all PCB isomers in the blood of healthy subjects, 12 men and 12 women between 25 and 45 years old, and reported the levels ranging from 54 to 228 ng/g-lipid with the mean of 112 ng/g-lipid. All the levels in breast milk or blood samples reported by Konishi, Nakano, or Hirai are at the levels around 100 ng/g-lipid.

Food. The background exposure to PCBs for citizens is mainly from food, and in Japan fish is the main source of PCB intake. Table 2.3 shows the 20-year trend of daily intake of PCBs from sea food (fish, mollusks, crustaceans, and their products) for a Japanese derived from total market basket studies. We can see the proportion of #153 in total PCBs is gradually increasing.

DISCUSSION

Shift in Isomeric Distributions among Media

PCB isomer patterns in air, water, or sediment are almost identical to the pattern in PCB products, except in some cases where #11 (3,3′-dichlorobiphenyl) shows a specific distribution. Also, PCB isomers found in environmental samples tend to shift from lower to higher chlorinated compounds in air, water, and sediment, respectively. In biological samples, the isomeric patterns differ from those of products reflecting the involvement of metabolic process in organisms. Water samples are known sometimes to contain di-chlorinated CBs at higher proportions and we can suggest the contribution from the pigment material, 3,3′-dichlorobenzidine. In biological samples, the shift to higher chlorinated isomers is further observed in the order of mussels, fish, and birds. In the birds, hexa-CBs are the most prominent of homologue groups, which is also the case in human samples such as breast milk and blood (fig. 2.2) (Nakano et al. 2004). This difference in PCB isomer distributions in environmental and human samples is clearly illustrated in figure 2.3 (Nakano et al. 2002).

Characteristics in Congener Profiles

Prominent congeners in human blood and breast milk are shown in tables 2.4 and 2.5, respectively. The three most prominent isomers, #153 (245-245), #138 (245-234) and #180 (245-2345), are common in blood and breast milk. The concentration ratio of #153/#138 was 1.7 in breast milk and 2.7 in blood, indicating marked reduction of #138 in blood.

Figure 2.4 illustrates the difference in PCB isomer profiles of breast milk and food (fish). Prominent isomers in PCB products, #66 (24-34) and #101 (245-25), are seen remaining in fish. The contribution of these isomers lessens in breast milk, suggesting different metabolic capacities in different animal groups.

While PCB isomer profiles are characteristic to each medium such as PCB products, environmental samples or human specimens, there is a dramatic difference in isomeric profiles between environmental media and human tissues. For example, relative concentrations of some isomer pairs become inverted between PCB products/environmental media and human samples. While in PCB products and environmental media #66 (24-34) is greater than #74 (245-4), in human samples #74 prevails. The same phenomena are observed in that in environmental media, the relation between concentrations are #99 (245-24) >#101 (245-25) and #146 (245-235) >#149 (245-236), which also are inverted in human samples (table 2.6).

Differential Reduction of Isomers Due to Biological Processes

Figure 2.5 illustrates the relation of #153 (245-245) and #138 (234-245) concentrations in animal food products over 20 years of time. A gradual increase in the concentration ratio of #153 (245-245) to #138 (234-245) is observed, suggesting that differential metabolism by organisms is causing the change in the occurrence of PCB isomers, which may appear as the transitions in PCB isomer profiles in environmental, biological, or food samples.

When the isomer profile of PCBs in breast milk is examined in relation to #153, the most prevalent isomer, although proportions of hepta-CBs and hexa-CBs do not show transition over time, those of penta-CBs and tetra-CBs declined greatly. The difference in isomer profiles in PCB products and breast milk is shown in figure 2.6. In the upper column, prevalent isomers in PCB products are shown and in the lower column, those in breast milk. Twelve isomers that are attributed with dioxin Toxic Equivalency Factors (TEFs) are low, except for #118 and #156. The prevailing isomer species are those that have their 2,4,5 positions chlorinated, such as #153 (245-245). PCB isomer profiles of environmental samples resemble closely those of PCB commercial products. However, in human samples, metabolism of the isomers results in more simplified isomeric patterns. Prevailing isomers that remain in human samples are those which have their 2,4,5 positions of phenyl rings chlorinated, such as #74 (2,4,4′,5-), #99 (2,2′,4,4′,5-), #118 (2,3′,4,4′,5-), #153 (2,2′,4,4′,5,5′-), #138 (2,2′,3,4,4′,5′-), #180 (2,2′, 3,4,4′,5,5′-), and #187 (2,2′,3,4′,5,5′,6-).

When relative concentrations of PCB isomers in breast milk of 1973 and 2000 are compared, #74 (245-4) and #118 (245-34), which reached around 80 percent of the level of #153 (245-245) in 1973, declined to about 20 percent of #153 in 2000. Similarly, #138 (234-245), which surpassed the level of #153 in 1973, dropped to approximately 80 percent of the level of #153 in 2000. On the other hand, highly chlorinated isomers such as #187 (2356-245) or #180 (2345-245) maintained almost the same concentration in relation to #153 over the period.

Table 2.3. Daily Intake of PCBs from Seafood (ng/Day) for Selected Years.

IUPAC No.	1982	1985	1986	1987	1988	1992	1993	1994	1995	1996	1997	1998	1999	2000	2001
31+28	79.6	68.4	68.1	32.3	55.0	31.9	15.8	24.2	20.9	21.5	24.2	30.5	8.9	71.8	14.4
52	82.1	104.6	74.2	35.6	62.6	27.8	18.1	25.0	26.5	19.7	35.3	44.7	15.1	46.2	16.0
49	64.6	113.0	68.7	27.7	40.3	26.9	12.3	29.7	21.0	19.1	39.5	48.8	11.7	42.3	15.6
74	49.8	62.8	44.0	28.8	29.9	16.4	9.3	16.7	15.4	11.0	24.3	32.9	9.5	17.9	10.1
70	76.6	63.3	58.4	36.6	47.8	21.8	12.9	19.4	20.3	15.0	29.5	31.1	13.1	18.9	11.7
66	87.7	134.8	70.9	37.3	45.5	33.6	15.6	25.8	25.2	21.8	48.1	60.4	16.2	30.0	19.9
95	77.5	102.0	66.3	40.2	50.6	30.0	19.0	25.2	18.7	19.0	36.0	40.9	19.8	18.9	16.9
(90+)101	153.4	235.8	163.2	89.0	116.5	63.6	45.0	72.9	61.6	49.9	104.3	104.6	49.7	50.2	48.8
99	82.1	172.4	111.7	57.2	68.6	47.3	26.1	70.1	45.5	39.3	90.2	103.3	31.6	35.9	38.8
87	50.5	64.9	48.4	32.8	42.2	19.4	14.6	18.1	15.6	13.3	24.2	24.2	14.9	15.5	13.1
110	102.7	131.6	96.3	65.6	74.4	38.6	28.0	33.9	32.7	27.2	52.2	50.1	28.3	28.8	27.7
118	123.3	205.6	139.1	95.1	108.8	59.9	36.4	69.4	54.8	47.2	93.8	97.0	44.4	48.7	47.4
105	41.0	62.5	41.4	30.2	34.1	17.6	11.3	18.3	16.3	13.2	23.3	25.5	13.1	15.0	15.0
151	42.4	85.2	45.9	27.0	33.6	20.5	10.1	27.4	23.1	15.0	35.6	44.0	15.9	15.6	15.5
149	127.3	220.6	142.2	92.8	96.6	65.0	36.7	79.2	67.6	47.5	99.2	126.0	52.8	47.3	52.8
146	31.0	75.9	42.6	24.4	26.8	20.9	10.3	33.9	25.7	18.3	43.0	58.2	17.7	18.7	19.2
153	215.5	393.2	276.4	167.5	183.0	130.3	68.6	200.8	160.5	114.7	255.1	342.0	111.9	115.6	118.0
138	194.9	341.2	228.0	161.3	169.6	101.8	60.9	130.6	113.0	87.2	170.6	192.4	88.1	92.9	86.0
128+167	27.7	44.2	32.7	25.1	25.2	14.5	8.4	17.5	14.3	11.7	23.3	24.2	11.7	13.5	11.8
187	56.4	157.1	70.8	43.2	44.5	36.3	15.8	61.8	56.5	35.8	74.5	127.2	38.9	33.1	35.9
180	64.3	123.0	64.8	46.9	52.0	31.2	16.5	43.6	56.0	30.4	67.6	83.8	38.8	28.7	28.3
170+190	27.6	52.8	29.4	21.5	23.5	14.1	7.3	18.8	23.1	13.5	28.9	35.9	17.5	13.8	13.3
Total (ng/day)	1858	3015	1984	1218	1431	869	499	1063	914	691	1423	1728	670	819	676

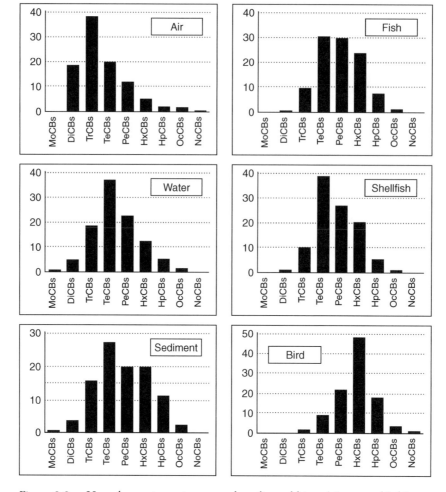

Figure 2.2.   Homologues in environmental media and biota (air, water, birds).

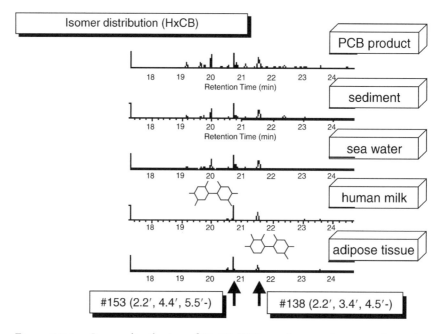

Figure 2.3.1.   Isomer distribution of HxCB (PCB product, environmental sample, human sample).

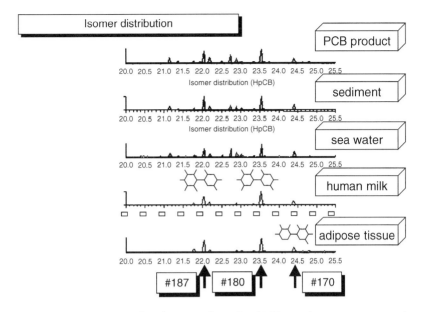

Figure 2.3.2. Isomer distribution of HpCB (PCB product, environmental sample, human sample).

A similar example of the change in isomer profiles through metabolism in biota is seen in the case of chlordane isomers in air (trans-chlordane, compound-5, cis-chlordane, trans-nonachlor, cis-nonachlor, oxy-chlordane). Twenty years ago, chlordane isomers found in air samples were the same as those in commercial products. Oxy-chlordane, which was not contained in products, was not detected in air. On the other hand, in biological samples, especially in birds, 20 years ago and today alike, a metabolite, oxy-chlordane, is the principal isomer, while trans-chlordane, compound-5, and cis-chlordane are found less often. Recently, the oxy-chlordane started to be detected in air, serving as evidence that the environmental load of the agrochemical isomers in the past has changed over time through animal metabolism and the isomer profiles in environmental media are influenced.

Metabolism similarly changes the profiles of environmental PCBs. Methyl sulfonyl metabolites have long been observed in wildlife and persistent metabolites are now well known in the blood of humans (Bergman, Klasson-Wehler, and Kuroki 1994; Letcher, Klasson-Wehler, and Bergman 2000). Multiple, nonpersistent, experimentally generated

Table 2.4. Levels of the Predominant PCB Congeners in Human Blood.

	Congener BZ No.	Mean (ng/g-Lipid)	2SD (%)	Existence Ratio[a] (%)	Cumulative
245-245	153	28.0	28.0	22.3	22.3
245-2345	180	14.9	16.6	11.8	34.1
245-234	138	10.4	9.2	8.3	42.4
245-2356	187	8.3	8.5	6.6	49.0
245-34	118	6.9	5.9	5.5	54.5
2356-34+	163, 164	6.4	6.2	5.1	59.6
236-345	-	-	-	-	-
245-24	99	4.7	4.2	3.8	63.4
245-4	74	4.6	2.6	3.6	67.0
245-235	146	4.1	3.7	3.3	70.3
2345-234	170	3.8	4.1	3.0	73.3
2345-34	156	2.8	3.0	2.3	75.6

[a]Mean is compared to total PCB level. Hirai et al. 2005.

Table 2.5. Levels of the Predominant PCB Congeners in Breast Milk.

BZ No.	Levels in Milk
CB-153	29.89
CB-138	17.72
CB-180	13.67
CB-118	9.64
CB-187, CB-182	9.34
CB-99	6.12
CB-170	5.11
CB-146	4.12
CB-74	3.86
CB-156	2.55
CB-183	2.4
CB-66	1.96
CB-177	1.79
CB-105	1.77
CB-178	1.76
CB-167	1.41
CB-201	1.27
CB-101	1.19
CB-196, CB-203	1.11
CB-28	1.02
CB-172	1.02

metabolites of tri-CBs, tetra-CBs, and penta-CBs are also known to accumulate in fish and invertebrates (Metcalf et al. 1975; Saghir, Koritz, and Hansen 1994). Numerous hydroxy tri- through penta-CB metabolites are also found in natural sediments and feral fish (Sakiyama et al. 2004).

Risk-Assessment Related Aspects in Analysis

Fluctuations in analytical data. Daily fluctuations of PCB levels in ambient air are considerable. In assessing risks from environmental PCBs, it is significant to take into account the PCB concentration fluctuations along with the meteorological conditions such as wind directions and velocity, temperatures, and rainfalls. When sampling air, we carried out 24-hour sampling using a high-volume air sampler together with two weeks sampling using a low-volume air sampler. To know the factors that affect concentrations at the monitoring site and to assess the risks from the chemicals in ambient air, it is essential that short-term monitoring is conducted in combination with long-term monitoring.

Aerial portions of plants are also good indicators of PCB levels in air. Accumulation is a function of air concentrations, so fluctuating air levels are averaged for the growing season, minimizing the daily variations which may affect risk assessments. Common golden rod is a good monitor species because it is plentiful, its surface area is large, and its

adsorption/absorption efficiency is good. Lipids in inert tree bark are also good long-term indicators of air levels. Care must be taken to remove adhering soil and dust particles prior to extraction of the samples.

Figure 2.8 shows the time trend of PCB concentration in air. In 1974, right after PCB use was banned, the air level was 100 ng/m^3 and dropped to about one-tenth of that within a few years. The air concentration declined to about 1 ng/m^3 in 1988 and has been fluctuating around this level since then (Nakano, Tsuji, and Okuno 1987, 1990b). We believe that this indicates that the local equilibrium between the environmental media (air, water, and soil) has been reached in Japan for the PCB loads to the nation's environment. Therefore, to aim at further reduction of PCB concentrations, we must bring forward what was ratified in the POPs protocols and propel PCB destructions that are proclaimed in our law.

Newly emerging isomers in the environment. There are some cases in which isomers from new sources influence the isomer profiles in environmental media. The DIN approach specifies that the sum of six PCB congeners (IUPAC 28, 52, 101, 138, 153, and 180) multiplied by a factor of five will approximate the total PCB concentration in a sample. However, this approach may underestimate total PCB in samples from unique sources such as the manufacture of cement, ink, chemicals, and so on, and from waste combustion. Combustion processes may result in the formation of, for example, PCBs 35, 37, 47, 51, 68, 77, 78, 79 and 81. Unique PCB congener profiles were reported in a dump site that received concrete blocks contaminated with KC-300 in capacitor paper (Yamamoto et al. 1992). The relative abundance of congener #54 (26-26) and #50 (246-2) was increased in the soil of the Japanese cypress forest around this dump site. Such an increase in the abundance of ortho-rich congeners not found in the PCB products is often based on anaerobic microbial dechlorination (Bedard and May 1996). Unusual tetra-, penta-, and hexa-CBs containing 24- and 246-chlorination patterns are uncommon in higher Aroclors and provided strong evidence for dechlorination of Aroclor 1260 in the sediments of Woods Pond. Relative to Aroclor 1260, the most extensively dechlorinated samples had lost 11–19 percent of the meta-chlorines and 2–7 percent of the para-chlorines (Bedard and May 1996).

PCB isomer #11 is not contained in PCB products. This isomer comes from 3,3'-dichlorobenzidin, which is a pigment material, and is detected at high concentrations in the Hudson River. In Japan, #11 and #35, which were not detected 30 years ago, are found in environmental media since about 10 years ago (fig. 2.9).

Contribution to dioxin toxic equivalent. In a three-day continuous nationwide monitoring, PCB concentrations in air

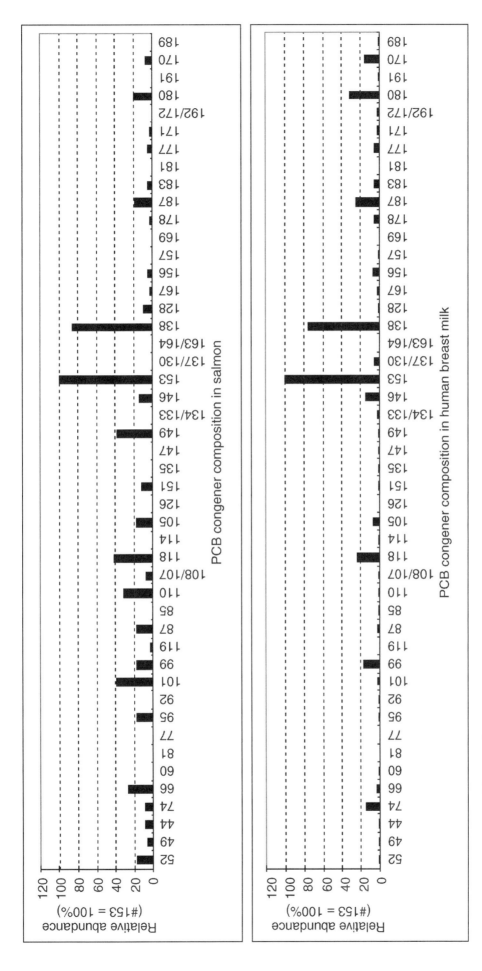

Figure 2.4. The salmon and breast milk comparison.

Table 2.6. PCB Isomeric Ratio Shifts between PCB
Product and Human Samples.

PCB Product	Environment	Human
#74 < #66	#74 < #66	#74 > #66
245-4 24-34	245-4 24-34	245-4 24-34
#99 < #101	#99 < #101	#99 > #101
245-24 245-25	245-24 245-25	245-24 245-25
#146 < #149	#146 < #149	#146 > #149
245-235 245-236	245-235 245-236	245-235 245-236

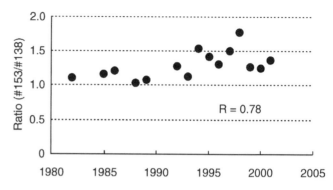

Figure 2.5. Time trend of the ratio (CB-153/CB-138) in food
(animal protein) (fish, shellfishes, meat, milk).

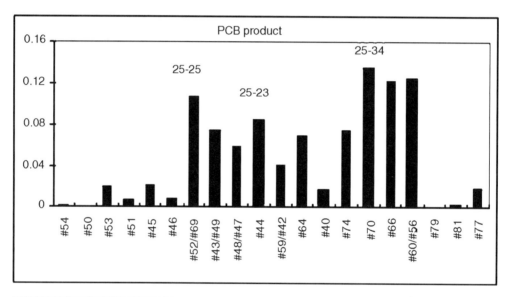

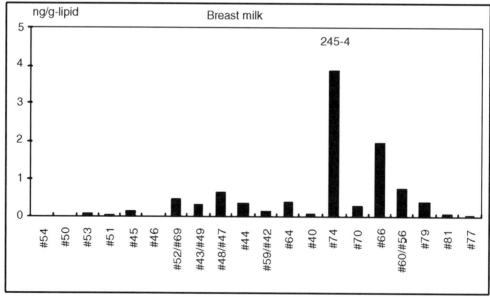

Figure 2.6.1. Comparisons of product versus breast milk for tetra-PCBs.

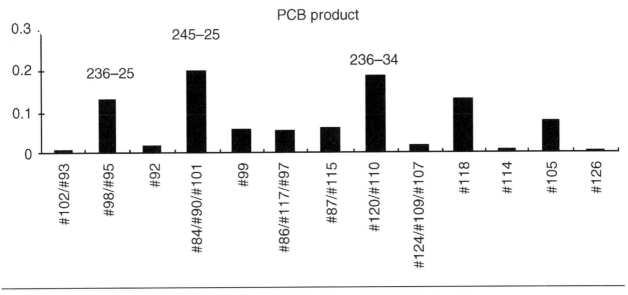

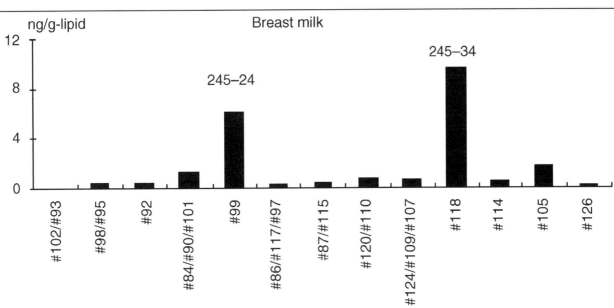

Figure 2.6.2.   Comparisons of product versus breast milk for penta-PCBs.

from different Japanese regions were surveyed (Ministry of the Environment, Environmental Health Division, in 1997). The contributions of each congener to the dioxin toxic equivalent (TEQ) from dioxin-like PCBs were large for #126 (Cl5), 89 percent and for #118 (Cl5), 4 percent, and small (1–2 %) for other congeners (#169 [Cl6], #105 [Cl5], #156 [Cl6], #114 [Cl5], and #77 [Cl4]). Of the PCBs, penta-chlorinated #126 contributes greatly to the dioxin TEQ, while seven isomers, #126, #118, #169, #105, #156, #114, and #77, constitute more than 99 percent of the dioxin TEQ due to dioxin-like PCBs. Of the total dioxin TEQ of

air samples, PCBs constitute an average of 0.01pg–TEQ/m3, which is less than a few percent of the TEQ of poly-chlorinated dibenzo-*p*-dioxins (PCDDs) and PCDFs, indi-cating the relatively small PCB dioxin-like health risk through air. On the other hand, the contribution of dioxin-like PCBs to the dioxin TEQ reaches over 50 percent of total TEQ in animals and food products. Food is a much more significant exposure route than air when risks of dioxin-like PCBs to human health are considered. Among food, con-taminated fish is the main source of PCB intake. PCB levels in fish are influenced by the levels in water and sediment.

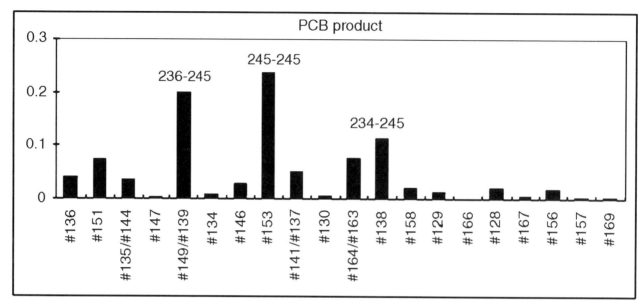

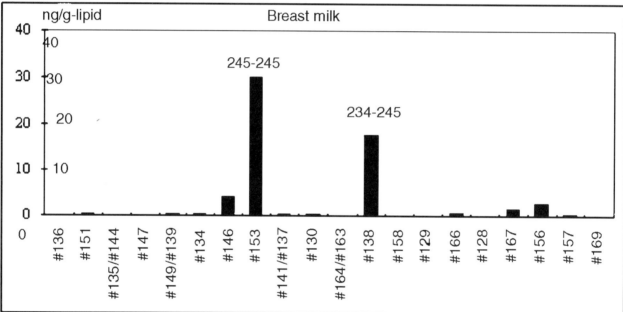

Figure 2.6.3.   Comparisons of product versus breast milk for hexa-PCBs.

Dioxin-like PCBs that are attributed with dioxin TEFs existed as components of PCB products; on the other hand, they also are newly formed unintentionally in combustion processes. The data on each PCB congener in the environmental media are indispensable in elucidating the sources of these congeners. Inclusion of dioxin-like PCBs in the risk assessment of dioxins has introduced a new perspective. Considering the amount of dioxin-like PCBs in products, we realize that the reduction of the risk from dioxin-like PCBs cannot be done only through regulating combustion. For example, 1 g of KC-500, the product used in trans-

formers, contains about 7×10^{-5}g of PCB #126 (TEF = 0.1), which results in 7×10^{-6}g–TEQ (7000 ng-TEQ). The needs for good PCB waste management and safe disposal along with dioxin reduction measures are obvious.

In Japan, since monitoring of dioxins is obligatory, a large database on dioxin pollution has accumulated. However, while TEF-attributed PCB isomers are regularly monitored as part of dioxin measures, monitoring of other PCB isomers without TEF values tends to be neglected. This situation is a great setback for the purpose of PCB risk assessment.

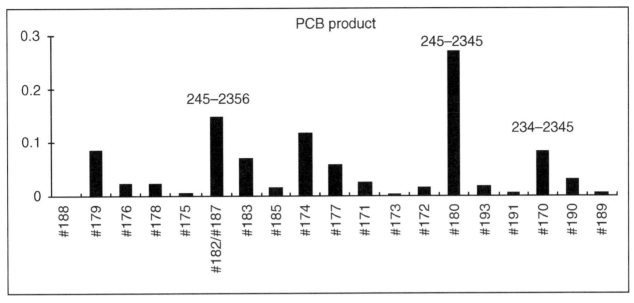

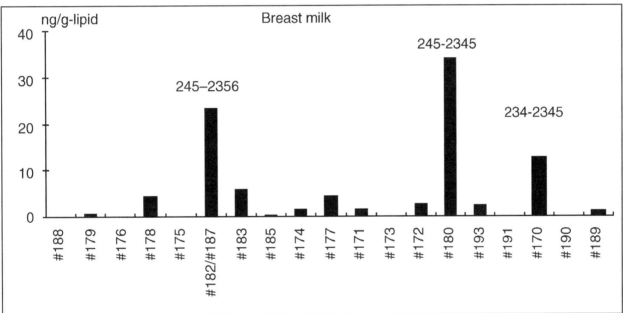

Figure 2.6.4. Comparisons of product versus breast milk for hepta-PCBs.

CONCLUSIONS

PCB Risks in Japan

In assessing PCB risks in Japan, information on PCB concentration levels and trends and isomer profiles are significant. Since toxicological studies using specific congeners instead of particular products are now taking place, and we now know that some toxicity such as in endocrine disruption is specific to certain congeners, we now realize that it is crucial to know PCB congener profiles in environmental media and metabolite analysis in organisms. We also realize that dioxin TEQ calculation alone is not enough for assessing the risks of PCBs. Not enough data have been available for the purpose of comprehensive PCB risk assessment. There are limited toxicity data on a limited number of PCB congeners. We do not know differences in responsiveness in different animal species. We have little data on PCB metabolism in organisms.

Environmental levels of dioxins are declining due to dioxin-reduction measures in Japan (fig. 2.10). On the other hand, we have almost no information on the risks of PCB released into the environment from aged transformers in use

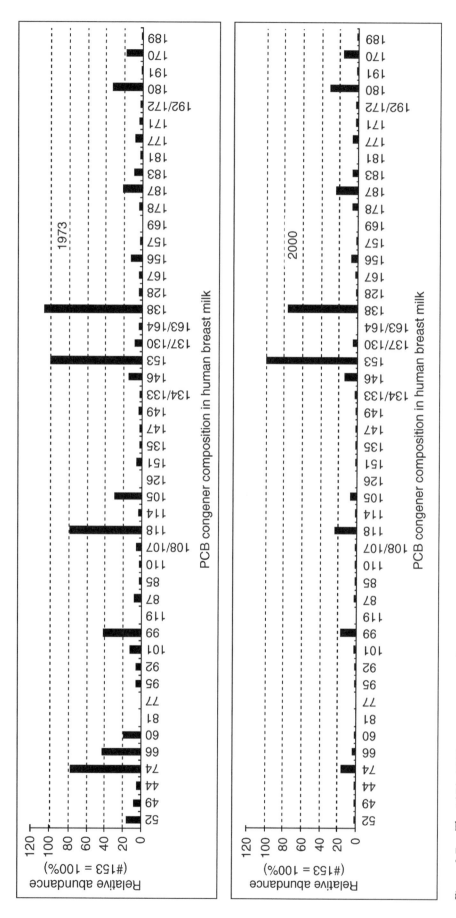

Figure 2.7. The 1973–2000 congener comparison.

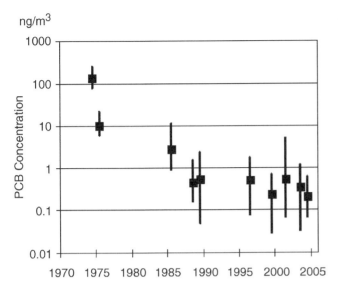

Figure 2.8.   Time trend of PCBs in ambient air.

or from stored equipment in the case of natural disasters such as earthquakes or floods. Swift progress in PCB destruction is believed to be the best way to reduce the risks from PCBs.

Importance of PCB Isomeric Analysis

Investigating the behaviors of certain isomers in the environment provides us with significant information on the emission sources or their origins and allows us to predict how isomers from different sources—such as from PCB products, combustion, chemical process, or from byproducts of chemical processes—influence the isomer profiles in environmental media, wildlife, food, or human milk/blood. In table 2.1, we marked the isomers we propose to be monitored with the symbol "#" and listed the reasons for their inclusions.

PCB isomeric analyses are cost- and time-consuming work. We propose that the PCB researchers have a common

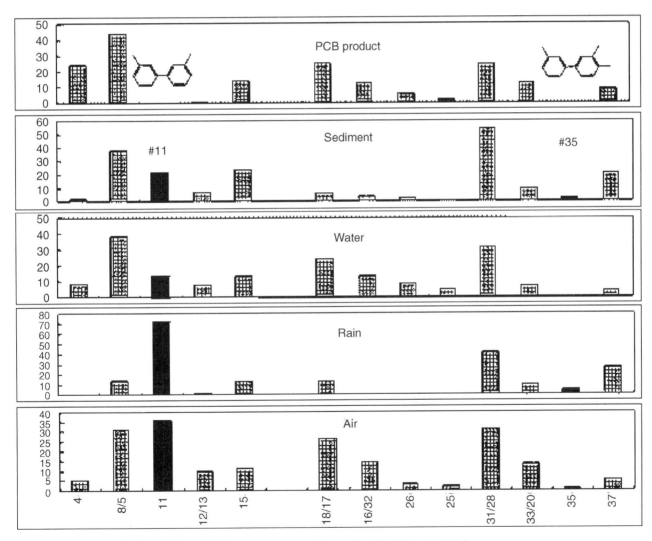

Figure 2.9.   CB-11(3-3) and CB-35(34-3) in environmental samples (di-PCBs, tri-PCBs).

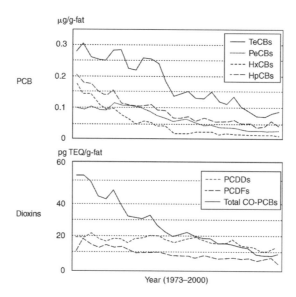

Figure 2.10.   Time trend of breast-milk levels.

understanding of the isomers they include in their analysis, knowing the reasons for their inclusion. In that way, we will have more isomeric data that are comparable among laboratories and researchers, and that will give us clearer pictures of PCB behaviors and lead to a reduction of the overall cost of analysis.

ACKNOWLEDGMENTS

We thank Professor Larry Hansen, who gave us the chance to take part in the 2004 PCB Workshop and later helped us through to the completion of this paper. Thanks are also due to Chisato Matsumura for work with PCB congener profiles; to Toshihiro Okuno, Keiichi Tanno, Yukie Majima, Hiroaki Kitamoto, Misao Gotoh, and Mayumi Kurata for breast milk work; and to Professor Chiharu Tohyama for useful discussions and encouragement.

REFERENCES

Bedard, D. L., and R. J. May. 1996. "Characterization of the Polychlorinated Biphenyls in the Sediments of Woods Pond: Evidence for Microbial Dechlorination of Aroclor 1260 in Situ." *Environ. Sci. Technol.* 30:237–41.

Bergman, Å., E. Klasson-Wehler, and H. Kuroki. 1994. "Selective Retention of Hydroxylated PCB Metabolites in Blood." *Environ. Health Perspect.* 102:464–69.

Hansen, L. G. 1999. *The Ortho Side of PCBs: Occurrence and Disposition.* Boston: Kluwer Academic Publishers.

Hirai, T., Y. Fujimine, S. Watanabe, and T. Nakano. 2005. "Congener-Specific Analysis of Polychlorinated Biphenyls in Human Blood from Japanese." *Environ. Geochemistry and Health* 27:65–73.

Konishi, Y., K. Kuwabara, and S. Hori. 2001. "Continuous Surveillance of Organochlorine Compounds in Human Breast Milk Samples from 1972 to 1998 in Osaka, Japan." *Arch. Environ. Contam. and Toxicol.* 40:571–78.

———. 2003. "Continuous Monitoring of PCB Isomers in Human Breast Milk from 1972 to 2000 in Osaka, Japan." *Organohalogen Compounds* 63:441–44.

Letcher, R. J., E. Klasson-Wehler, and Å. Bergman. 2000. "Methyl Sulfone and Hydroxylated Metabolites of Polychlorinated Biphenyls." In *New Types of Persistent Halogenated Compounds,* vol. 3, edited by J. Paasivirta, 315–57. Berlin: Springer-Verlag.

Metcalf, R. L., J. R. Sanborn, P. Y. Lu, and D. Nye. 1975. "Laboratory Model Ecosystem Studies of the Degradation and Fate of Radiolabeled Tri-, Tetra-, and Pentachlorobiphenyl Compared with DDE." *Arch. Environ. Contam. Toxicol.* 3:151–65.

Nakano, T., and S. Yukizo. 1979. "Sediment Contamination of the Organic Chlorinated Compounds (Polychlorinated Biphenyls and Polychlorinated Terphenyls) in Harima-nada and Osaka Bay." *Hyogo-kenritsu Kogai Kenkyusho Kenkyu Hokoku.* 11:33–9.

Nakano, T., M. Tsuji, and T. Okuno. 1987. "Level of Chlorinated Compounds in Atmosphere." *Chemosphere* 16:1781–86.

———. 1990a. "Characteristics of Isomeric Pattern for Organic Chlorinated Compounds in Environment." *International Association on Water Pollution Research and Control,* 203–6.

———. 1990b. "Distribution of PCDDs, PCDFs and PCBs in the Atmosphere." *Atmospheric Environment* 24A (6): 1361–68.

Nakano, T., K. Tanno, H. Kitamoto, C. Matsumura, M. Goto, Y. Majima, R. Masho, C. Tohyama, and T. Okuno. 2002. "Congener Specific Analysis of Polychlorinated Biphenyls in the Environment and Human Samples." *Organohalogen Compounds* 55:339–42.

Nakano, T., M. Fukushima, Y. Shibata, N. Suzuki, Y. Takazawa, Y. Yoshida, N. Nakajima, Y. Enomoto, S. Tanabe, and M. Morita. 2004. "POPs Monitoring in Japan—Fate and Behavior of POPs." *Organohalogen Compounds* 66:1490–96.

Oki, N., T. Nakano, T. Okuno, M. Tsuji, and A. Yasuhara. 1990. "Emission of Volatile Chlorinated Organic Compounds by Combustion of Waste Organochlorine Material." *Chemosphere* 21 (6): 761–70.

Saghir, S. A., G. D. Koritz, and L. G. Hansen. 1994. "Toxicokinetics of 2,2′,4,4′- and 3,3′,4,4′-Tetrachlorobiphenyl in House Flies Following Topical Administration." *Pesticide Biochemistry and Physiology* 49:94–113.

Sakiyama T., T. Okumura, N. Kakutani, and Y. Mori. 2004. "Hydroxy Polychlorinated Biphenyls in Fishes and Sediments." *13th Symposium on Env. Chem. Programs and Abstracts,* 556–57.

Tsuji, M., T. Nakano, and T. Okuno. 1987. "Measurement of Combustion Products from Liquid PCB Waste Incinerator." *Chemosphere* 16:1889–94.

Yamamoto, K., T. Nakano, M. Tsuji, and T. Okuno. 1992. "Isomers Isomer-Specific Analysis of PCB in Plants as Indicator of PCB in Air." *International Conference on Biological Mass Spectrometry,* 226–27.

3

Chiral Polychlorinated Biphenyls and Their Metabolites

Charles S. Wong, *University of Alberta*

ABBREVIATIONS

BPDM   Benzphetamine N-demethylase
CE   Capillary electrophoresis
CYP   Cytochrome P-450
E1   First-eluting enantiomer
E2   Second-eluting enantiomer
ECD   Electron capture detection
ECNI   Electron capture negative ionization
ee   Enantiomeric excess
EF   Enantiomer fraction
EI-MS   Electron impact mass spectrometry
ER   Enantiomer ratio
EROD   Ethoxyresorufin-O-deethylase
GC   Gas chromatography
GCxGC   Comprehensive two-dimensional gas chromatography
GSH   Glutathione
HPLC   High-performance liquid chromatography
MDGC   Multidimensional gas chromatography
MEKC   Micellar electrokinetic chromatography
$MeSO_2$-PCB   Methyl sulfonyl polychlorinated biphenyl
MS   Mass spectrometry
NIST   National Institute of Standards and Technology
PCB   Polychlorinated biphenyl
PM-CD   Permethyl-β-cyclodextrin
SRM   NIST Standard Reference Material

This chapter outlines the current state of knowledge about chiral polychlorinated biphenyls and their methylsulfonyl metabolites. Much is known about the environmental chemistry and effects of chiral PCBs, but little is known about their metabolites. In particular, little is known about the stereoisomers of those PCB compounds that are atropisomeric and are chiral at ambient temperature, despite predictions of the existence of these enantiomers decades ago. Many chiral chromatographic techniques have been developed to resolve and measure PCB and metabolite enantiomers in environmental samples. The enantiomer composition of these compounds in environmental biota is often non-racemic, indicating stereoselective biological processes such as biotransformation have occurred. The occurrence and fate processes of chiral PCB compounds in soils and sediments, aquatic organisms, and birds and mammals are summarized. Current results suggest that chiral analysis provides enhanced insights into biologically mediated environmental processes that would otherwise remain hidden, but understanding of factors controlling these processes is poor.

INTRODUCTION

Polychlorinated biphenyls (PCBs) are legacy pollutants that were banned decades ago but are present ubiquitously in the environment. As a result, they remain a potential health hazard due to their persistence and bioaccumulative ability. The environmental chemistry, fate, and toxicology of these chemicals have been studied extensively over this time period. However, little remains known about the behavior and toxicological effects of those PCB congeners which exhibit chirality, even though the existence of chiral PCB congeners was predicted thirty years ago (Kaiser 1974), or almost as long as PCBs have been an environmental problem.

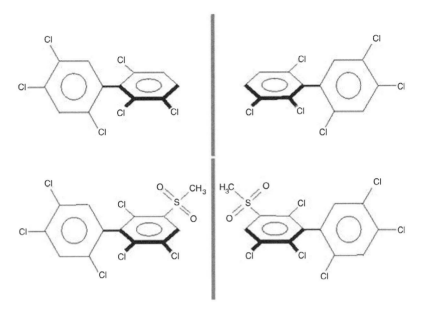

Figure 3.1.   Atropisomers of (top) PCB 149 and (bottom) 3-methylsulfonyl
PCB 149 (3-MeSO$_2$-PCB 149), drawn in their nonplanar configurations.

There are 209 possible PCB congeners, based on number and position of chlorine atoms on the biphenyl rings. Substituted biphenyls such as PCBs can be chiral (fig. 3.1) if two criteria are met. First, they must be asymmetrically substituted about the long axis of the molecule, and thus chiral in their energetically favored nonplanar conformation. This type of chirality is known as atropisomerism, and of the 209 PCB congeners, 78 satisfy this first criterion. Second, the two phenyl rings must not be free to rotate around the central C-C bond. Of the 78 chiral PCBs, 19 have three or four *ortho*-chlorine atoms, which are large enough to provide sufficient steric hindrance to rotation. The rotational energy barrier (DG‡) is on the order of 177 to 186 kJ/mol for tri-*ortho* substituted PCBs, and 246 kJ/mol for tetra-*ortho* congeners (Harju and Haglund 1999; Schurig, Glausch, and Fluck 1995). The presence of a buttressing *meta*-chlorine atom in tri-*ortho* substituted PCBs adds approximately another 6.4 kJ/mol to the rotational energy barrier (Harju and Haglund 1999). The energy barrier for the 19 stable chiral PCBs is equivalent to a racemization half-life of more than 200 billion years at 37°C. Thus these atropisomers are not only stable to racemization under environmental conditions but also stable at the elevated temperatures needed to resolve such stereoisomers by chiral gas chromatography (GC). Of the 19 stable chiral PCBs, seven have been classified as environmentally relevant, as they are present in Aroclors (Frame 1997) and thus in the environment in significant proportions (table 3.1). In addition, a number of other less prevalent chiral PCB congeners have also been studied in the environment.

Table 3.1. PCBs Forming Stable Atropisomers and Their Chlorine Substitution Patterns.

PCB Congener	Substitution on First Ring	Substitution on Second Ring
45	2,3,6	2
84	2,3,6	2,3
88	2,3,4,6	2
91	2,3,6	2,4
95	2,3,6	2,5
131	2,3,4,6	2,3
132	2,3,6	2,3,4
135	2,3,6	2,3,5
136	2,3,6	2,3,6
139	2,3,4,6	2,4
144	2,3,4,6	2,5
149	2,3,6	2,4,5
171	2,3,4,6	2,3,4
174	2,3,4,5	2,3,6
175	2,3,4,6	2,3,5
176	2,3,4,6	2,3,6
183	2,3,4,6	2,4,5
196	2,3,4,6	2,3,4,5
197	2,3,4,6	2,3,4,6

Metabolites of PCBs, such as hydroxylated, methylthio, and methylsulfonyl derivatives, can also be chiral. Cytochrome P-450 (CYP) attack on PCBs produces arene oxide intermediates, which can then further react to form hydroxylated PCBs. PCB arene oxide intermediates may also

Table 3.2. Environmentally Relevant Methylsulfonyl PCBs (MeSO$_2$-PCBs) Forming Stable Atropisomers and Their Chlorine and MeSO$_2$ Substitution Patterns.

MeSO$_2$-PCB Congener	MeSO$_2$ Substitution on First Ring	Chlorine Substitution on First Ring	Chlorine Substitution on Second Ring
3-MeSO$_2$-PCB 91	5	2,3,6	2,4
4-MeSO$_2$-PCB 91	4	2,3,6	2,4
3-MeSO$_2$-PCB 95	5	2,3,6	2,5
4-MeSO$_2$-PCB 95	4	2,3,6	2,5
3-MeSO$_2$-PCB 132	5	2,3,6	2,3,4
4-MeSO$_2$-PCB 132	4	2,3,6	2,3,4
3-MeSO$_2$-PCB 149	5	2,3,6	2,4,5
4-MeSO$_2$-PCB 149	4	2,3,6	2,4,5
3-MeSO$_2$-PCB 174	5	2,3,4,5	2,3,6
4-MeSO$_2$-PCB 174	4	2,3,4,5	2,3,6

conjugate with glutathione (GSH) either spontaneously or via glutathione-S-transferase enzymes. After dehydration, the PCB-GSH conjugate can undergo a series of reactions through the mercapturic acid pathway (MAP) (Bakke, Bergman, and Larsen 1982) to form methylthio PCBs, which can then further react to form PCB methyl sulfones (MeSO$_2$-PCB, fig. 3.1). Calculations based on quantum chemical methods predict that asymmetrically substituted methylthio and methylsulfonyl PCBs also form atropisomers (Nezel and Müller-Plathe 1997); one would imagine that hydroxylated PCBs can be chiral as well. Indeed, chiral metabolites can theoretically be produced by an achiral (i.e., prochiral) parent PCB. Of the 837 possible MeSO$_2$-PCBs, 456 are chiral, and 170 with tri- or tetra-*ortho* substitution are environmentally stable (Nezel and Müller-Plathe 1997). In practice, only about 60 MeSO$_2$ congeners have been observed in the environment, as only parent PCBs with 2,5- or 2,3,6-chlorine substitution are most likely to undergo MAP metabolism to form persistent MeSO$_2$-PCBs, although some exceptions exist (Letcher et al. 2000). Of the environmentally relevant MeSO$_2$-PCBs, ten are atropisomeric and stable at ambient temperatures (table 3.2). The environmental chemistry and toxicology of hydroxylated and MeSO$_2$-PCBs was recently reviewed (Letcher et al. 2000). In this chapter, we will limit our discussion of chiral PCB metabolites to the methyl sulfones, as nothing is known about the environmental fate and occurrence of chiral hydroxylated and methylthio PCBs.

Chiral PCBs are significant for several reasons. First, the stereoisomers of a chiral compound may have different biological and toxicological effects from each other and from the racemic mixture, because the individual enantiomers can interact differentially with other chiral compounds such as enzymes in living systems. Some of the earliest research on chiral PCBs has shown that CYP induction can be stereo-

selective (Püttmann et al. 1989, 1990; Rodman et al. 1991). Such enantioselectivity can result in inaccurate predictions of contaminant risks (Magrans, Alonso-Prados, and García-Baudín 2002; Williams 1996) based on the properties of the racemate, which is typically the form in which chiral chemicals (including PCBs) are introduced to the environment. Little is known about enantiomer-specific effects of chiral chemicals in the environment, but these effects can be significant enough to warrant registration of single-enantiomer pesticides such as S-metolachlor, which provide maximum benefit due to application of the enantiomers with the highest biological activity while minimizing application rates and hence environmental impact (Williams 1996).

In addition, chiral compounds are also useful markers of biological activity affecting such chemicals in the environment. Because the enantiomers of a chiral compound have identical physical-chemical properties, environmental abiotic processes such as air-water exchange, sorption, and abiotic transformation proceed at identical rates for both stereoisomers. However, biological processes such as biotransformation can be enantioselective, and proceed at different rates for each enantiomer. This differential activity may change the relative amounts of the stereoisomers. Thus, enantiomer compositions of chiral chemicals are tracers of biological processes affecting such compounds in the environment. Moreover, this tracer is insensitive to abiotic processes, as the many physical and chemical processes that can affect concentrations and fluxes of the compound will generally not change the enantiomer composition. This feature of chiral compounds has been extensively studied in recent years to understand enantioselective fate processes of such pollutants, including PCBs, in the environment, as highlighted in some recent reviews (Hühnerfuss 2000; Vetter and Schurig 1997).

This chapter details the current state of knowledge of chiral PCBs in the environment, including known studies through May 2005. We will discuss analytical chemistry and separation of chiral PCB compounds, environmental occurrence and fate processes of these chemicals, and the little that is known about stereoselective biological and toxicological effects of atropisomeric PCB compounds.

ANALYTICAL SEPARATION OF ATROPISOMERS OF PCB COMPOUNDS

Two tasks must be accomplished in order to measure the stereoisomer composition of a chiral compound in environmental matrices: the enantiomers must be separated through their differential interactions with a chiral selector compound and they must be separated and/or distinguished from other compounds in the sample that may interfere with the measurement. To this end, analytical separations of chiral PCBs and metabolites have been accomplished using several different techniques: chiral gas chromatography (GC), multidimensional gas chromatography (MDGC) and comprehensive two-dimensional gas chromatography (GCxGC), high-performance liquid chromatography (HPLC), and capillary electrophoresis/micellar electrokinetic chromatography (CE/MEKC). For environmental analysis, chromatographic separation techniques are typically coupled to detection by mass spectrometry (MS) to provide the selectivity required for samples with complex analyte mixtures. Because the analytical separation of PCB atropisomers has been extensively reviewed (Lehmler and Robertson 2001), this discussion will focus particularly on developments taking place after that review was published.

Gas Chromatography

Gas chromatographic separation of chiral PCB compounds has typically used cyclodextrins as chiral selectors, ever since the first report of PCB atropisomer separation by GC over ten years ago (König et al. 1993). A number of different substituted cyclodextrins (Schurig 2001) have been utilized in both GC columns, both custom made (Hardt et al. 1994; Jin and Jin 1994; König et al. 1993) and commercially available (Schurig and Glausch 1993; Vetter et al. 1997b, 1999). These columns can separate the stereoisomers of all 19 stable chiral PCB congeners, at least partially (Wong and Garrison 2000). One-dimensional chiral GC can quantify at least nine congeners in environmental samples (Wong and Garrison 2000) if retention time calibration of all 209 congeners is done to identify enantiomers separated from achiral congeners and if electron impact ionization mass spectrometry (EI-MS) is used to distinguish atropisomers from coeluting congeners with different molecular masses (Wong et al.

2002a). This method has the advantage of simplicity, as it can be used with commonly available analytical GC/MS instrumentation. A disadvantage is the inability to separate some chiral PCBs (e.g., PCBs 45, 84, 131, 135, 171, 175) due to coelutions with congeners of the same molecular mass. In addition, one-dimensional GC/EI-MS for chiral PCB analysis offers less sensitivity than electron capture detection (ECD) typically used for trace PCB analysis, because of the large number of interfering coelutions of achiral congeners of differing molecular mass that can be distinguished by EI-MS, but not by ECD (Wong and Garrison 2000). Electron capture negative ionization MS (ECNI-MS) can also be used instead of EI-MS, but congeners with fewer than six chlorine atoms have poor response factors using this ionization method.

One-dimensional GC/MS methods have also been developed to measure stereoisomers of PCB methyl sulfones. The stereoisomers of 3- and 4-MeSO$_2$-PCB 91, 3- and 4-MeSO$_2$-PCB 132, 3-MeSO$_2$-PCB 149, and 3- and 4-MeSO$_2$-PCB 174 were separated using a BGB-172 chiral column by Wiberg and colleagues (1998), who quantified the PCB metabolites using ion trap GC/MS/MS and GC/ECNI-MS. Enantiomers of eight MeSO$_2$-PCBs were resolved using a custom column (Ellerichmann et al. 1998). Meta-substituted MeSO$_2$-PCBs 95, 132, 149, and 174 were separated on a partially ethylated γ-cyclodextrin column, while the corresponding 4-MeSO$_2$-PCB atropisomers could not be separated (Jaus and Oehme 2000). A Chirasil-Dex column with a 3-m achiral DLB-XLB precolumn upstream of it was found to resolve the stereoisomers of 3-MeSO$_2$-PCB 132, 3- and 4-MeSO$_2$-PCB 149, and 3- and 4-MeSO$_2$-PCB 174 (Chu et al. 2003a). However, long run times were necessary (>300 min.), resulting in peak broadening and poor detection limits. The authors also warned of shifts in enantioselectivity over time on BGB-172 as columns age and wear out from loss of the stationary phase over time through column bleed. This phenomenon has been noted by other authors (Wong and Garrison 2000), who suggested periodic recalibration of retention times for chiral columns, which are usually unbonded and more fragile compared to achiral columns.

The mechanisms for chiral separation by GC are not well understood (Schurig 2001). There are no a priori methods by which to predict enantiomer separations. Nonetheless, some progress has been made to understand chiral GC separations with respect to PCBs. With the exception of PCB 45, congeners with 2,3,6-chlorine substitution patterns were separated on the commercially available stationary phase Chirasil-Dex (Haglund and Wiberg 1996). Those with 2,3,4,6-substitution were not separated, with the exception of PCB 176, which has both substitution patterns. Variable purities in some commonly used randomly substituted cyclodextrins result in widely varying column performance. Pure permethyl-β-cyclodextrin (PM-CD) resolved more chiral compounds and separated their enantiomers better than PM-CD batches with more impurities (Jaus and Oehme 2001). However, columns with randomly silyated heptakis (6-*O-tert*-butyldimethylsilyl-2,3-di-*O*-methyl)-β-cyclodextrin, a commonly used chiral stationary phase, resolved enantiomers of more chiral compounds tested than columns with less heterogeneous substitution (Ruppe et al. 2000), perhaps due to the different cyclodextrin derivatives serving as a mixed set of chiral selectors, each with its own specificity. The use of GC columns with several substituted cyclodextrin chiral selectors within the stationary phase resulted in improved PCB atropisomer selectivity (Magnusson et al. 2000), but at the expense of extremely long and impractical analysis times (more than two hours). A cyclodextrin derivative synthesized with the substituents of two "parent" cyclodextrins was highly enantioselective. It had not only the combined separation power of both sets of "parents" but also greater separation power than the two-"parent" cyclodextrins mixed in one stationary phase (Junge and König 2003). In summary, the development of new chiral selectors and the elucidation of factors controlling enantiomer separation is an ongoing area of research that should continue to yield improvements in chiral analysis.

Multidimensional Gas Chromatography

Multidimensional GC techniques such as MDGC and GCxGC are highly suitable for chiral analysis in environmental matrices. In MDGC, a conventional achiral column is connected in series to a chiral column for enantiomer separation. Analytes eluting from the first column are selectively heart-cut into a cold trap. After achiral separation is complete, the cold trap is heated to reinject trapped analytes into the second, chiral column for enantiomer separation. The use of heart-cutting to remove interferences results in longer column life, as well as greater selectivity compared to one-dimensional GC. Thus MDGC can be coupled with ECD for high sensitivity and selectivity. This technique was first demonstrated for chiral PCBs in commercial PCB mixtures (Glausch et al. 1994) and has been used to quantify PCB atropisomers in human milk (Blanch, Glausch, and Schurig 1999; Glausch, Hahn, and Schurig 1995), foodstuffs (Bordajandi et al. 2005), biota (Blanch et al. 1996), and sediments (Benická et al. 1996; Glausch, Blanch, and Schurig 1996). Multidimensional GC/MS has also been used to verify enantiomer compositions measured by one-dimensional chiral GC/MS (Wong et al. 2002a). The disadvantage of MDGC is that the requisite instrumentation is expensive and not widely available.

Comprehensive GCxGC is a relatively new and promising technique for analysis of extremely complex matrices, including environmental residues of PCBs. Like MDGC, this technique also uses two columns with different stationary phases in series. However, the effluent from the first column enters a modulator, which uses thermal or cryogenic trapping to focus small portions of eluate prior to injection into a short (ca. 1 m × 0.1 mm film thickness) second column for further separation. Separations on the second column are fast (e.g., 5–10 seconds), so that sequential fractions do not interfere with one another. Thus, unlike MDGC, the entire sample is subjected to separation on both stationary phases and high peak capacities and separations are possible. Comprehensive GCxGC to resolve eight chiral PCBs in commercial mixtures in a single run with m-ECD detection was achieved using a narrow-bore Chirasil-Dex chiral column in the first dimension and a liquid crystal column in the second dimension (Harju and Haglund 2001). This system, as well as a GCxGC system coupled to time-of-flight MS for additional selectivity, was successfully used to determine PCB enantiomer compositions in extracts of gray seal tissues (Harju et al. 2003). Comprehensive GCxGC has also been used to quantify enantiomers of all nine chiral PCB congeners separated by Chirasil-Dex, when this column was used in the first dimension in conjunction with a combination of two different polar stationary phase columns in the second dimension (Bordajandi et al. 2005). As with MDGC, the major disadvantages of comprehensive GCxGC are the expense and availability of analytical instrumentation, and the operator expertise needed to run such systems effectively. However, the use of GCxGC is currently in its infancy and should become a significant asset in complex sample analysis, including chiral analysis.

Liquid Chromatography and Capillary Electrophoresis

There are two major uses for liquid chromatography in chiral PCB analysis. First, HPLC has been used to separate enantiomers of PCB compounds. Separation of PCBs 84, 131, 132, 135, 136, 174, 175, 176, and 196 by reversed-phase chiral HPLC was done using permethylated β-cyclodextrin derivatized silica (Haglund 1996a, 1996b). In addition, enantiomers of some di-*ortho* substituted PCBs, which racemize under ambient conditions, were separated at subambient temperatures (0°C). Enantiomer separation of twelve PCBs was reported using three different HPLC stationary phases (Reich and Schurig 1999). All chiral PCBs except for 171 and 197 have been separated by HPLC. Liquid chromatographic separation was used to isolate single enantiomers of PCB congeners (Haglund 1996b; Mannschreck et al. 1985) for determining elution order of (+) and (–) stereoisomers (Haglund and Wiberg 1996), and for studies of physical and chemical properties of stereoisomers (Schurig, Glausch, and Fluck 1995), including the determination of the absolute configuration of 3- and 4-MeSO$_2$-PCB 132, 3- and 4-MeSO$_2$-PCB 149, and 3-MeSO$_2$-PCB 174 by vibrational circular dichroism (Döbler et al. 2002; Pham-Tuan et al. 2005). The absolute configurations of other PCB compound enantiomers are currently unknown. Semipreparative scale isolation (Pham-Tuan et al. 2005) can also be used to obtain stereoisomers for enantiomer-specific biological studies (Lehmler et al. 2005). No research has been done to date to use HPLC to quantify PCB compound enantiomers in environmental matrices. The use of liquid chromatography for this purpose is promising, given recent development of achiral HPLC/MS-based methods to measure hydroxylated PCBs (Berger, Herzke, and Sandanger 2004).

Liquid chromatography has also been used to fractionate samples to simplify analysis. PCBs in environmental extracts were fractionated based on *ortho* chlorine substitution using a 2-(1-pyrenyl) ethyldimethylsilyated silica HPLC column (Ramos, Hernández, and González 1999) to reduce the number of analytes in each fraction compared to the raw extract. Fractions containing tri- and tetra-*ortho* PCBs could then be analyzed by GC/ECD with minimal interferences, resulting in greater sensitivity at the expense of more sample preparation. This method has been used to quantify PCB enantiomer compositions in tissues of cetaceans (Chu et al. 2003c; Jiménez, Jiménez, and Gonzalez 2000; Reich et al. 1999; Serrano et al. 2000) and humans (Chu, Covaci, and Schepens 2003b). Collected fractions were also useful in quantifying non- and mono-*ortho* coplanar PCBs that are most responsible for dioxin-like toxicity (Jiménez, Jiménez, and Gonzalez 2000; Reich et al. 1999; Serrano et al. 2000).

Capillary electrophoresis/micellular electrokinetic chromatography is another promising method by which to separate chemical stereoisomers. In CE, charged molecules in a buffered polar mobile phase are separated through the application of an electrical potential in a narrow capillary tube. Neutral compounds such as PCBs are separated by MEKC through the addition of surfactants, which form micelles into which the neutral analytes partition. Stereoisomers are separated with the addition of a chiral selector. Advantages of CE/MEKC include more efficient separations than HPLC or GC, minimal sample preparation, and small amounts of sample needed. The first report of CE/MEKC for chiral PCBs separated the enantiomers of 12 congeners in about 35 minutes (compared to about two hours or more for GC analysis), using β- and γ-cyclodextrin in a separation buffer containing 2-(N-cyclohexylamino) ethanesulfonic acid with urea and sodium dodecyl sulfate micelles (Marina et al. 1996a,

1996b). Since then, researchers (Crego, García, and Marina 2000; Edwards and Shamsi 2002; García-Ruiz et al. 2001; García-Ruiz, Crego, and Marina 2003; Lin, Chang, and Kuei 1999) have experimented with a variety of buffers, organic modifiers, and cyclodextrin and polymeric chiral selectors to resolve all chiral PCB stereoisomers except for those of PCB 174. Microbial degradation of chiral PCBs was monitored using CE/MEKC in one study (García-Ruiz et al. 2002). To date, UV absorption has been used for detection, which results in poor sensitivity due to the short path length of the capillary. The potential of CE/MEKC for measuring chiral PCBs in complex environmental samples, by coupling to selective detectors such as MS, has not yet been assessed.

Quantification of Enantiomer Composition

Enantiomer compositions for environmental contaminants are reported in several common formats: the enantiomer ratio (ER), enantiomer fraction (EF) and enantiomeric excess (ee).

The enantiomer ratio (ER) is the concentration ratio of the (+)-enantiomer over that of the (–)-enantiomer. Chromatographic data for calculating concentrations, such as peak area or height, are also employed in computing ERs. If the elution order of the enantiomers is unknown on the chiral column used for the analysis, then the ER is generally reported as the ratio of the first-eluting enantiomer (E1) to the second-eluting one (E2):

$$ER = (+)/(-) \tag{1}$$
$$ER = (E1)/(E2) \tag{2}$$

The range of ER is from zero to infinity, with a racemic mixture having an ER of 1. Enantiomer ratios were typically used in the earlier literature on chiral environmental analysis (Faller et al. 1991).

Enantiomer fractions (EFs) are a more recent descriptor of enantiomer signatures (de Geus et al. 2000; Harner, Wiberg, and Norstrom 2000), in which the range in EF is from zero to unity, and the racemate has an EF of 0.5:

$$EF = (+)/[(+) + (-)] \tag{3}$$
$$or \ EF = (E1)/[(E1) + (E2)] \tag{4}$$

Enantiomer fractions, which are used throughout this paper, are preferred to ERs. Enantiomer fractions are based on a bounded, linear additive scale and are symmetric about the equivalency (racemic) value of 0.5. Enantiomer ratios are on a multiplicative scale, and thus produce skewed data that are inappropriate for statistical summaries such as sample mean and standard errors (Ulrich, Helsel, and Foreman 2003). As a result, EFs are more amenable compared to ERs for graphical representations of data, as well as for mathematical expressions and environmental modeling (Harner, Wiberg, and Norstrom 2000; Ulrich, Helsel, and Foreman 2003). Individual ER and EF measurements can be converted as follows (de Geus et al. 2000; Harner, Wiberg, and Norstrom 2000):

$$EF = ER/(ER + 1) = 1/[(1 + (1/ER))] \qquad (5)$$

Conversion of summarized values (e.g., mean ER ± single-value standard deviation to EF) can lead to substantial discrepancies and should be avoided (Ulrich, Helsel, and Foreman 2003). It is also important to keep in mind that conventions used in describing ERs and EFs may differ between studies and analytical methods. For example, chiral separations on different stationary phases may result in reversal of elution order and lead to different values if elution orders are not known. In this chapter, EF is defined using equation 3 for those congeners in which elution order is known (Haglund and Wiberg 1996), and using equation 4 on the specified column if not, unless otherwise indicated.

A third metric of enantiomer composition is the ee, which is the percentage difference of the more abundant enantiomer:

$$ee = \%(+) - \%(-) \text{ (assuming } [+] \text{ enantiomer in excess)} \quad (6)$$

Although ee is commonly used in the organic synthesis and pharmaceutical fields, it is less commonly used in chiral environmental chemistry than EFs and ERs.

ENANTIOMERS OF PCB COMPOUNDS IN THE ENVIRONMENT

Soil and Sediment

Soils and sediments are a major environmental sink for PCBs, which are hydrophobic and thus sorb to natural organic carbon in soil and sediment particles. PCBs in these matrices may be biotransformed through microbial action by way of both aerobic (Bedard et al. 1987) and anaerobic (Bedard and Quensen 1995) pathways, which may be enantioselective. Studies on the enantiomer composition of chiral PCBs in sediments and soils to date have provided insight into the different microbial populations and enzymes involved in these reactions, and highlighted the utility of chiral compounds as tracers of environmental microbial biotransformation.

Field studies have found site-specific PCB enantiomer compositions. Racemic amounts of PCBs 95, 132, and 149

were measured in sediments of the Elsenz River in southern Germany (Glausch, Blanch, and Schurig 1996) and provided no evidence for microbial degradation of PCBs at this site. Similar results for PCBs 91, 95, 136, 149, 174, 176, and 183 were observed in Environment Canada Certified Reference Material EC-5 sediments, taken from the mouth of the Humber River in Toronto (Wong et al. 2002a). In contrast, sediment SRM 1939 (National Institute of Standards and Technology, or NIST) taken from the Hudson River in New York state, had significantly nonracemic amounts of PCB 95 (Benická et al. 1996). Hudson River sediments are heavily contaminated with PCBs, which are known there to undergo microbially mediated anaerobic reductive dechlorination (Bedard and Quensen 1995) and aerobic oxygenase and dioxygenase transformation (Bedard et al. 1987). Thus Benická et al. (1996) provided the first confirmation that microbial PCB biotransformation can be stereoselective. A more extensive survey found nonracemic signatures for PCBs 91, 95, 132, 136, 149, 174, 176, and 183 in river sediments throughout the U.S. (Wong, Garrison, and Foreman 2001), particularly those of the Hudson (consistent with the results of Benická et al. 1996) and the Housatonic in Connecticut. Patterns in EFs among congeners were consistent with known reductive dechlorination patterns (Bedard and Quensen 1995) in both these river basins. In addition, the enantiomer signature of PCB 91 was reversed between the Hudson (EF >0.5 using equation 4 on Chirasil-Dex) and the Housatonic River (EF <0.5) sediments (Wong, Garrison, and Foreman 2001). This observation implies that the two sites have different PCB biotransformation processes with different enantiomer preferences, consistent with known differences in microbial consortia and dechlorination patterns (Bedard and Quensen 1995). Reversals in PCB enantioselectivity have also been observed at different depths within individual dated sediment cores of Lake Hartwell, a contaminated reservoir in South Carolina. This observation suggests that microbial consortia capable of biotransforming PCBs in different ways may also be active at the same site, although perhaps at different points in time.

Controlled laboratory studies of aerobic microbial degradation of chiral PCBs have found that some pathways are not stereoselective, while others can be highly enantioselective. The soil bacterial isolate *Jonibacter* sp. strain MS3-02 did not demonstrate any evidence of enantioselectivity in the degradation of PCBs 45, 88, 91, 95, 136, 144, and 149 (García-Ruiz et al. 2002). In another study (Singer, Wong, and Crowley 2002), PCBs 45, 84, 91, and 95 were exposed in the presence of various cosubstrates to five different bacterial strains: gram-negative *Ralstonia euthrophus* H850 and *Burkholderia cepacia* LB 400; and gram-positive *Arthrobacter*

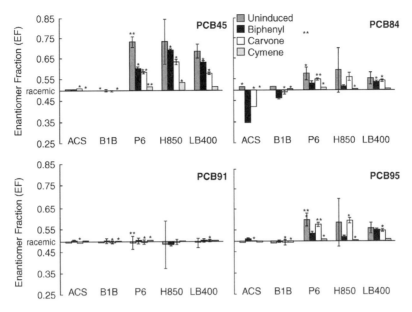

Figure 3.2.   Enantiomer fractions for PCBs 45, 84, 91, and 95 after exposure to aerobic gram-positive bacterial isolates *Rhodococcus* sp. strain ACS, *Arthrobacter* sp. strain B1B, and *Rhodococcus globerulus* P6; and gram-negative isolates *Ralstonia euthrophus* H850 and *Burkholderia cepacia* LB 400 with different inducing compounds. Treatments with significant PCB degradation are noted with an asterisk ($P < 0.05$) or double asterisk ($P < 0.01$). Adapted from Singer et al. (2002). Reprinted with permission from Appl Environ Microbiol (2002) 68:5756–5759. ©2002 American Society for Microbiology.

sp. strain B1B, *Rhodococcus* sp. strain ACS, and *Rhodococcus globerulus* P6. Enantioselective biodegradation was observed, with the gram-negative strains possessing similar enantiomer selectivity and the gram-positive strains a completely different stereoisomer pattern (fig. 3.2). The exception was P6, which had the gram-negative enantiomer pattern and is known to be genetically similar to the gram-negative strains with regards to biphenyl dioxygenase enzymes (Asturias, Diaz, and Timmis 1995). The authors (Singer, Wong, and Crowley 2002) concluded that dissimilar PCB biotransformation enzymes and pathways between the various bacterial strains were responsible for the observed results, and suggested that chiral analysis can be used to lend insight into biotransformation mechanisms.

Anaerobic reductive dechlorination of chiral PCBs was observed in laboratory microcosms of sediments from Lake Hartwell (Pakdeesusuk et al. 2003). Previous field measurements had observed nonracemic PCBs buried deep within anaerobic bed sediments of this lake (Wong, Garrison, and Foreman 2001), suggesting *in situ* reductive dechlorination. Live microcosms spiked with racemic PCB 132 underwent non-enantioselective *meta* dechlorination to racemic PCB 91, which in turn was *meta* dechlorinated to achiral PCB 51 in a highly stereoselective manner (fig. 3.3). Similarly, PCB

149 was nonstereoselectively *para* dechlorinated to PCB 95, which was then *meta* dechlorinated to achiral PCB 53 enantioselectively. These results confirmed field studies suggesting reductive dechlorination as a possible stereoselective fate process for chiral PCBs, and suggested that these compounds might be useful as markers for *in situ* biotransformation. However, little is known about the microbial strains and enzymes involved in anaerobic reductive dechlorination, or the factors that control their stereospecificity.

It is possible that microbial degradation of PCBs takes place at concentrations far below threshold concentrations of about 1 mg/g, as suggested by achiral measurements (Rhee et al. 1993), at least in some systems. Chiral PCB EFs in rural topsoils in the United Kingdom had nonracemic amounts of PCBs 95, 136, and 149 (Robson and Harrad 2004) at concentrations on the order of pg/g dry weight (fig. 3.4). Similar results for these congeners were also noted in forested soils of the greater Toronto area (Wong et al. 2003). These observations suggest that natural attenuation of PCBs at trace background concentrations by soil microbes occurred slowly (e.g., on the order of years or decades) but appreciably over time, and that chiral analysis is the only feasible method to date to detect such biodegradation. On the other hand, nearly racemic profiles of PCBs in Lake Ontario

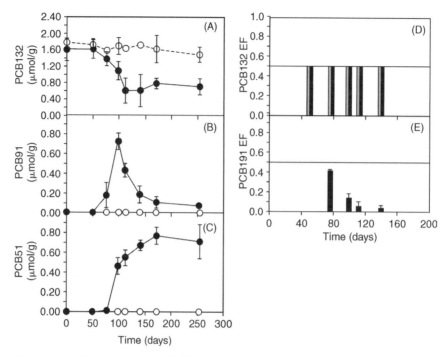

Figure 3.3.   Concentrations of PCB 132 and its major dechlorination products (A–C) and EFs for PCB 132 and PCB 91 (D, E) in racemic PCB 132–spiked Lake Hartwell sediment microcosms: open circles, crosshatched bars are autoclaved controls with PCB 132; filled circles, filled bars are live treatments with PCB 132. Racemic value (EF = 0.5) indicated with dashed line (D, E). Error bars represent standard deviations for triplicates; where not shown, the deviation was smaller than the size of the symbols (Pakdeesusuk et al. 2003). Reprinted with permission from *Environ. Sci. Technol.* (2003) 37:1100–1107. ©2003 American Chemical Society.

sediments, with total PCB concentrations below 1 mg/g, suggest that stereoselective reductive dechlorination is not an important process at low concentrations, at least in this lake.

Chiral PCBs in sediments and soils are also useful in providing insight into fate processes affecting these compounds, which in turn may influence regulatory decision making in risk assessment. The UK study that found nonracemic soil PCBs at trace concentrations also saw racemic PCBs in air samples (fig. 3.4) taken simultaneously at those sites (Robson and Harrad 2004). This observation suggests that atmospheric PCBs were from primary racemic sources, at least at the sites investigated, rather than from volatilization of biologically weathered soil residues. This conclusion is contrary to results from similar studies (Bidleman et al. 1998; Mattina et al. 2002) with chiral organochlorine pesticides, and suggests that control of primary PCB emissions sources (e.g., transformers, capacitors, elastic construction sealants still in use) may be effective in minimizing atmospheric PCB loads. Chiral signatures can be also used for estimating *in situ* biotransformation rates. Enantiomer compositions of PCBs in Lake Hartwell sediments suggested that the half-

life of reductive dechlorination was approximately 30 years, on the same order of decrease as that estimated for natural recovery via burial (Brenner et al. 2004).

Aquatic Organisms

Aquatic invertebrates and fish bioaccumulate PCBs and other recalcitrant organic pollutants through the food web (i.e., uptake through prey), resulting in higher concentrations of such compounds in organisms at higher trophic levels. Invertebrates and fish are also generally considered to have poor biotransformation capability toward persistent organic pollutants because they have lower levels and activities of CYP 1A and 2B enzymes that attack xenobiotic compounds entering the body (Kleinow, Melancon, and Lech 1987; Stegeman and Klopper-Sams 1987). As a result, aquatic organisms are readily impacted by persistent contaminants such as PCBs, which may bioaccumulate to concentrations sufficient to cause deleterious effects and lead to consumption advisories. Chiral analysis can thus provide a way to understand biotransformation as an elimination mechanism in biota. A specific tracer for biotransformation processes is vital for elucidating biological processes affect-

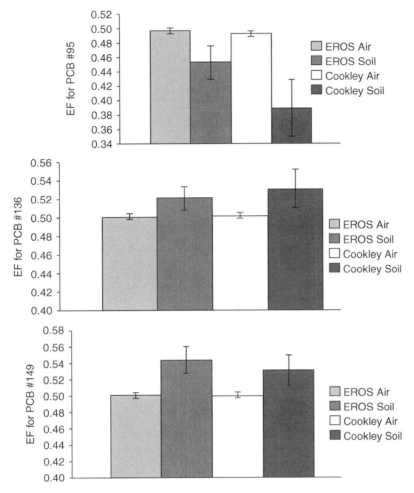

Figure 3.4.   Comparison of average EF values for PCB 95, 136, and 149 in paired air and soil samples taken in the UK. Error bars are s_{n-1}. Reprinted with permission from *Environ. Sci. Technol.* (2004) 38:1662–66. ©2004 American Chemical Society.

ing contaminants in food webs, given the many parameters that exist which may affect contaminant concentrations in biota, such as lipid content (Kucklick and Baker 1998), trophic position (Kidd et al. 1998), changes in diet (Stow et al. 1995), and food chain length (Rasmussen et al. 1990), as well as the intrinsic natural variability that may be associated with them.

Chiral analysis suggests that most aquatic invertebrates lack significant biotransformation capability toward PCBs, and that nonracemic residues in such organisms are most likely due to uptake from their environment. Near-racemic amounts of PCB 149 (EF = 0.50–0.55 using equation 4 on the custom column) were observed in blue mussels (*Mytilus edulis*) of the German Bight (Hühnerfuss et al. 1995), suggesting only weak enzymatic degradation of this congener in this species. Similarly, racemic or near-racemic EFs were observed for PCBs 136, 149, 174, and 176 in fresh-

water bivalves (*Corbicula* sp.) in U.S. streams (Wong et al. 2001). However, PCBs 95, 136, and 183 were more non-racemic, with EFs deviating 0.1–0.2 units away from the racemic value of 0.5. Analysis of the food web was not done in either study, so it is not clear if the nonracemic residues in the bivalves were due to *in vivo* biotransformation or from plankton and other organic detritus consumed by these filter feeders. Crayfish (*Procambaraus* sp.) in Lake Hartwell also had significantly nonracemic PCBs (Wong et al. 2001), but the known sediment microbial biotransformation of PCBs at this site suggests that the crayfish residues were due to uptake from sediments. The marine zooplankton *Calanus* sp. in the Arctic had racemic PCBs (Hoekstra et al. 2002; Warner et al. 2005). These observations, in conjunction with findings of chiral chlordane and α-hexachlorocyclohexane enantiomer compositions in *Calanus* similar to that in water (Hoekstra et al. 2003; Moisey et al. 2001),

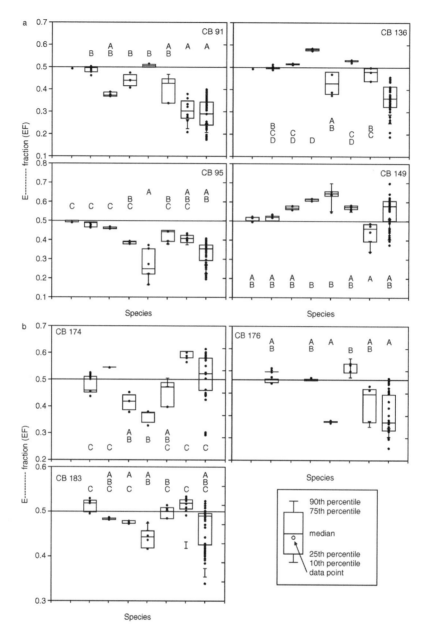

Figure 3.5.   Box plots of EFs for chiral penta- and hexachlorobiphenyls (A) and heptachlorobiphenyls (B) measured in Lake Superior biota. The horizontal line across each panel indicated racemic EF of 0.5. "Phyto" = phytoplankton, "Zoo" = zooplankton. Distributions with the same letter are not statistically different (Wong et al. 2004a). Reprinted with permission from *Environ. Sci. Technol.* (2004) 38:84–92. ©2004 American Chemical Society.

suggest that *Calanus* does not biotransform persistent organic compounds, at least not enantioselectively.

Recent results suggest that a few aquatic invertebrates may have some capability to biotransform PCBs, although the evidence is inconclusive at this point. Nonracemic PCBs were observed for the amphipod *Diporeia hoyi* and the opossum shrimp *Mysis relicta* in Lake Superior (fig. 3.5), while phytoplankton and zooplankton prey were racemic (Wong et al. 2004a). This observation suggests that these macrozooplankton species may be capable of biotransforming PCBs. Alternatively, these signatures may be due to uptake of nonracemic PCBs in bed sediments and particulate organic matter, upon which both *Diporeia* and *Mysis* also feed. If this is true, then microbial biotransformation to produce such nonracemic PCBs may be occurring in Lake Superior at sediment concentrations (ng/g for total PCBs) far below the threshold levels believed necessary to support such activity. The observation of nonracemic PCBs at similar concentrations in soils (Robson and Harrad 2004) suggests the latter hypothesis may be true. However, more research is needed to understand biotransformation dynamics of PCBs in aquatic invertebrates.

Studies with fish have shown species-specific trends for PCB enantiomer compositions, indicating that chiral signatures in fish are due to a combination of species-dependent food web interactions and *in vivo* biotransformation. Data for marine fish are sparse to date, but suggest that the few species studied so far do not process PCBs stereoselectively to any significant extent. Livers of the Portuguese dogfish shark (*Centroscymnus coelolepis*) had racemic amounts of PCBs 95 and 149 (Blanch et al. 1996), but slightly nonracemic amounts of PCB 132 (EF = 0.53–0.57, defined with equation 3). Racemic or near-racemic EFs were observed for PCBs 95, 136, 149, and 174 in livers of groupers (*Epinephelus marginatus*) from the Atlantic; however, one liver sample had a PCB 132 EF of 0.61 with the remaining five samples racemic (Serrano et al. 2000). The source of nonracemic PCBs is not clear in these organisms. Arctic cod (*Boreogadus saida*) also had near-racemic amounts of PCBs 95 and 149 similar to planktonic prey (Warner et al. 2005), suggesting nonenantioselective processes controlled PCB levels in this species.

Freshwater fish, which are more extensively studied than marine fish, also appear to have more nonracemic PCB residues, and thus may degrade these congeners slowly but significantly over the course of their lifetimes. Largemouth bass (*Micropterus salmoides*) and bluegill sunfish (*Lepomis macrochirus*) from the heavily contaminated Lake Hartwell had highly nonracemic PCBs 91, 95, 136, and 149 (Wong et al. 2001). Species-specific differences were observed; for example, EFs for PCB 95 ranged from 0.4 to 0.6 in bass, but ranged from 0.45 to 0.9 in bluegills. This observation indicates differences in *in vivo* metabolism and/or accumulation from prey. Similar species-dependent EFs were observed for PCBs 91, 95, 136, 149, 174, and 183 in suckers (*Catostomus* sp.) and sculpins (*Cottus* sp.) in U.S. rivers (Wong et al. 2001). The nonracemic EFs in sculpins are consistent with achiral observations of MeSO$_2$-PCB metabolites in these fish (Bright, Grundy, and Reimer 1995; Stapleton, Letcher, and Baker 2001) that are lacking in other benthic fish in the same habitat (Stapleton, Letcher, and Baker 2001), suggesting that sculpins can metabolize PCBs to some extent. However, stereoisomers of chiral PCB metabolites have not yet been studied to date in fish. Sculpins in Lake Superior had nonracemic PCBs 95, 136, and 174 significantly different than their *Diporeia* and *Mysis* prey, consistent with *in vivo* enantioselective biotransformation of these congeners

(Wong et al. 2004a). Similar results were also observed for Lake Superior fish, such as lake herring (*Coregonus artedii*) for PCB 91, and lake trout (*Salvelinus namaycush*) for PCB 136. Enantiomer mass balances of chiral PCBs in Lake Superior fish, based on concentrations and EFs between predator and prey, suggest that biotransformation half-lives of metabolizable PCBs were on similar time scales as the lifetime of the organisms (Wong et al. 2004a). Estimation of *in situ* biotransformation rates using enantiomers in field studies has not been possible to this point, except by curve-fitting in modeling studies. Although crude, the rates calculated suggest that biotransformation of persistent contaminants by aquatic biota may be more extensive than previously assumed.

The ability of fish to eliminate PCBs stereoselectively, inferred by chiral analysis in field studies, has been demonstrated experimentally. Rainbow trout (*Oncorhynchus mykiss*) exposed to food spiked with mg/g concentrations of PCBs 95 and 136 maintained racemic residues of PCB 95 but preferentially eliminated (+)-PCB 136 compared to its antipode (fig. 3.6) with a minimum biotransformation half-life of approximately one year (Wong et al. 2002b). This enantiomer preference agreed with PCB 136 EFs in lake trout, a related salmonid species (Wong et al. 2004a). Forage fish also have some ability to eliminate PCBs stereoselectively, as fathead minnows (*Pimephales promelas*) maintained nonracemic amounts of PCBs 84, 136, and 149 when fed racemic levels in food, while PCBs 91, 95, and 174 remained racemic (Wong et al. 2004b). The similarities in enantiomer preferences for PCB 136 among the trout species and fathead minnows suggest that similar enzymes with the same enantioselectivity may be responsible for producing the enantiomer composition observed.

The occurrence of nonracemic PCBs in aquatic organisms is consistent with known PCB metabolic processes. However, little is known about how such processes are enantioselective, and chiral analysis does reveal some unexpected surprises. With the exception of PCB 183, all the chiral congeners studied in freshwater fish have at least one set of vicinal *meta, para* hydrogen atoms, a structure amenable to attack by CYP2B (Boon et al. 1989). Thus the PCB stereoselectivity observed in freshwater fish is consistent with known detoxification mechanisms. However, the occurrence of CYP2B in fish has not been demonstrated (Kleinow, Melancon, and Lech 1987; Stegeman and Klopper-Sams 1987), even though proteins in fish similar to CYP2B in mammals (Kleinow et al. 1990; Stegeman, Woodin, and Waxman 1990) but with different regulatory mechanisms (Iwata et al. 2002) have been isolated. Thus it is not clear at this point what enzymes and biochemical pathways act upon PCBs stereoselectively. In addition, these processes appear

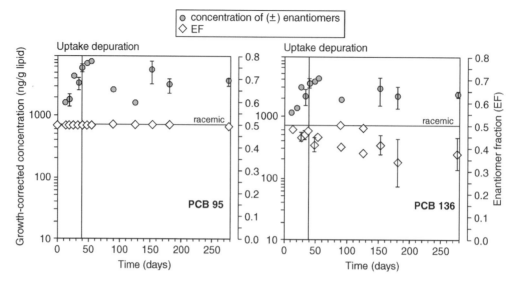

Figure 3.6.   Growth- and lipid-normalized concentrations (sum of both enantiomers) and EFs of PCB 95 and 136 in rainbow trout exposed to contaminated food over time. Each point represents mean and standard deviation (if larger than symbol used) of fish sampled at that time point (Wong et al. 2002b). Reprinted with permission from *Environ. Sci. Technol.* (2002) 36: 1257–62. ©2002 American Chemical Society.

to be regioselective as well as stereoselective. PCBs 95 and 136 differ in structure by only one *ortho* chlorine atom (table 3.1). That difference is enough to have resulted in preferential elimination of (+)-PCB 136 but no stereoselectivity in PCB 95 in fish (Wong et al. 2002b, 2004b). Factors controlling PCB stereoselectivity are essentially unknown, as EFs in aquatic biota do not appear to be related to concentration, organism age, sex, or other physiological factors. More research needs to be done to understand PCB enantiomer dynamics in aquatic organisms. Regardless, it is important to recognize that chiral analysis can detect and can provide insight into biologically mediated processing of PCBs and other optically active compounds that would otherwise be obscured by the many achiral processes affecting contaminant concentrations in the environment.

Birds and Mammals

The toxicokinetics of chiral PCBs in mammals and birds is more complex than in lower organisms for several reasons. First, mammals and birds have much higher CYP activities than lower organisms, resulting in more extensive detoxification and biotransformation of metabolizable xenobiotic compounds such as PCBs. In addition, mammals and birds tend to occupy the highest trophic levels in food webs. This high biomagnification potential results in the buildup of recalcitrant contaminants to concentrations more likely to induce CYP detoxification activity than in lower organisms.

The sparse data on chiral PCBs in birds suggests that PCB metabolism in avians is highly stereoselective. Barn swallows (*Hirundo rustica*) from Lake Hartwell had highly nonracemic amounts of PCB 91 (Wong et al. 2001) as well as near-racemic amounts of PCBs 95, 136, and 149. However, the source of these enantiomer compositions was unknown. Highly nonracemic PCBs 91, 95, 149, and 183 were found in seven seabird species in the Northwater Polynya between Greenland and the Canadian archipelago (Warner et al. 2005), with average EFs as high as 0.84 for PCB 95 (defined using equation 4 on Chirasil-Dex) in ivory gulls (*Pagophila eburnea*) and as low as 0.15 for PCB 149 in thick-billed murres (*Uria lomvia*). Food web analysis indicated that the seabird species biotransformed PCBs stereoselectively in a species-specific manner, consistent with similar observations with aquatic biota. This biotransformation is consistent with findings of hydroxyl-PCB and MeSO$_2$-PCB metabolites in albatrosses (Klasson-Wehler et al. 1998), as well as observations of lower biomagnification and higher elimination rates of congeners amenable to CYP attack in American kestrels (Drouillard and Norstrom 2000).

Cetaceans have lower CYP activities than other mammals and thus accumulate persistent organic pollutants such as PCBs to a greater extent. However, chiral analysis has shown that cetaceans possess some capacity for metabolizing PCBs. Blubber and livers of striped dolphins (*Stenella coeruleoalba*) found dead in the Mediterranean had nonracemic amounts of PCBs 95, 132, 135, 149, and 176, while PCBs

136 and 174 were racemic (Reich et al. 1999). Similar results were found in four other cetacean species in the Mediterranean (Jiménez, Jiménez, and Gonzalez 2000), and in pilot whales (*Globicephala melas*) of NIST SRM 1945 and Marine Mammal Quality Assurance Exercise Control Material IV (Wong et al. 2002a). In addition, highly non-racemic PCB 91 (EF = 0.12, defined using equation 4 on Chirasil-Dex) and PCB 84 (EF = 0.60, defined using equation 3) were observed in tissues of bottlenose dolphins (*Tursiops truncatus*) (Jiménez, Jiménez, and Gonzalez 2000). However, it is not clear in these studies if these residues were due to *in vivo* biotransformation or from uptake of non-racemic PCBs in prey. Bowhead whales (*Balaena mysticetus*) harvested off the Alaskan coast had nonracemic PCB 91, 135, 149, 174, 176, and 183, while their main prey, *Calanus*, was racemic (Hoekstra et al. 2002). This observation suggests *in vivo* biotransformation of this congener by bowhead whales. Enantioselectivity was both sex and age dependent, as PCB 91 EFs decreased with increasing body length (correlated to age) in male bowheads, but was racemic in females regardless of size (fig. 3.7). Age-dependent EFs were also observed in livers of harbor porpoises (*Phocoena phocoena*) from the southern North Sea (Chu et al. 2003c), as PCBs 95, 132, and 149 were near-racemic in most of the juvenile specimens sampled, but were markedly nonracemic in all adults studied (PCB 95 EF = 0.57–0.72, PCB 132 EF = 0.52–0.69, PCB 149 EF = 0.36–0.46). Enantiomer fractions correlated with the ratio of PCB 153, a recalcitrant congener, to PCB 101, a congener labile to CYP attack. This observation suggested that EFs may reflect the proportion of the metabolized congener. Methyl sulfonyl PCB metabolites (3-MeSO$_2$-PCB 132, 3- and 4-MeSO$_2$-PCB 149, and 3- and 4-MeSO$_2$-PCB 174) were also highly non-racemic in the harbor porpoise samples (EF > 0.73 or EF < 0.23, all defined with equation 4 on Chirasil-Dex), suggesting that MeSO$_2$-PCBs may have been preferentially formed, or were retained in a highly stereoselective manner (Chu et al. 2003a).

Enantiomer compositions of PCB compounds in seals and polar bears are more nonracemic than in cetaceans, consistent with their higher xenobiotic detoxification capability. The first reports of chiral PCBs in pinnipeds found PCB 149 EFs (defined using equation 4 on the chiral column used) of 0.58–0.71 and 0.60–0.70 in blubber of harbor seals (*Phoca vitulina*) and gray seals (*Halichoerus grypus*), respectively, from Iceland (Klobes et al. 1998; Vetter et al. 1997a). Gray seals from the Baltic Sea had essentially only the second-eluting enantiomer (on Chirasil-Dex) of PCB 91 (EF < 0.01), while other chiral PCBs were almost as nonracemic (PCB 95 EFs = 0.65–0.8, PCB 132 EFs = 0.1–0.2) (Harju et al. 2003). In contrast, ringed seals (*Phoca hispida*)

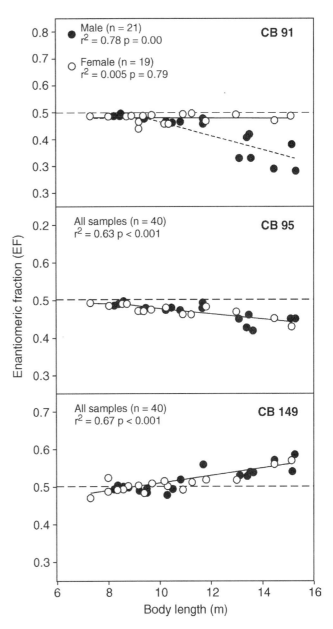

Figure 3.7.   Relationship between EFs of PCBs 91, 95, and 149 in blubber from male and female bowhead whales with body length. Dashed line represents racemic EF = 0.5. Solid lines provide first-order linear regression of EF values for each gender. The EF values for PCB 95 (EF = –0.0064[length] + 0.539) and PCB 149 (EF = 0.01[length] + 0.408) in males and females were pooled since the correlation with gender was not significant (Hoekstra et al. 2002). Reprinted with permission from *Environ. Sci. Technol.* (2002) 36:1419–25. ©2002 American Chemical Society.

from the Northwater Polynya had nearly racemic PCB 149 (average EF = 0.49) but highly nonracemic PCB 95 (average EF = 0.85) that was constant for specimens between 4 and 40 years old (Warner et al. 2005). These findings suggest species-dependent differences in diet and/or metabolism in ringed seals compared to the other pinniped species. It is not clear if the constant PCB 95 EFs in ringed seals arose from accumulation over the organisms' lifetimes, or if the enantiomer signature observed came from placental transfer from the mother before birth and/or absorption from the mother via milk when the seals were young. Ringed seals also formed highly nonracemic 3-MeSO$_2$-PCB 149 (EF = 0.24) which was not detected in their major prey, arctic cod. This observation indicated that 3-MeSO$_2$-PCB 149 arose from enantioselective sulfone formation and/or clearance in the ringed seal (Wiberg et al. 1998). This MeSO$_2$-PCB congener was almost single enantiomer (EF < 0.09) in polar bears (*Ursa maritimus*). The authors concluded, based on enantiomer analysis, that at least a portion of the 3- and 4-MeSO$_2$-PCB 91 and MeSO2-PCB 149 residue in polar bears was due to *in vivo* metabolism, while 3- and 4-MeSO$_2$-PCB 132 residues were due to accumulation from seal consumption (Wiberg et al. 1998). Similar results were also observed in gray seals (*Halichoerus grypus*) from the Baltic Sea, with liver tissues containing much higher MeSO$_2$-PCB concentrations than lung and blubber tissues, at nearly single-enantiomer composition (EF < 0.14 or EF > 0.94) in all three tissues (Larsson et al. 2004).

As with aquatic mammals, terrestrial mammals also appear to have a similar capacity to biotransform PCB compounds, although factors affecting stereoselectivity are complex and not well understood. Livers of wolverines (*Gulo gulo*) from the Canadian Arctic had significantly but slightly nonracemic amounts of PCBs 95, 136, and 149 (EFs deviating to 0.41) (Hoekstra et al. 2003). Biomagnification factors of PCBs suggested that the wolverine population studied may have been scavenging marine mammals, and had a PCB congener profile consisting of only a handful of recalcitrant congeners. These observations suggest that wolverines may biotransform PCBs, but stereoselectivity is minor in this species. Mice dosed with racemic PCB 84 preferentially eliminated the (–)-enantiomer within 6 days (Lehmler et al. 2003). Tissue-specific enantiomer preferences were observed, with brain tissue having the most (+)-PCB 84 (EF = 0.60), while spleen tissue was nearly racemic (EF = 0.52). These observations suggest that mice stereoselectively bioprocess this congener, and that different tissues may exhibit different binding affinities for PCB 84 atropisomers. Rats dosed with the technical PCB mixture Clophen A50 also produced highly stereoselective *meta* and *para* PCB methyl sulfones (Larsson et al. 2002). Tissue retention was organ-specific and enantioselective, with lung tissues showing reversed enantiomer preferences compared to adipose and liver tissue for 4-MeSO$_2$-PCB 149 (fig. 3.8). Parent PCB compounds were not analyzed in this study. An *in vitro* experiment showed that rat liver hepatocytes degraded PCB 149 nonstereoselec-

tively, but transformed the *S*-enantiomer of 3-MeSO$_2$-PCB 149 in a highly enantioselective manner (Hühnerfuss et al. 2003), consistent with the *in vivo* results (Larsson et al. 2002). Human milk tissue from Germany was found to have racemic PCB 149, slightly enantioenriched PCB 95 (EF = 0.50–0.54), and nonracemic PCB 132 (EF = 0.50–0.71) (Blanch, Glausch, and Schurig 1999; Glausch, Hahn, and Schurig 1995). Human liver tissues had high enantiomeric excesses of 3-MeSO$_2$-PCB 132 and 3-MeSO$_2$-PCB 149 (Ellerichmann et al. 1998). A Belgian study (Chu, Covaci, and Schepens 2003b) found near-racemic amounts of PCBs 95, 132, and 149 in muscle, kidney, and brain samples of cadavers (EFs ranging from 0.48 to 0.57) but enantioenrichment of these congeners in liver tissue (mean EFs of 0.63, 0.43, and 0.57, respectively), suggesting *in vivo* metabolism in humans.

As with aquatic organisms, it is important to bear in mind that the biochemical pathways and factors controlling the observed atropisomer residues for birds and mammals are also essentially unknown.

BIOLOGICAL EFFECTS OF CHIRAL PCB COMPOUNDS

There is almost no literature to describe stereoselective biological and toxicological effects of chiral PCBs and their metabolites. In chick embryo hepatocytes (Rodman et al. 1991), both PCB 88 stereoisomers were equally potent as inducers of total CYP, while ethoxyresorufin-O-deethylase (EROD) activity was induced more by (+)-PCB 88. The (–)-enantiomer was inactive in EROD induction, but was more active than (+)-PCB 88 in induction of benzphetamine N-demethylase (BPDM) activity. The (+)-enantiomer of PCB 139 was more effective than the (–)-enantiomer in induction of CYP, EROD, and BPDM. In BPDM induction, (–)-PCB 197 was more effective than (+)-PCB 197, but the opposite was true for EROD induction. In male Sprague-Dawley rats (Püttmann et al. 1989), racemic PCB 197 and its stereoisomers only weakly induced CYP2B, with no enantiomer preferences. However, racemic PCB 139 was a potent CYP2B inducer. In addition, (+)-PCB 139 was more effective than (–)-PCB 139 in enhancing aminopyrene N-demethylase, aldrin epoxidase, and total CYP content (Püttmann et al. 1989), as well as induction of morphine UDP-glucoronosyltransferase (Püttmann et al. 1990). Recently, both enantiomers of PCB 84 have been shown to increase phorobol ester binding in rat cerebellar granule cells, with the (–)-enantiomer being more potent (Lehmler et al. 2005). This same study also found both PCB 84 stereoisomers to be equally effective in inhibiting microsomal calcium ion uptake by these cells. Interestingly enough, the PCB 84 racemate was significantly

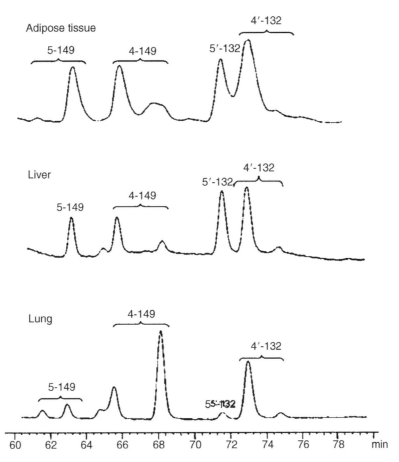

Figure 3.8.   Separation of atropisomers of MeSO$_2$-PCBs in adipose, liver, and lung tissue from rats dosed with the technical PCB mixture Clophen A50 (Larsson et al. 2002). Reprinted with permission from *Environ. Sci. Technol.* (2002) 36:2833–38. ©2002 American Chemical Society.

more potent for both neurotoxicity endpoints (Lehmler et al. 2005), suggesting that enantioselective biotransformation of PCB 84 may lower these neurotoxic effects.

The PCB congeners assessed for biochemical effects stereoselectivity are minor constituents in technical and environmental PCB mixtures (Frame 1997). With one exception (Lehmler et al. 2005), no research has been published to date to understand such effects, if any, from enantiomers of environmentally relevant PCBs, nor has any research been done to date on stereoselective effects on atropisomeric PCB metabolites. Until recently, toxic effects from multiple *ortho* substituted PCBs has attracted little attention. The main mode of action for PCBs was regarded to be dioxin-like toxicity of coplanar congeners, agonists of the Ah receptor (Safe 1990). However, recent research has shown that multiple *ortho* substituted PCBs have a host of adverse effects, such as neurotoxicity (Schantz and Widholm 2001), promotion of liver carcinogenicity (Glauert, Robertson, and Silberhorn 2001), antiestrogenicity (Gierthy et al. 2001), and calcium

signaling effects (Pessah and Wong 2001). Similarly, MeSO$_2$-PCBs are known to bind to proteins, induce CYP, and have endocrine-related effects (Letcher et al. 2000). The stereoselectivity of these effects is generally unknown. It is clear that much work must be done to understand stereoselectivity in biological and toxic effects of chiral PCBs and their metabolites.

REFERENCES

Asturias, J. A., E. Diaz, and K. N. Timmis. 1995. "The Evolutionary Relationship of Biphenyl Dioxygenase from Gram-Positive *Rhodococcus globerulus* P6 to Multicomponent Dioxygenases from Gram-Negative Bacteria." *Gene* 156:11–18.

Bakke, J. E., A. L. Bergman, and G. L. Larsen. 1982. "Metabolism of 2,4′,5-Trichlorobiphenyl by the Mercapturic Acid Pathway." *Science* 217:645–47.

Bedard, D. L., and J. F. Quensen III. 1995. "Microbial Reductive Dechlorination of Polychlorinated Biphenyls." In *Microbial Transformation and Degradation of Toxic Organic Contaminants,* edited by L. Y. Young and C. E. Cerniglia, 127–216. New York: Wiley-Liss.

Bedard, D. L., R. E. Wagner, M. J. Brennan, M. L. Haberl, and J. F. Brown Jr. 1987. "Extensive Degradation of Aroclors and Environmentally Transformed Polychlorinated Biphenyls by *Alcaligenes eutrophus* H850." *Appl. Environ. Microbiol.* 53:1094–1102.

Benická, E., R. Novakovsky, J. Hrouzek, and J. Krupčík. 1996. "Multidimensional Gas Chromatographic Separation of Selected PCB Atropisomers in Technical Formulations and Sediments." *J. High Resol. Chromatogr.* 19:95–98.

Berger, U., D. Herzke, and T. M. Sandanger. 2004. "Two Trace Analytical Methods for Determination of Hydroxylated PCBs and Other Halogenated Phenolic Compounds in Eggs from Norwegian Birds of Prey." *Anal. Chem.* 76:441–52.

Bidleman, T. F., L. M. M. Jantunen, K. Wiberg, T. Harner, K. A. Brice, K. Su, et al. 1998. "Soil as a Source of Atmospheric Heptachlor Epoxide." *Environ. Sci. Technol.* 32:1546–48.

Blanch, G. P., A. Glausch, and V. Schurig. 1999. "Determination of the Enantiomeric Ratios of Chiral PCB 95 and 149 in Human Milk Samples by Multidimensional Gas Chromatography with ECD and MS(SIM) Detection." *Eur. Food Res. Technol.* 209:294–96.

Blanch, G. P., A. Glausch, V. Schurig, and M. J. Gonzalez. 1996. "Quantification and Determination of Enantiomeric Ratios of Chiral PCB 95, PCB 132, and PCB 149 in Shark Liver Samples (*C. coelolepis*) from the Atlantic Ocean." *J. High Resol. Chromatogr.* 19:392–96.

Boon, J. P., F. Eijgenraam, J. M. Everaarts, and J. C. Duinker. 1989. "A Structure-Activity Relationship (SAR) Approach towards Metabolism of PCBs in Marine Animals from Different Trophic Levels." *Mar. Environ. Res.* 27:159–76.

Bordajandi, L. R., P. Korytar, J. de Boer, and M. J. Gonzalez. 2005. "Enantiomeric Separation of Chiral Polychlorinated Biphenyls on β-Cyclodextrin Capillary Columns by Means of Heart-Cut Multidimensional Gas Chromatography and Comprehensive Two-Dimensional Gas Chromatography: Application to Food Samples." *J. Sep. Sci.* 28:163–71.

Brenner, R. C., V. S. Magar, J. A. Ickes, E. A. Foote, J. E. Abbott, L. S. Bingler, et al. 2004. "Long-Term Recovery of PCB-Contaminated Surface Sediments at the Sangamo-Weston/ Twelvemile Creek/Lake Hartwell Superfund Site." *Environ. Sci. Technol.* 38:2328–37.

Bright, D. A., S. L. Grundy, and K. J. Reimer. 1995. "Differential Bioaccumulation of Non-Ortho-Substituted and Other PCB Congeners in Coastal Arctic Invertebrates and Fish." *Environ. Sci. Technol.* 29:2504–12.

Chu, S., A. Covaci, K. Haraguchi, S. Voorspoels, K. Van de Vijver, K. Das, et al. 2003a. "Levels and Enatiomeric Signatures of Methyl Sulfonyl PCB and DDE Metabolites in Liver of Harbor Porpoises (*Phocoena phocoena*) from the Southern North Sea." *Environ. Sci. Technol.* 37:4573–78.

Chu, S., A. Covaci, and P. Schepens. 2003b. "Levels and Chiral Signatures of Persistent Organochlorine Pollutants in Human Tissues from Belgium." *Environ. Res.* 93:167–76.

Chu, S., A. Covaci, K. Van de Vijver, W. De Coen, R. Blust, and P. Schepens. 2003c. "Enantiomeric Signatures of Chiral Polychlorinated Biphenyl Atropisomers in Livers of Harbor Porpoises (*Phocoena phocoena*) from the Southern North Sea." *J. Environ. Monit.* 5:521–26.

Crego, A. L., M. A. García, and M. L. Marina. 2000. "Enantiomeric Separation of Chiral Polychlorinated Biphenyls by Micellar Electrokinetic Chromatography Using Mixtures of Bile Salts and Sodium Dodecyl Sulphate with and without γ-Cyclodextrin in the Separation Buffer." *J. Microcol. Sep.* 12:33–40.

de Geus, H. J., P. G. Wester, J. de Boer, and U. A. Th. Brinkman. 2000. "Enantiomer Fractions Instead of Enantiomer Ratios." *Chemosphere* 41:725–27.

Döbler, J., N. Peters, C. Larsson, Å Bergman, E. Geidel, and H. Hühnerfuss. 2002. "The Absolute Structures of Separated PCB-Methylsulfone Enantiomers Determined by Vibrational Circular Dichroism and Quantum Mechanical Calculations." *J. Molec. Struct. Theochem.* 586:159–66.

Drouillard, K. G., and R. J. Norstrom. 2000. "Dietary Absorption Efficiencies and Toxicokinetics of Polychlorinated Biphenyls in Ring Doves Following Exposures to Aroclor Mixtures." *Environ. Toxicol. Chem.* 19:2707–14.

Edwards, S. H., and S. A. Shamsi. 2002. "Chiral Separation of Polychlorinated Biphenyls Using a Combination of Hydroxy-propyl-γ-Cyclodextrin and a Polymeric Chiral Surfactant." *Electrophoresis* 23:1320–27.

Ellerichmann, T., Å Bergman, S. Franke, H. Hühnerfuss, E. Jakobsson, W. A. König et al. 1998. "Gas Chromatographic Enantiomer Separations of Chiral PCB Methyl Sulfons and Identification of Selectively Retained Enantiomers in Human Liver." *Fresenius Environ. Bull.* 7:244–57.

Faller, J., H. Hühnerfuss, W. A. König, R. Krebber, and P. Ludwig. 1991. "Do Marine Bacteria Degrade α-Hexachloro Cyclohexane Stereoselectively?" *Environ. Sci. Technol.* 25:676–78.

Frame, G. M. 1997. "A Collaborative Study of 209 PCB Congeners and 6 Aroclors on 20 Different HRGC Columns: 2, Semi-Quantitative Aroclor Congener Distributions." *Fresenius J. Anal. Chem.* 357:714–22.

García-Ruiz, C., A. L. Crego, and M. L. Marina. 2003. "Comparison of Charged Cyclodextrin Derivatives for the Chiral Separation of Atropisomeric Polychlorinated Biphenyls by Capillary Electrophoresis." *Electrophoresis* 24:2657–64.

García-Ruiz, C., Y. Martín-Biosca, A. L. Crego, and M. L. Marina. 2001. "Rapid Enantiomeric Separation of Polychlorinated Biphenyls by Electrokinetic Chromatography Using Mixtures of Neutral and Charged Cyclodextrin Derivatives." *J. Chromatogr.* A 910:157–64.

García-Ruiz, C., R. Andrés, J. L. Valera, F. Laborda, and M. L. Marina. 2002. "Monitoring the Stereoselectivity of Biodegradation of Chiral Polychlorinated Biphenyls Using Electrokinetic Chromatography." *J. Sep. Sci.* 25:17–22.

Gierthy, J. F., K. F. Arcaro, D. D. Vakharia, and Y. Yang. 2001. "Use of Estrogen-Induced Postconfluent Cell Proliferation and Focus Development in Human MCF-7 Breast Cells as an Assay to Characterize PCB Estrogen Modulation." In *PCBs: Recent Advances in Environmental Toxicology and Health Effects,* edited by L. W. Robertson and L. G. Hansen, 281–84. Lexington: University Press of Kentucky.

Glauert, H. P., L. W. Robertson, and E. M. Silberhorn. 2001. "PCBs and Tumor Promotion." In *PCBs: Recent Advances in Environmental Toxicology and Health Effects,* edited by L. W. Robertson and L. G. Hansen, 355–72. Lexington: University Press of Kentucky.

Glausch, A., G. P. Blanch, and V. Schurig. 1996. "Enantioselective Analysis of Chiral Polychlorinated Biphenyls in Sediment Samples by Multidimensional Gas Chromatography-Electron-Capture Detection after Steam Distillation-Solvent Extraction and Sulfur Removal." *J. Chromatgr. A.* 723:399–404.

Glausch, A., J. Hahn, V. Schurig. 1995. "Enantioselective Determination of Chiral 2,2′,3,3′,4,6′-Hexachlorobiphenyl (PCB 132) in Human Milk Samples by Multidimensional Gas Chromatography/Electron Capture Detection and by Mass Spectrometry." *Chemosphere* 30:2079–85.

Glausch, A., G. J. Nicholson, M. Fluck, and V. Schurig. 1994. "Separation of the Enantiomers of Stable Atropisomeric Polychlorinated Biphenyls (PCBs) by Multidimensional Gas Chromatography on Chirasil-Dex." *J. High Resol. Chromatogr.* 17:347–49.

Haglund, P. 1996a. "Enantioselective Separation of Polychlorinated Biphenyl Atropisomers Using Chiral High-Performance Liquid Chromatography." *J. Chromatogr. A.* 724:219–28.

———. 1996b. "Isolation and Characterization of Polychlorinated Biphenyl (PCB) Atropisomers." *Chemosphere* 32:2133–40.

Haglund, P., and K. Wiberg. 1996. "Determination of the Gas Chromatographic Elution Sequences of the (+)- and (–)-Enantiomers of Stable Atropisomeric PCBs on Chirasil-Dex." *J. High Resol. Chromatogr.* 19:373–76.

Hardt, I. H., C. Wolf, B. Gehrcke, D. H. Hochmuth, B. Pfaffenberger, H. Hühnerfuss, et al. 1994. "Gas Chromatographic Enantiomer Separation of Agrochemicals and Polychlorinated Biphenyls (PCBs) Using Modified Cyclodextrins." *J. High Resol. Chromatogr.* 17:859–64.

Harju, M. T., and P. Haglund. 1999. "Determination of the Rotational Energy Barriers of Atropisomeric Polychlorinated Biphenyls." *Fresenius J. Anal. Chem.* 364:219–23.

———. 2001. "Comprehensive Two-Dimensional Gas Chromatography (GCxGC) of Atropisomeric PCBs, Combining a Narrow Bore β-Cyclodextrin Column and a Liquid Crystal Column." *J. Microcol. Sep.* 13:300–305.

Harju, M., Å. Bergman, M. Olsson, A. Roos, and P. Haglund. 2003. "Determination of Atropisomeric and Planar Polychlorinated Biphenyls, Their Enantiomeric Fractions and Tissue Distribution in Gray Seals Using Comprehnsive 2D Gas Chromatography." *J. Chromatogr. A.* 1019:127–42.

Harner, T., K. Wiberg, and R. Norstrom. 2000. "Enantiomer Fractions Are Preferred to Enantiomer Ratios for Describing Chiral Signatures in Environmental Analysis." *Environ. Sci. Technol.* 34:218–20.

Hoekstra, P. F., C. S. Wong, T. M. O'Hara, K. R. Solomon, S. A. Mabury, and D. C. G. Muir. 2002. "Enantiomer-Specific Accumulation of PCB Atropisomers in the Bowhead Whale (*Balaena mysticetus*)." *Environ. Sci. Technol.* 36:1419–25.

Hoekstra, P. F., T. M. O'Hara, H. Karlsson, K. R. Solomon, and D. C. G. Muir. 2003. "Enantiomer-Specific Biomagnification of α-Hexachlorocyclohexane and Selected Chiral Chlordane-Related Compounds within an Arctic Marine Food Web." *Environ. Toxicol. Chem.* 22:2482–91.

Hühnerfuss, H. 2000. "Chromatographic Enantiomer Separation of Chiral Xenobiotics and Their Metabolites: A Versatile Tool for Process Studies in Marine and Terrestrial Ecosystems." *Chemosphere* 40:913–19.

Hühnerfuss, H., B. Pfaffenberger, B. Gehrcke, L. Karbe, W. A. König, and O. Landgraff. 1995. "Stereochemical Effects of PCBs in the Marine Environment: Seasonal Variation of Coplanar and Atropisomeric PCBs in Blue Mussels (*Mytilus edulis* L.) of the German Bight." *Mar. Poll. Bull.* 30: 332–40.

Hühnerfuss, H, Å Bergman, C. Larsson, N. Peters, and J. Westendorf. 2003. "Enantioselective Transformation of Atropisomeric PCBs or of Their Methylsulfonyl Metabolites by Rat Hepatocytes?" *Organohalogen Compounds* 62:265–68.

Iwata, H., K. Yoshinari, M. Negishi, and J. J. Stegeman. 2002. "Species-Specific Responses for Constitutively Active Receptor (CAR)-CYP2B Coupling: Lack of CYP2B Inducer-Responsive Nuclear Translocation of CAR in Marine Teleost, Scup (*Stenotomus chrysops*)." *Comp. Biochem. Physiol.* 131: 501–10.

Jaus, A., and M. Oehme. 2000. "Gas Chromatographic Separation of Atropisomeric Polychlorinated Biphenyls and Methylsulfonylated Derivatives with Partially Ethylated γ-Cyclodextrin." *Chromatographia* 52:242–44.

———. 2001. "Consequences of Variable Purity of Heptakis (2,3,6-tri-O-methyl)-β-Cyclodextrin Determined by Liquid Chromatography–Mass Spectrometry on the Enantioselective Separation of Polychlorinated Compounds." *J. Chromatogr. A.* 905:59–67.

Jiménez, O., B. Jiménez, and M. J. Gonzalez. 2000. "Isomer-Specific Polychlorinated Biphenyl Determination in Cetaceans from the Mediterranean Sea: Enantioselective Occurrence of Chiral Polychlorinated Biphenyl Congeners." *Environ. Toxicol. Chem.* 19:2653–60.

Jin, Z., and H. L. Jin. 1994. "2,6-di-o-Pentyl-3-o-Propionyl-γ-Cyclodextrin as an Enantiomeric Stationary Phase for Capillary Gas Chromatography." *Chromatographia* 38:22–28.

Junge, M., and W. A. König. 2003. "Selectivity Tuning of Cyclodextrin Derivatives by Specific Substitution." *J. Sep. Sci.* 26:1607–14.

Kaiser, K. L. E. 1974. "On the Optical Activity of Polychlorinated Biphenyls." *Environ. Poll.* 7:93–101.

Kidd, K. A., D. W. Schindler, R. H. Hesslein, and D. C. G. Muir. 1998. "Effects of Trophic Position and Lipid on Organochlorine Concentrations in Fishes from Subarctic Lakes in Yukon Territory." *Can. J. Fish Aquat. Sci.* 55:869–81.

Klasson-Wehler, E., Å Bergman, M. Athanasiadou, J. P. Ludwig, H. J. Auman, K. Kannan et al. 1998. "Hydroxylated and Methylsulfonyl Polychlorinated Biphenyl Metabolites in Albatrosses from Midway Atoll, North Pacific Ocean." *Environ. Toxicol. Chem.* 17:1620–25.

Kleinow, K. M., M. J. Melancon, and J. J. Lech. 1987. "Biotransformation and Induction: Implications for Toxicity, Bioaccumulation, and Monitoring of Environmental Xenobiotics in Fish." *Environ. Health Perspect.* 71:105–19.

Kleinow, K. M., M. L. Haasch, D. E. Williams, and J. J. Lech. 1990. "A Comparison of Hepatic P450 Induction in Rat and Trout (*Oncorhynchus mykiss*): Delineation of the Site of Resistance of Fish to Phenobarbital-Type Inducers." *Comp. Biochem. Physiol.* 96:259–70.

Klobes, U., W. Vetter, B. Luckas, K. Skirnisson, and J. Plötz. 1998. "Levels and Enantiomeric Ratios of α-HCH, Oxychlordane, and PCB 149 in Blubber of Harbour Seals (*Phoca vitulina*) and Grey Seals (*Halichoerus grypus*) from Iceland and Further Species." *Chemosphere* 37:2501–2512.

König, W. A., B. Gehrcke, T. Runge, and C. Wolf. 1993. "Gas Chromatographic Separation of Atropisomeric Alkylated and Polychlorinated Biphenyls Using Modified Cyclodextrins." *J. High Resol. Chromatogr.* 16:376–78.

Kucklick, J. R., and J. E. Baker. 1998. "Organochlorines in Lake Superior's Food Web." *Environ. Sci. Technol.* 32:1192–98.

Larsson, C., T. Ellerichmann, H. Hühnerfuss, and Å. Bergman. 2002. "Chiral PCB Methyl Sulfones in Rat Tissues after Exposure to Technical PCBs." *Environ. Sci. Technol.* 36:2833–38.

Larsson, C., K. Norström, I. Athanansiadis, A. Bignert, W. A. König, and Å. Bergman. 2004. "Enantiomeric Specificity of Methylsulfonyl-PCBs and Distribution of Bis(4-Chlorophenyl) Sulfone, PCB, and DDE Methyl Sulfones in Grey Seal Tissues." *Environ. Sci. Technol.* 38:4950–55.

Lehmler, H., and L. W. Robertson. 2001. "Atropisomers of PCBs." In *PCBs: Recent Advances in Environmental Toxicology and Health Effects,* edited by L. W. Robertson and L. G. Hansen, 61–65. Lexington: University Press of Kentucky.

Lehmler, H., D. J. Price, A. W. Garrison, W. J. Birge, and L. W. Robertson. 2003. "Distribution of PCB 84 Enantiomers in C57Bl/6 Mice." *Fresenius Environ. Bull.* 12:254–60.

Lehmler, H., L. W. Robertson, A. W. Garrison, and P. R. S. Kodavanti. 2005. "Effects of PCB 84 Enantiomers on [^{3}H]-Phorbol Ester Binding in Rat Cerebellar Granule Cells and ^{45}Ca^{2+}-Uptake in Rat Cerebellum." *Toxicol. Lett.* 156:391–400.

Letcher, R. J., R. J. Norstrom, D. C. G. Muir, C. D. Sandau, K. Koczanski, R. Michaud, et al. 2000. "Methylsulfone Polychlorinated Biphenyl and 2,2-Bis(Chlorophyenyl)-1,1-Dichloroethylene Metabolites in Beluga Whale (*Delphinapterus leucas*) from the St. Lawrence River Estuary and Western Hudson Bay, Canada." *Environ. Toxicol. Chem.* 19: 1378–88.

Lin, W., F. Chang, and C. Kuei. 1999. "Separation of Atropisomeric Polychlorinated Biphenyls by Cyclodextrin Modified Micellar Electrokinetic Chromatography." *J. Microcol. Sep.* 11:231–38.

Magnusson, J., L. G. Blomberg, S. Claude, R. Tabacchi, A. Saxer, and S. Schürch. 2000. "Gas Chromatographic Enantiomer Separation of Atropisomeric PCBs Using Modified Cyclodextrins as Chiral Phases." *J. High Resol. Chromatogr.* 23:619–27.

Magrans, J. O., J. L. Alonso-Prados, and J. M. García-Baudín. 2002. "Importance of Considering Pesticide Stereoisomerism —Proposal of a Scheme to Apply Directive 91/414/EEC Framework to Pesticide Active Substances Manufactured as Isomeric Mixtures." *Chemosphere* 49:461–69.

Mannschreck, A., N. Pustet, L. W. Robertson, F. Oesch, and M. Püttmann. 1985. "Enantiomers of Polychlorinated Biphenyls Semipreparative Enrichment by Liquid Chromatography." *Liebigs Ann. Chem.,* 2101–3.

Marina, M. L., I. Benito, J. C. Díez-Masa, and M. J. González. 1996a. "Chiral Separation of Polychlorinated Biphenyls by Micellar Electrokinetic Chromatography with γ-Cyclodextrin as Modifier in the Separation Buffer." *Chromatographia* 42: 269–72.

———. 1996b. "Separation of Chiral Polychlorinated Biphenyls by Micellar Electrokinetic Chromatography Using β- and γ-Cyclodextrin Mixtures in the Separation Buffer." *J. Chromatogr. A.* 752:265–70.

Mattina, M. I., J. White, B. Eitzer, and W. Iannucci-Berger. 2002. "Cycling of Weathered Chlordane Residues in the Environment: Compositional and Chiral Profiles in Contiguous Soil, Vegetation, and Air Compartments." *Environ. Toxicol. Chem.* 21:281–88.

Moisey, J., A. T. Fisk, K. A. Hobson, and R. J. Norstrom. 2001. "Hexachlorocyclohexane (HCH) Isomers and Chiral Signatures of α-HCH in the Arctic Marine Food Web of the Northwater Polynya." *Environ. Sci. Technol.* 35:1920–27.

Nezel, T., and F. Müller-Plathe, M. D. Müller, and H.-R. Buser. 1997. "Theoretical Considerations about Chiral PCBs and Their Methylthio and Methylsulfonyl Metabolites Being Possibly Present as Stable Enantiomers." *Chemosphere* 35: 1895–1906.

Pakdeesusuk, U., W. J. Jones, C. M. Lee, A. W. Garrison, W. L. O'Neill, J. T. Coates, et al. 2003. "Changes in Enantiomeric Fraction (EF) during Microbial Reductive Dechlorination of PCB 132, PCB 149, and Aroclor 1254 in Lake Hartwell Sediment Microcosms." *Environ. Sci. Technol.* 37:1100–1107.

Pessah, I. N., and P. W. Wong. 2001. "Etiology of PCB Neurotoxicity: From Molecules to Cellular Dysfunction." In *PCBs: Recent Advances in Environmental Toxicology and Health Effects,* edited by L. W. Robertson and L. G. Hansen, 179–85. Lexington: University Press of Kentucky.

Pham-Tuan, H., C. Larsson, F. Hoffmann, Å Bergman, M. Fröba, and H. Hühnerfuss. 2005. "Enantioselective Semipreparative HPLC Separation of PCB Metabolites and Their Absolute Structure Elucidation Using Electronic and Vibrational Circular Dichroism." *Chirality* 17:266–80.

Püttmann, M., A. Mannschreck, F. Oesch, and L. Robertson. 1989. "Chiral Effects in the Induction of Drug-Metabolizing Enzymes Using Synthetic Atropisomers of Polychlorinated Biphenyls (PCBs)." *Biochem. Pharmacol.* 38:1345–52.

Püttmann, M., M. Arand, F. Oesch, A. Mannschreck, and L. W. Robertson. 1990. "Chirality and the Induction of Xenobiotic-Metabolizing Enzymes: Effects of the Atropisomers of the Polychlorinated Biphenyl 2,2',3,4,4',6-Hexachlorobiphenyl." In *Chirality and Biological Activity*, edited by B. Holmstedt, H. Frank, and B. Testa, 177–84. New York: Liss.

Ramos, L., L. M. Hernández, and M. J. González. 1999. "Simultaneous Separation of Coplanar and Chiral Polychlorinated Biphenyls by Off-Line Pyrenyl-Silica Liquid Chromatography and Gas Chromatography: Enantiomeric Ratios of Chiral Congeners." *Anal. Chem.* 71:70–77.

Rasmussen, J. B., D. J. Rowan, D. R. S. Lean, and J. H. Carey. 1990. "Food-Chain Structure in Ontario Lakes Determines PCB Levels in Lake Trout (*Salvelinus namaycush*) and Other Pelagic Fish." *Can. J. Fish Aquat. Sci.* 47:2030–38.

Reich, S., and V. Schurig. 1999. "Enantiomerentrennung Atropisomer PCB Mittels HPLC." *GIT Spezial* 1:15–16.

Reich, S., B. Jiminez, L. Marsili, L. M. Hernández, V. Schurig, and M. J. González. 1999. "Congener Specific Determination and Enantiomeric Ratios of Chiral Polychlorinated Biphenyls in Striped Dolphins (*Stenella coeruleoalba*) from the Mediterranean Sea." *Environ. Sci. Technol.* 33:1787–93.

Rhee, G. Y., B. Bush, C. M. Bethoney, A. DeNucci, H. M. Oh, and R. C. Sokol. 1993. "Reductive Dechlorination of Aroclor 1242 in Anaerobic Sediments: Pattern, Rate, and Concentration Dependence." *Environ. Toxicol. Chem.* 12:1025–32.

Robson, M., and S. Harrad. 2004. "Chiral PCB Signatures in Air and Soil: Implications for Atmospheric Source Apportionment." *Environ. Sci. Technol.* 38:1662–66.

Rodman, L. E., S. I. Shedlofsky, A. Mannschreck, M. Püttmann, A. T. Swim, and L. W. Robertson. 1991. "Differential Potency of Atropisomers of Polychlorinated Biphenyls on Cytochrome P450 Induction and Uroporphyrin Accumulation in the Chick Embryo Hepatocyte Culture." *Biochem. Pharmacol.* 41:915–22.

Ruppe, S., W. Vetter, B. Luckas, and G. Hottinger. 2000. "Application of Well-Defined β-Cyclodextrins for the Enantioseparation of Compounds of Technical Toxaphene and Further Organochlorines." *J. Microcol. Sep.* 12:541–49.

Safe, S. 1990. "Polychlorinated Biphenyls (PCBs), Dibenzo-*p*-Dioxins (PCDDs), Dibenzofurans (PCDFs), and Related Compounds: Environmental and Mechanistic Considerations Which Support the Development of Toxic Equivalency Factors (TEFs)." *Crit. Rev. Toxicol.* 21:51–88.

Schantz, S. L., and J. J. Widholm. 2001. "Effects of PCB Exposure on Neurobehavorial Function in Animal Models." In *PCBs: Recent Advances in Environmental Toxicology and Health Effects*, edited by L. W. Robertson and L. G. Hansen, 221–40. Lexington: University Press of Kentucky.

Schurig, V. 2001. "Separation of Enantiomers by Gas Chromatography." *J. Chromatogr. A.* 906:275–99.

Schurig, V., and A. Glausch. 1993. "Enantiomer Separation of Atropisomeric Polychlorinated Biphenyls (PCBs) by Gas Chromatography on Chirasil-Dex." *Naturwissenschaften* 80:468–69.

Schurig, V., A. Glausch, and M. Fluck. 1995. "On the Enantiomerization Barrier of Atropisomeric 2,2',3,3',4,6'-Hexachlorobiphenyl (PCB 132)." *Tetrahedron Asymmetry* 6:2161–64.

Serrano, R., M. Fernández, R. Rabanal, M. Hernández, and M. J. Gonzalez. 2000. "Congener-Specific Determination of Polychlorinated Biphenyls in Shark and Grouper Livers from the Northwest African Atlantic Ocean." *Arch. Environ. Contam. Toxicol.* 38:217–24.

Singer, A. C., C. S. Wong, and D. E. Crowley. 2002. "Differential Enantioselective Transformation of Atropisomeric Polychlorinated Biphenyls by Multiple Bacterial Strains with Differing Inducing Compounds." *Appl. Environ. Microbiol.* 68:5756–59.

Stapleton, H. M., R. J. Letcher, and J. E. Baker. 2001. "Metabolism of PCBs by the Deepwater Sculpin (*Myoxocephalus thompsoni*)." *Environ. Sci. Technol.* 35:4747–52.

Stegeman, J. J., and P. J. Klopper-Sams. 1987. "Cytochrome P-450 Isozymes and Monooxygenase Activity in Aquatic Animals." *Environ. Health Perspect.* 71:87–95.

Stegeman, J. J., B. R. Woodin, and D. J. Waxman. 1990. "Structural Relatedness of Mammalian Cytochromes P450 IIB and Cytochrome P450 B from the Marine Fish Scup (*Stenotomus chrysops*)." *FASEB J.* 4:A739.

Stow, C. A., S. R. Carpenter, L. A. Eby, J. F. Amrhein, and R. J. Hesselberg. 1995. "Evidence That PCBs Are Approaching Stable Concentrations in Lake Michigan Fishes." *Ecol. Appl.* 5:248–60.

Ulrich, E. M., D. R. Helsel, and W. T. Foreman. 2003. "Complications with Using Ratios for Environmental Data: Comparing Enantiomeric Ratios (ERs) and Enantiomer Fractions (EFs)." *Chemosphere* 53:531–38.

Vetter, W., and V. Schurig. 1997. "Enantioselective Determination of Chiral Organochlorine Compounds in Biota by Gas Chromatography on Modified Cyclodextrins." *J. Chromatogr. A.* 774:143–75.

Vetter, W., U. Klobes, K. Hummert, and B. Luckas. 1997a. "Gas Chromatographic Separation of Chiral Organochlorines on Modified Cyclodextrin Phases and Results of Marine Biota Samples." *J. High Resol. Chromatogr.* 20:85–93.

Vetter, W., U. Klobes, B. Luckas, and G. Hottinger. 1997b. "Enantiomer Separation of Selected Atropisomeric Polychlorinated Biphenyls Including PCB 144 on *Tert*-Butyldimethylsilylated β-Cyclodextrin." *J. Chromatogr. A.* 769:247–52.

———. 1999. "Use of *6-O-Tert.*-Butyldimethylsilylated β-Cyclodextrins for the Enantioseparation of Chiral Organochlorine Compounds." *J. Chromatogr. A.* 846:375–81.

Warner, N. A., R. J. Norstrom, C. S. Wong, and A. T. Fisk. 2005. "Enantiomeric Fractions of Chiral PCBs Provide Insights on Biotransformation Capacity of Arctic Biota." *Environ. Toxicol. Chem.* 24:2763–67.

Wiberg, K., R. Letcher, C. Sandau, J. Duffe, R. Norstrom, P. Haglund, et al. 1998. "Enantioselective Gas Chromatography/Mass Spectrometry of Methylsulfonyl PCBs with Application to Arctic Marine Mammals." *Anal. Chem.* 70:3845–52.

Williams, A. 1996. "Opportunities for Chiral Agrochemicals." *Pestic. Sci.* 46:3–9.

Wong, C. S., and A. W. Garrison. 2000. "Enantiomer Separation of Polychlorinated Biphenyl Atropisomers and Polychlorinated Biphenyl Retention Behavior on Modified Cyclodextrin Capillary Gas Chromatography Columns." *J. Chromatogr. A.* 866:213–20.

Wong, C. S., A. W. Garrison, and W. T. Foreman. 2001. "Enantiomeric Composition of Chiral Polychlorinated Biphenyl Atropisomers in Aquatic Bed Sediment." *Environ. Sci. Technol.* 35:33–39.

Wong, C. S., A. W. Garrison, P. D. Smith, and W. T. Foreman. 2001. "Enantiomeric Composition of Chiral Polychlorinated Biphenyl Atropisomers in Aquatic and Riparian Biota." *Environ. Sci. Technol.* 35:2448–54.

Wong, C. S., P. F. Hoekstra, H. Karlsson, S. M. Backus, S. A. Mabury, and D. C. G. Muir. 2002a. "Enantiomer Fractions of Chiral Organochlorine Pesticides and Polychlorinated Biphenyls in Standard and Certified Reference Materials." *Chemosphere* 49:1339–47.

Wong, C. S., F. Lau, M. Clark, S. A. Mabury, and D. C. G. Muir. 2002b. "Rainbow Trout (*Oncorhynchus mykiss*) Can Elimi-

nate Chiral Organochlorine Compounds Enantioselectively." *Environ. Sci. Technol.* 36:1257–62.

Wong, F., M. Diamond, J. Truong, M. Robson, and S. Harrad. 2003. "Chirality as Indication for PCB Accumulation in Forest Soils along an Urban-Rural Gradient." *Organohalogen Compounds* 62:301–4.

Wong, C. S., S. A. Mabury, D. M. Whittle, S. M. Backus, C. Teixeira, D. S. DeVault, et al. 2004a. "Polychlorinated Biphenyls in Lake Superior: Chiral Congeners and Biotransformation in the Aquatic Food Web." *Environ. Sci. Technol.* 38:84–92.

Wong, C. S., E. Z. Shao, C. Darling, A. Baitz, P. F. Hoekstra, M. Clark, et al. 2004b. "Bioaccumulation and Elimination of Chiral Organochlorine Compounds in Low Trophic Level Organisms." *Organohalogen Compounds* 62:269–72.

Wong, C. S., U. Pakdeesusuk, J. A. Morrissey, C. M. Lee, J. T. Coates, A. W. Garrison, et al. 2005. "Enantiomeric Composition of Chiral Polychlorinated Biphenyl Atropisomers in Dated Sediment Cores." *Environ. Toxicol. Chem.* 26: 254–63.

4

The Importance of Atmospheric Interactions to PCB Cycling in the Hudson and Delaware River Estuaries

Lisa A. Totten, *Rutgers University*

ABBREVIATIONS

DRE   Delaware River estuary
HE   New York–New Jersey harbor estuary
MW   Molecular weight
NJADN New Jersey Atmospheric Deposition Network
PCB   Polychlorinated biphenyl
TMDL   Total maximum daily load
WQS   Water-quality standards

Recent efforts to establish total maximum daily loads (TMDLs) for polychlorinated biphenyls (PCBs) in the Hudson River and Delaware River estuaries have vastly improved our understanding of the cycling of PCBs in these systems and have increased awareness of the importance of the atmosphere to PCB cycling. While the importance of the atmosphere as a sink for PCBs (via air-water exchange and volatilization) is relatively well understood, the atmosphere as a source of PCBs is of significant concern in achieving the 1,000- to 10,000-fold reductions in PCBs loads required by the TMDLs. Various research efforts have been conducted to assess atmospheric cycling of PCBs in these systems, including the New Jersey Atmospheric Deposition Network (NJADN), which included seven monitoring locations in the Delaware River estuary. Results indicate that while the atmosphere acts as net sink for PCBs with <5 chlorines via volatilization from the water column, the atmosphere is a net source of PCBs containing >5 chlorines. The size of the PCB load from the atmosphere is orders of magnitude greater than the TMDL recently established in the Delaware River, a situation that is likely to be repeated in watersheds throughout the United States. Furthermore, since a substantial fraction of PCBs in stormwater and tributary loads arise from indirect atmospheric deposition, efforts to track down atmospheric PCB sources should be included in plans to implement PCB TMDLs nationwide.

INTRODUCTION

In the United States, 794 watersheds are on the list mandated by section 303(d) of the Clean Water Act for impairment due to polychlorinated biphenyls (PCBs) (EPA 2004). Development of TMDLs will be required for a large proportion of these impaired waterways. A TMDL for PCBs in the Delaware River estuary (DRE) was adopted on December 15, 2003, after a massive effort to gather data on the sources and sinks of PCBs in the tidal portion of the river south of Trenton, New Jersey (Fikslin and Suk 2003). A similar effort is underway in the tidal Hudson River, more correctly known as the New York–New Jersey harbor estuary (HE). In both systems, PCBs were measured in the operationally defined dissolved phase, suspended particles, surface sediment, and regional atmosphere. Atmospheric PCB levels were characterized via the establishment of monitoring stations associated with the New Jersey Atmospheric Deposition Network (NJADN), which is operated by researchers at Rutgers University. Operating in various stages from 1997 to the present, the NJADN included 13 sites in New Jersey, Pennsylvania, and Delaware (fig. 4.1). Additionally, simultaneous air and water samples were collected for PCB analysis in the HE and DRE for the purpose of calculating air-water exchange fluxes. These efforts have clarified the role of the atmosphere as both a source and sink for PCBs in these estuaries.

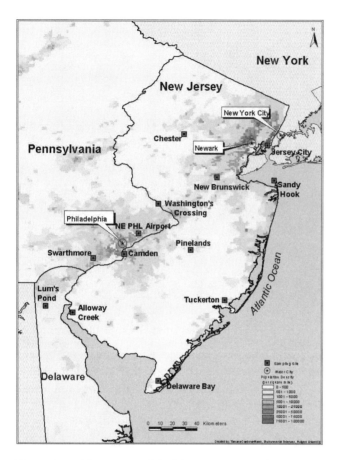

Figure 4.1. The 13 sites of the New Jersey Atmospheric Deposition Network (squares).

These two estuaries are excellent model systems since they are well characterized and are typical of most urban estuaries in the United States. The DRE is one of the first watersheds in the United States to adopt a TMDL for PCBs based on a state-of-the-art water-quality model, and it therefore serves as a test case for the rest of the nation. This chapter will discuss the role of the atmosphere in supporting and controlling PCB concentrations in these estuaries based on data from NJADN and other sources. The conclusions drawn will often be applicable to surface waters throughout the United States.

PCB LEVELS IN THE DRE AND HE

The Hudson River has been heavily contaminated by PCBs from a variety of sources, including wastewater treatment plant effluents, stormwater, and industrial discharges. The most prominent source of PCBs in the upper Hudson River (above Troy, New York) is the former General Electric manufacturing facilities that released between 95,000 and 603,000 kg to the river from about 1947 to 1977 (EPA 2001). PCBs from the upper Hudson represent about half

of the PCBs entering the HE (Totten 2005; Farley et al. 1999; TAMS 1997). In the northern portions of the HE in the Hudson River itself, dissolved ΣPCB concentrations are typically 20–30 ng/L (Farley et al. 1999), while in the southern portion of the estuary in Raritan Bay, large volumes of seawater lowers the concentrations to about 2 ng/L (Totten et al. 2001). In the DRE, dissolved ΣPCBs are generally less than 1 ng/L, while particle-bound ΣPCBs are generally less than 2 ng/L (Fikslin and Suk 2003). The sources of PCBs in the DRE are dominated by urban runoff and inputs from contaminated sites. Point discharges, tributaries, combined sewer overflows, and atmospheric deposition are also important sources of PCBs in the DRE. Therefore the DRE is perhaps more representative than the HE of a generic water body that has been impacted by urban and industrial PCB sources.

Atmospheric PCB levels display a strong correlation with urbanization. For example, PCB concentrations in Camden, New Jersey, are about 20 times higher than those in remote areas such as the New Jersey Pinelands (Totten et al. 2004). In great portions of the NJADN area, which encompass forested and coastal environments as well as suburban areas, gas-phase ΣPCB concentrations are essentially the same (averaging 150–220 pg m^{-3}), representing a regional background PCB concentration. This spatial trend suggests that atmospheric PCBs arise from highly localized, urban sources that influence atmospheric concentrations and deposition fluxes over a distance of a few tens of km. Thus while the urban PCB signal in the atmosphere is diluted to background concentrations within about 40 km, the urban signal in rivers may persist for hundreds of km before becoming diluted by clean water.

EQUILIBRIUM CONSIDERATIONS

The levels of PCBs in the water columns of both rivers greatly exceed established water-quality standards (WQS). The New York state WQS for ΣPCBs is 1 pg/L (New York 2000), while a WQS of 7.9 pg/L drives the TMDL for PCBs in the DRE (Fikslin and Suk 2003). In the next section, atmospheric deposition of PCBs to the HE and DRE will be quantified by applying deposition velocities to atmospheric concentrations in order to determine the size of the atmospheric load to these estuaries. An even simpler calculation may be performed, however, to determine the concentration of PCBs in the water column in equilibrium with the atmosphere. The assumption of equilibrium is justifiable because the characteristic time for air-water exchange in these systems is about 10 days, while the residence times of water within the DRE and HE are about 30–70 days. The characteristic time for air-water exchange is obtained by dividing the depth of the water body (about 5 m for these estuaries) by the mass transfer coefficient for air-water exchange of PCBs (about 0.5 m/d; Farley et al. 1999). The equilibrium partitioning of a chemical between air and water is described by the Henry's Law Constant (K_{aw}):

$$K_{aw} = \frac{C_a}{C_w} \qquad (1)$$

where C_a and C_w represent the concentrations of the chemical in air and water, respectively. C_w was calculated using the K_{aw} values of Bamford et al. (2000, 2002) and adjusting them for the average year-round temperature in the DRE/HE region (about 15°C). In equilibrium with the regional background concentration of PCBs, the dissolved-phase PCB concentration would be 29 pg/L, in excess of established WQSs. The standards apply to whole-water concentrations, however, thereby including PCBs in the colloidal and suspended particle phases. The *whole-water* concentration of PCBs in equilibrium with the atmosphere at Washington's Crossing may be much higher, on the order of 80 pg/L or more under the conditions observed in the DRE and HE, where total suspended matter may be 5 mg/L or higher. In addition, atmospheric concentrations of PCBs in urban areas such as Camden and Jersey City, New Jersey, are as much as 20 times higher than regional background, leading to a corresponding increase in the whole-water concentration at equilibrium. Therefore neither the HE nor the DRE can meet current WQS due to atmospheric inputs alone. This is generally true of water bodies throughout the United States wherever average gas-phase PCB concentrations exceed about 3 pg/m^3.

DIRECT ATMOSPHERIC DEPOSITION

The equilibrium calculations above suggest that the atmospheric inputs to the DRE exceed the TMDL there. These atmospheric inputs can be more accurately quantified by modeling the NJADN data on atmospheric PCB concentrations. PCBs may enter a water body from the atmosphere via three pathways. First, they may be present on atmospheric particles that deposit to the water surface (dry particle deposition). Second, they may be present in rain or snow that falls on the water surface (wet deposition). And third, they may be in the gas phase and may absorb into the water phase (gaseous deposition). This third process, unlike the first two, is reversible, such that PCBs in the dissolved phase in water may volatilize, that is, leave the water column and enter the atmosphere as gases. This reversible process is referred to as "air-water exchange." The atmosphere may therefore act as either a net source, depositing PCBs into the water

column, or a net sink, removing PCBs from the water column. Whether the atmosphere acts as a net source or sink for PCBs depends on the size and direction of the imbalance between the atmospheric and water column concentrations of PCBs, a factor that can change in time and space within the system. All three processes are characterized by calculating fluxes, which represent movement of mass across a unit of surface area in a unit of time. The fluxes are then multiplied by surface area to obtain yearly loads (or losses). The area of the HE, from the Newburgh Bridge south, is about 811 km^2 (Totten 2005), whereas the area of the Delaware River from Trenton south is about 2,000 km^2 (Fikslin and Suk 2003).

The dry deposition flux (F_{dry}) is the concentration of PCBs in each particle size fraction (C_i) multiplied by the deposition velocity of that size fraction ($V_{d,i}$), summed over all size fractions:

$$F = \sum_{i=1}^{n} C_i V_{d,i} \qquad (2)$$

Because it is difficult and expensive to collect size-fractionated particle samples, NJADN used high-volume air samplers to collect a bulk atmospheric particle sample. The dry deposition flux is calculated from the resulting data by applying a single deposition velocity (V_d) to the bulk particle phase:

$$F_{dry} = V_d C_{part} \qquad (3)$$

where F_{dry} is the flux in ng m^{-2} d^{-1} and C_{part} is the average particle concentration of the chemical in ng m^{-3}. A value for the V_d of 0.5 cm s^{-1} is selected since it reflects the disproportionate influence that large particles have on atmospheric deposition, especially in urbanized and industrialized regions (Franz, Eisenreich, and Holsen 1998; Pirrone, Keeler, and Holsen 1995a and 1995b). Other studies have employed a deposition velocity of 0.2 cm s^{-1} (Hillery et al. 1998; Miller et al. 2001). Thus the choice of V_d represents by far the largest source of error in the calculation of dry deposition flux. Moreover, V_d may differ between sites due to changes in particle characteristics and local meteorology.

Wet deposition fluxes (F_{wet}) were estimated at each site as follows:

$$F_{wet} = C_{VWM} P \qquad (4)$$

where C_{VWM} is the volume-weighted mean concentration of the chemical in precipitation, and P is the precipitation intensity (m y^{-1}).

Calculations of absorptive gas fluxes ($F_{gas,abs}$) are described in previous NJADN publications (Totten et al. 2001, 2004) and will be summarized here. The modified two-layer model used assumes that the rate of gas transfer is controlled by the compound's ability to diffuse across the water and air layer on either side of the air-water interface. The net flux calculation is

$$F_{gas,\,net} = K_{OL} \left(C_w - \frac{C_a}{K_{aw}} \right) \qquad (5)$$

where $F_{gas,net}$ is the net flux (ng m^{-2} d^{-1}), K_{OL} (m d^{-1}) is the overall mass transfer coefficient, and ($C_w - C_a / K_{aw}$) describes the concentration gradient (ng m^{-3}); C_w (ng m^{-3}) is the dissolved phase concentration of the compound in water; C_a (ng m^{-3}) is the gas phase concentration of the compound in air. For PCBs, values for K_{aw} and its temperature dependence (DH$_{aw}$) were taken from Bamford et al. (2000, 2002). K_{aw} also depends on the salinity of the water. The maximum salinity in the HE (~0.35 M) occurs in Raritan Bay. At this level, the correction for salinity causes about a 30 percent increase in the K_{aw} for PCBs (Totten et al. 2001). The net flux is divided into volatilization (F$_{gas,vol}$) and absorption (F$_{gas,abs}$) terms as follows:

$$F_{gas,vol} = K_{OL} C_w \qquad (6)$$
$$F_{gas,abs} = K_{OL} C_a / K_{aw} \qquad (7)$$

Calculation of these fluxes requires knowledge of air and water temperature and wind speed. An increase in K_{aw} due to salinity will decrease the gas absorption flux directly due to the presence of K_{aw} in equation 7 and indirectly as a result of a smaller increase in K_{OL}. It is important to remember that only truly dissolved PCBs can volatilize. Once sorbed to particles or colloids, PCB molecules are no longer available for air-water exchange.

In the atmosphere, about 90 percent of the ΣPCB burden is in the gas phase, but this percentage varies by homolog. At New Brunswick, for example, the percentage in the particle phase for the 2, 3, 4, 5, 6, 7, 8, and 9 homolog groups is 2, 3, 6, 13, 22, 37, and 66 percent respectively. As a result, gas absorption is the most important mechanism for deposition of low MW PCBs to the water column from the atmosphere (fig. 4.2). For heavier homologs, dry deposition is increasingly important. Because most of the PCBs in rain and snow derive from particles scavenged from the atmosphere, the importance of wet deposition also increases for the heavy homologs.

Data from the NJADN have been used to estimate atmospheric deposition fluxes to the HE and DRE in support of their respective TMDL models. The TMDL model for the DRE focused on penta-chlorinated PCBs. Loadings and partitioning processes for the penta-PCBs were used to gen-

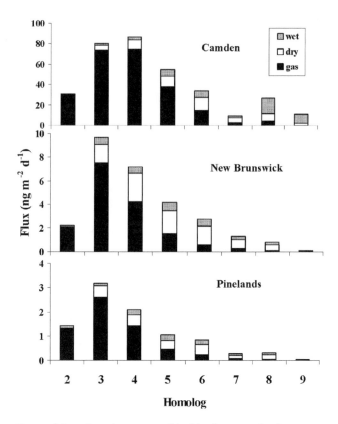

Figure 4.2. Gas absorption (black), dry particle deposition (white), and wet deposition (striped) fluxes of PCB homologues at three NJADN sites: Camden (top), New Brunswick (middle), and Pinelands (bottom). Note difference in the y-axis scales. From Totten et al. 2004, used by permission.

erate a penta-TMDL, which was then scaled (multiplied by 4) to calculate a TMDL for ΣPCBs (see Fikslin and Suk 2003 for a full description). The NJADN data were used to estimate the wet and dry atmospheric deposition of penta-PCBs to the estuary. Seven monitoring sites were located on or near the DRE. The water-quality model of the river included 87 model segments, each of which was assigned a dry deposition flux and a precipitation PCB concentration based on measurements at the closest monitoring station. Dry deposition fluxes were calculated seasonally, while the precipitation concentration was an annual volume-weighted mean average. Dry deposition fluxes ranged from 0.26 ng m^{-2} d^{-1} at Washington's Crossing in summer to 16.2 ng m^{-2} d^{-1} at Camden in winter (Fikslin and Suk 2003). Wet deposition was applied to the model according to the meteorological data. That is, when it rained, the rain contained from 0.11 to 1.28 ng L^{-1} penta-PCBs. The atmospheric load over the 577-day model calibration period was 2.34 kg, or 4.06 g/day^{-1}. This is in excess of the TMDL of 379.96 mg/day^{-1} (Fikslin and Suk 2003), and it does not include wet and dry deposition of the other homologs, nor does it include gas absorption. When these are included, the total atmospheric inputs (wet plus dry plus gaseous deposition) of ΣPCB to the DRE are estimated to be about 16 g/day^{-1}.

In the HE, the situation is similar. Although the HE does not yet have a TMDL for PCBs, it is clear that the TMDL will be on the order of a few hundred milligrams per day. Currently the total atmospheric inputs (wet plus dry plus gaseous deposition) of ΣPCB to the estuary are estimated to be 13–40 kg/year, or 36–110 g/day (Totten 2005). In general, surface waters proximate to urban areas are likely to receive atmospheric inputs of PCBs that, by themselves, exceed potential TMDL levels.

In order for the DRE to meet the TMDL, atmospheric concentrations of PCBs must decline by a factor of about 40. Hillery et al. (1997) measured long-term declines in PCB concentrations with half-lives of about six years at sites surrounding the Great Lakes as part of the Integrated Atmospheric Deposition Network. If this rate of decline is occurring in the DRE region as well, then the time required for atmospheric PCB concentrations to decline to levels at which the TMDL might be met is about 30 years. At New Brunswick, the NJADN has measured a long-term decline in gas-phase PCB concentrations with a half-life of 4.2 years, with 95 percent confidence limits ranging from 3.0 to 7.2 years. Given these half-lives, in background areas 18–45 years will be required, while in urban areas such as Camden, 30–75 years must pass before atmospheric PCB concentrations will decline to levels at which the TMDL may be met.

VOLATILIZATION

So far this analysis has considered atmospheric deposition only as a source of PCBs to the water column. However, volatilization has been recognized as an important process by which PCBs are eliminated from the water column. Eisenreich, Hornbuckle, and Achman (1997) estimated that it is responsible for 46–87 percent of the removal of ΣPCBs from the Great Lakes. Similarly, volatilization was found to be the most important loss process for PCBs containing 2–6 chlorines in the HE, responsible for about half of all the losses from the system in the 1992 model year (Farley et al. 1999). The importance of volatilization does not necessarily imply, however, that the atmosphere acts as a net sink for PCBs. In Lake Michigan, for example, volatilization is the most important loss process, but gas absorption is also an important loading process. As a result, volatilization and gas absorption are closely balanced (Totten et al. 2003) and the lake is close to equilibrium with respect to air-water exchange. In contrast, the atmosphere does act as a net sink for ΣPCBs exiting the water column in both the DRE and the HE (Totten et al. 2001; Yan 2003; Fikslin and Suk 2003). The

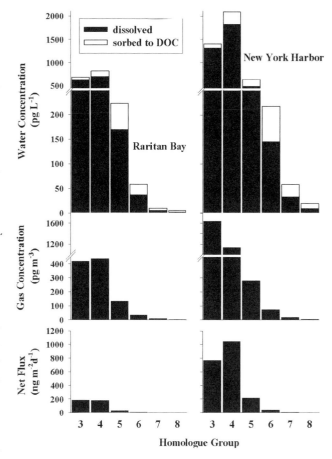

Figure 4.3.   Water column concentrations, gas-phase concentrations, and calculated net air-water exchange fluxes for PCB homologues in the Raritan Bay and New York Harbor during July 5–10, 1998. From Totten et al. 2001, used with permission.

volatilization of ΣPCBs greatly exceeds their gaseous absorption into the water column, as well as their wet and dry deposition. These observations are somewhat misleading, however, because they address the situation in terms of the sum of all PCB congeners. When examined on a homolog basis, it is clear that volatilization is important for PCBs containing 5 or fewer chlorines. Heavier PCBs do not have substantial volatilization fluxes because their truly dissolved concentrations are very low and their mass transfer coefficients (K_{OL}) are small. As figure 4.3 demonstrates, the net flux of heavy PCBs (those with 6 or more chlorines) from sites within the HE (Raritan Bay and New York Harbor) is not measurably different from zero (Totten et al. 2001). Therefore while volatilization is by far the most important removal mechanism for low MW PCBs in both the DRE and HE, it is unimportant for heavier PCBs. Temporal changes, especially in atmospheric PCB concentrations, may reverse the direction of the net air-water exchange fluxes over short periods of time. However, over longer time frames (months to years), we may make some general conclusions. In heavily impacted urban estuaries, volatilization of PCBs containing 1–5 chlorines virtually always exceeds their atmospheric deposition. In contrast, for PCBs with 6 or more chlorine atoms, volatilization is unimportant and the atmosphere acts as a net source via wet, dry, and gaseous deposition.

INDIRECT ATMOSPHERIC DEPOSITION

The previous discussion has focused on the direct interactions between the atmosphere and the water column. The atmosphere also, however, deposits PCBs to land surfaces, which may be remobilized via rain and runoff, subsequently entering surface waters. This "indirect" atmospheric deposition delivers PCBs to surface waters via stormwater runoff ("nonpoint sources") and tributary inputs. Storm water and tributaries accumulate PCBs from sources other than indirect atmospheric deposition, such as point discharges and contaminated sites. In calculating PCB loads and in considering ways in which those loads may be reduced, it is important to consider two issues. First, how large are tributary and stormwater loads to surface waters proximate to urban areas? And second, what fraction of the PCBs in those loads derive from indirect atmospheric deposition?

Figure 4.4 displays the penta-chloro biphenyl loading estimate for the DRE over the 577-day model calibration period (Fikslin and Suk 2003). Figure 4.5 displays the annual loadings of ΣPCBs to the HE (Totten 2005). Stormwater (nonpoint sources) is the second or third largest loading category, responsible for about 20 percent of the total loadings to both the DRE and the HE. Because the Hudson River load to the HE consists primarily of PCBs leached from known contaminated sites (i.e., the General Electric plants), it is reasonable to consider it to be not a tributary load, but a contaminated site load. Contaminated sites are therefore the biggest loading category to both estuaries. In the DRE, the contaminated sites are numerous and widely scattered throughout the estuary. When all tributaries are added together, tributary loads to the DRE are the largest loading category, comprising about 25 percent of the total load. Tributary loads (excluding the Hudson River) were only partially evaluated in the 2005 loading estimate (Totten 2005). They are expected to be small, however, less than about 30 kg/year.

What fractions of the stormwater and tributary PCB loads come from indirect atmospheric deposition? To construct loading estimates for stormwater in the DRE, literature data were used to construct an event mean concentration of PCBs in a typical urban stormwater of 62 ng L^{-1} ΣPCBs (Fikslin and Suk 2003). The volume-weighted mean concentration of ΣPCBs in rainwater in Camden, New Jersey, has been measured by the NJADN at 13 ng L^{-1} (Totten et al. 2004), or potentially 20 percent of the stormwater PCB concentration. Furthermore, NJADN measurements suggest that the dry deposition flux of ΣPCBs at Camden is about 20 ug m^{-2} y^{-1} (Totten et al. 2004). Assuming 100 percent runoff efficiency of both the rain (about 1 m y^{-1}) and the atmospheric particles, this dry deposition would add another 20 ng L^{-1} ΣPCBs to the stormwater. Atmospheric deposition may therefore contribute >50 percent of PCBs detected in a typical urban stormwater sample.

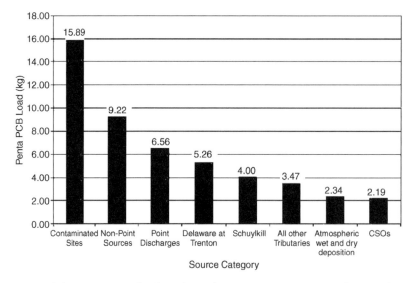

Figure 4.4.   Penta-PCB loads to the Delaware River Estuary over the 577-day model calibration period (Fikslin and Suk 2003; used by permission).

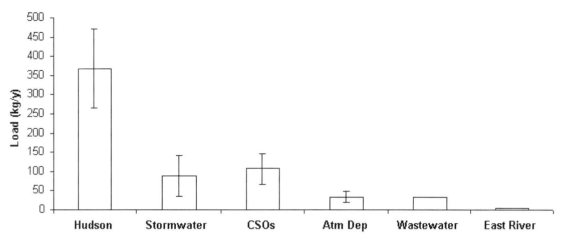

Figure 4.5.   Estimated annual loads of ΣPCBs to the NY/NJ Harbor Estuary (data from Totten 2005). CSOs is an abbreviation for contained sewer overflows.

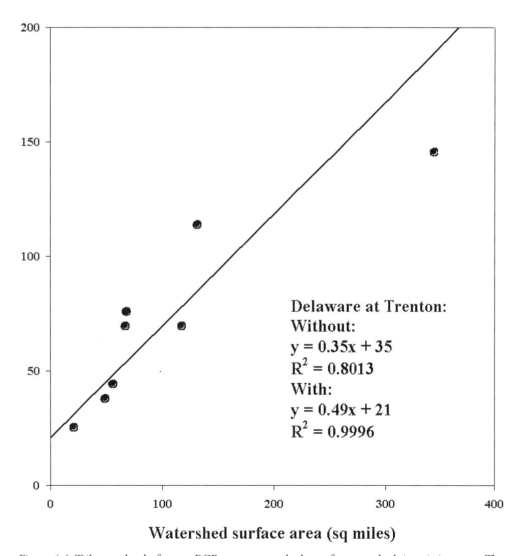

Delaware at Trenton:
Without:
$y = 0.35x + 35$
$R^2 = 0.8013$
With:
$y = 0.49x + 21$
$R^2 = 0.9996$

Figure 4.6. Tributary load of penta-PCB versus watershed area for watersheds in pristine areas. The data point for the Delaware at Trenton is not shown for clarity.

Estimating the importance of indirect atmospheric deposition to tributary PCB loads is more difficult. Penta-PCB loads from 22 tributaries to the DRE were estimated from water column measurements (Fikslin and Suk 2003). Some tributaries, such as the Schuylkill, obviously receive PCBs from sources other than indirect atmospheric deposition. In analyzing the tributary loading data from the DRE PCB TMDL, there is a significant ($p < 0.05$) correlation between watershed area and penta-PCB load for 9 tributaries whose watersheds lie in relatively pristine areas of southern New Jersey and Pennsylvania (fig. 4.6). This type of relationship would be expected for watersheds in which atmospheric deposition is the major source of PCBs. This relationship is significant with or without the Delaware at Trenton, which dominates the regression due to its large area. The slope of this correlation suggests that these watersheds contribute about 0.4 g of penta-PCB per square mile per year to the DRE, or 0.4 ng m^{-2} d^{-1}. If this flux applies throughout the watershed of the DRE, then indirect atmospheric deposition contributes about 3 kg y^{-1} of penta-PCB to the estuary, or about one-third of the total of all the tributary loads. These calculations suggest that indirect atmospheric deposition is an important part of tributary and stormwater PCB loads to the DRE. Indirect atmospheric deposition should be investigated as a potentially important loading process for surface waters throughout the United States.

CONCLUSIONS

The major PCB loads to the DRE and HE are contaminated sites, tributaries, and stormwater. Direct atmospheric deposition is much smaller than these other inputs, but is large enough to exceed TMDLs by more than an order of magnitude. In addition to *direct* atmospheric deposition, atmospheric PCBs may reach surface waters via *indirect* atmospheric deposition, a process that may be responsible for a substantial fraction of the PCBs in stormwater and tributary inputs. The implementation of TMDLs nationwide will therefore require that atmospheric PCB sources be addressed. Although the atmosphere acts as a net sink for low MW PCBs volatilizing from these estuaries, it is a net source for high MW PCBs (those with 6 or more chlorines). The importance of volatilization should not be viewed as a rationale for ignoring the inputs of PCBs to surface waters from the atmosphere.

ACKNOWLEDGMENTS

Funding was provided by the Hudson River Foundation, the New Jersey Department of Environmental Protection, and the Delaware River Basin Commission. Thanks to Amy Rowe, Cari Gigliotti, Steve Wall, and Steve Eisenreich for helpful discussions.

REFERENCES

Bamford, H. A., D. L. Poster, and J. E. Baker. 2000. "Henry's Law Constants of Polychlorinated Biphenyl Congeners and Their Variation with Temperature." *J. Chem. Eng. Data* 45: 1069–74.

Bamford, H. A., D. L. Poster, R. Huie, and J. E. Baker. 2002. "Using Extrathermodynamic Relationships to Model the Temperature Dependence of Henry's Law Constants of 209 PCB Congeners." *Environ. Sci. Technol.* 36:4395–4402.

Eisenreich, S. J., K. C. Hornbuckle, and D. R. Achman. 1997. "Air-Water Exchange of Semivolatile Organic Chemicals in the Great Lakes." In *Atmospheric Deposition of Contaminants to the Great Lakes and Coastal Waters,* edited by J. E. Baker. Pensacola, Fla.: SETAC Press.

Environmental Protection Agency (EPA). 2001. "Hudson River PCBs Site New York Record of Decision." Available at http://www.epa.gov/hudson/d_rod.htm#record (accessed June 1, 2007).

———. 2004. "National TMDL Report." Available at http://www.epa.gov/owow/tmdl (accessed July 15, 2004).

Farley, K. J., R. V. Thomann, T. F. I. Cooney, D. R. Damiani, and J. R. Wands. 1999. "An Integrated Model of Organic Chemical Fate and Bioaccumulation in the Hudson River Estuary." Final report to the Hudson River Foundation. Available at http://www.hudsonriver.org/ls/reports/Farley_016_95A_final_report.pdf (accessed June 1, 2007).

Fikslin, T. J., and N. Suk. 2003. "Total Maximum Daily Loads for Polychlorinated Biphenyls (PCBs) for Zones 2–5 of the Tidal Delaware River." Report to the USEPA, Regions II and III. December 15.

Franz, T. P., S. J. Eisenreich, and T. M. Holsen. 1998. "Dry Deposition of Particulate Polychlorinated Biphenyls and Polycyclic Aromatic Hydrocarbons to Lake Michigan." *Environ. Sci. Technol.* 32:3681–88.

Hillery, B. R., I. Basu, C. W. Sweet, and R. A. Hites. 1997. "Temporal and Spatial Trends in a Long-Term Study of Gas-Phase PCB Concentrations near the Great Lakes." *Environ. Sci. Technol.* 31:1811–16.

Hillery, B. R., M. F. Simcik, I. Basu, R. M. Hoff, W. M. J. Strachan, D. Burniston, C. H. Chan, K. A. Brice, C. W. Sweet, and R. A. Hites. 1998. "Atmospheric Deposition of Toxic Pollutants to the Great Lakes as Measured by the Integrated Atmospheric Deposition Network." *Environ. Sci. Technol.* 32:2216–21.

Miller, S. M., M. L. Green, J. V. DePinto, and K. C. Hornbuckle. 2001. "Results from the Lake Michigan Mass Balance Study: Concentrations and Fluxes of Atmospheric Polychlorinated Biphenyls and Trans-Nonachlor." *Environ. Sci. Technol.* 35: 278–85.

New York, state of. 2000. "Codes, Rules and Regulations: Water Quality Regulations Surface Water and Groundwater Classi-

fications and Standards." Available at http://www.epa.gov/
ost/standards/wqslibrary/ny/ny_2_water_quality_reg.pdf
(accessed June 1, 2007).

Pirrone, N., G. J. Keeler, and T. M. Holsen. 1995a. "Dry Deposition of Semivolatile Organic Compounds to Lake Michigan."
Environ. Sci. Technol. 29:2123–32.

———. 1995b. "Dry Deposition of Trace Elements to Lake Michigan: A Hybrid-Receptor Deposition Modeling Approach."
Environ. Sci. Technol. 29:2112–22.

TAMS Consultants, TCG, Inc., and the Gradient Corporation.
1997. "Phase 2 Report—Further Site Characterization and
Analysis Volume 2C—Data Evaluation and Interpretation
Report Hudson River PCBs Reassessment RI/FS." USEPA
and U.S. Army Corps of Engineers. February 13.

Totten, L. A. 2005. "Present-Day Sources and Sinks for Polychlorinated Biphenyls (PCBs) in the Lower Hudson River
Estuary." In *Pollution Prevention and Management Strategies
for Polychlorinated Biphenyls in the New York/New Jersey Harbor,* edited by M. Panero, S. Boehme, and G. Muñoz, 84–96.
New York: New York Academy of Sciences. http://www.nyas
.org/programs/harbor.asp (accessed June 1, 2007).

Totten, L. A., P. A. Brunciak, C. L. Gigliotti, J. Dachs IV, E. D.
Nelson, and S. J. Eisenreich. 2001. "Dynamic Air-Water
Exchange of Polychlorinated Biphenyls in the NY-NJ Har-
bor Estuary." *Environ. Sci. Technol.* 35:3834–40.

Totten, L. A., C. L. Gigliotti, J. H. Offenberg, J. E. Baker, and
S. J. Eisenreich. 2003. "Reevaluation of Air-Water Exchange
Fluxes of PCBs in Green Bay and Southern Lake Michigan."
Environ. Sci. Technol. 37:1739–43.

Totten, L. A., C. L. Gigliotti, D. A. VanRy, J. H. Offenberg, E. D.
Nelson, J. Dachs, J. R. Reinfelder, and S. J. Eisenreich. 2004.
"Atmospheric Concentrations and Deposition of PCBs to the
Hudson River Estuary." *Environ. Sci. Technol.*

Yan, S. 2003. "Air-Water Exchange Controls Phytoplankton
Concentrations of Polychlorinated Biphenyls in the Hudson
River Estuary." Master's thesis, Rutgers University.

5

Human and Environmental Contamination with Polychlorinated Biphenyls in Anniston, Alabama

Kenneth G. Orloff, *Agency for Toxic Substances and Disease Registry*
Steve Dearwent, *Agency for Toxic Substances and Disease Registry*
Susan Metcalf, *Agency for Toxic Substances and Disease Registry*
Wayman Turner, *National Center for Environmental Health*

Blood serum concentrations of polychlorinated biphenyls (PCBs) were measured in 80 residents of Anniston, Alabama, who were living near a chemical manufacturing plant that formerly produced PCBs. The highest blood PCB concentrations were detected in older adults who had been long-term residents of the area. Both age and length of residency in the area were independently correlated with blood PCB concentrations. Blood concentrations of PCBs did not correlate with the concentration of PCBs in soil or house-dust samples collected from the residents' properties. PCB congeners 153, 138/158, 180, 118, and 187 contributed more than 60 percent of the total PCBs in both adults and children. PCB congener 209, which has a long biological half-life, was detected in 19 percent of the blood samples and in 8 of the 9 blood samples with a total PCB concentration in excess of 10 parts per billion (ppb). PCB 209 was detected in soil samples from all of the properties of residents who had detectable blood concentrations of PCB 209. However, PCB 209 concentrations in soil and blood were not correlated. PCB congeners 44 and 52, which have short biological half-lives, were detected in 6 percent of the blood samples. PCB 44 and 52 were not detected in any of the blood samples that contained more than 5 ppb total PCBs. Past exposures to PCBs may have been a significant contributor to the elevated concentrations of PCBs detected in some residents.

INTRODUCTION

A chemical company produced polychlorinated biphenyls (PCBs) at a plant in Anniston, Alabama, from 1935 to 1971. During this time, the company buried PCB and other chemical wastes in two unlined landfills adjacent to the production facility. Surface water runoff from the landfills carried PCBs into drainage ditches in the surrounding residential communities. Environmental studies have documented the presence of PCB contamination in sediment in drainage ditches and in surface soils on nearby residential properties. The company has purchased numerous private properties near the plant and has remediated off-site contaminated areas. Nevertheless, some off-site environmental PCB contamination remains.

In order to assess human exposure to PCBs in the residential areas surrounding the plant, the Agency for Toxic Substances and Disease Registry (ATSDR) conducted an exposure investigation. The ATSDR recruited families with young children who lived near the plant to participate in this investigation so that PCB exposures in different age groups could be assessed. The agency collected blood samples from all family members and environmental samples (soil and house dust) from their homes and analyzed them for PCB congeners.

MATERIALS AND METHODS

Target Population

The ATSDR invited families who lived within a half-mile radius from the facility to participate in this investigation. In order to be eligible, at least one family member had to be a child between one and seven years old. A total of 18 families fully participated in this investigation. Agency staff collected surface-soil and house-dust samples from the 18 homes and blood samples from 78 residents of these homes. Blood samples were taken from two other residents who lived in the target area, although environmental samples were not collected from their homes.

Before blood samples were collected, written informed consent or assent was obtained from all adult and minor participants. Informed consent for environmental testing was also obtained from an adult resident of each house prior to testing.

Participants in this investigation ranged from 1 to 89 years old and consisted of 43 adults and 37 children (aged 16 years or less). The racial distribution of the participants was 81 percent African American and 19 percent white; the gender distribution was 56 percent female and 44 percent male.

Biological Sampling and Analyses

A licensed phlebotomist collected a 7-milliliter blood sample from each participant using a Vacutainer tube with no anti-coagulant. After collection, the blood samples were allowed to clot for two hours at room temperature. The tubes were then stored on ice until they were delivered to the National Center for Environmental Health laboratory at the Centers for Disease Control and Prevention in Atlanta, Georgia, for analyses.

Blood serum samples were analyzed for 37 PCB congeners using high-resolution gas chromatography isotope-dilution high-resolution mass spectrometry (HRGC/ID-HRMS). Details of the analytical methodology have been previously published (Turner et al. 1997; Barr et al. 2003). Results were reported as concentrations of individual PCB congeners in blood serum (micrograms/liter) or parts per billion (ppb), and as lipid-based concentrations (micrograms/gram serum lipid). Individual congeners were added together to yield total PCB concentrations.

Environmental Sampling and Analyses

ATSDR staff used a metal trowel to collect a surface-soil sample (0–3 inches deep) from an area of the yard identified as a frequent play area by the parent or child. In addition, ATSDR staff used a Nilfisk hepta vacuum cleaner to collect an indoor house-dust sample from a room in the house frequented by family members; this was typically the living room of the house. The environmental samples were analyzed for PCB congeners by the Midwest Research Institute in Kansas City, Missouri, using standard EPA methods.

RESULTS AND DISCUSSION

The concentrations of PCBs were measured in blood serum samples from 37 children (aged 16 years or less) and from 43 adults. The detection level for individual PCB congeners varied by congener and among samples, but the detection

Table 5.1. Blood Serum PCBs: Summary Statistics.

	Adults	Children (<16 Years)
Number with detectable PCBs/total samples	35/43 (81%)	10/37 (27%)
Concentration range (ppb)	ND–210	ND–4.6
Mean concentration (ppb)[a]	14.3	0.37
Mean concentration (ppb)[b]	14.6	1.59
Median concentration (ppb)	2.2	ND
Mean concentration (μg/g serum lipid)[a]	2,543	61.5

[a] Nondetect assumed to be zero.

[b] Nondetect assumed to be one-half of detection level.

Note: ND = not detected.

level was generally less than 0.05 parts per billion (ppb). In adults, the total PCB concentrations in blood ranged from nondetected to 210 ppb. The mean total PCB concentration in adults was 14.3 ppb, and the median concentration was 2.2 ppb.

The concentrations of PCBs in blood serum samples from children were considerably less, ranging from nondetected to 4.6 ppb. The mean total PCB concentration in children was 0.37 ppb, and the median concentration was nondetected. The summary statistics for PCB concentrations in adults and children are presented in table 5.1. A more complete description of these test results has been previously published (Orloff et al. 2001).

Blood PCB concentrations were strongly correlated with age, as has been reported for other populations (Miller et al. 1991; Gerstenberger et al. 1997). The Spearman rank correlation coefficient (r_s) between age and total PCB concentration in blood was 0.729 (p < 0.001). The participants' length of residency at the current address was also correlated with the blood PCB concentration (r_s = 0.645, p < 0.001). By calculating the Spearman partial correlation coefficient, it was determined that even with controlling for age, there was still a significant correlation between the blood PCB concentration and length of residency (r_s = 0.310, p = 0.0054). The results of these statistical analyses support the hypothesis that people were exposed to PCBs while living near the facility, although it is not known when the exposures occurred or what the source of PCB exposure was.

Environmental sampling at the participants' homes indicated the presence of PCB contamination in soil and house-dust samples from most of the homes. This contamination poses a potential source of exposure, since children may intentionally (pica behavior) or inadvertently ingest soil and house dust (Calabrese et al. 1989; Stanek and Calabrese 1995). Soil and house-dust ingestion by adults has not been well studied, but it is likely to be less than for children (Calabrese et al. 1990). However, some adults may intentionally ingest clay or soil (geophagia) (Vermeer and Frate 1979; Feldman 1986). If such behavior occurs, it can be a significant source of exposure to contaminants in clay or soil.

There was a significant correlation between the concentrations of PCBs in soil and in house-dust samples from the same home ($r_s = 0.628$, $p < 0.0052$). This correlation is expected because soil typically constitutes about 50 percent of the mass of house dust (Paustenbach, Finley, and Long 1997). However, further statistical analyses failed to demonstrate a significant correlation between the concentration of PCBs in blood and the concentration of PCBs in either soil ($r_s = -0.128$, $p = 0.26$), house dust ($r_s = 0.078$, $p = 0.51$), or the house-dust surface loading concentration ($\mu g/m^2$) ($r_s = 0.046$, $p = 0.70$). Further analyses in which the test population was divided into adults and children also failed to demonstrate any significant correlations between blood and environmental PCB concentrations. The absence of a correlation between PCB concentrations in blood and those in soil or house dust suggests that recent exposures to PCBs in soil and house dust are not a major determinant of blood PCB concentrations.

Congener Specific Analyses

An examination of the congener profiles of the blood PCB data revealed that PCB congeners 153, 138/158, 180, 118, and 187 accounted for about 61 percent of the total PCB concentrations in adults. In children, the same six PCB congeners accounted for 67 percent of the total blood PCB concentrations. Total PCB concentrations in children were generally lower than for adults. Furthermore, the children typically lacked detectable quantities of the minor PCB congeners that were detected at low concentrations in adults. The absence of detectable quantities of these minor PCB conge-

ners in children may account for the small increase in the relative percent of the major congeners.

PCB Congener 209

PCB congener 209 is the only fully chlorinated PCB (decachlorobiphenyl). PCB 209 is a significant component (4.8%) of Aroclor 1268, which was formerly produced in small quantities at this facility. PCB 209 is present at trace quantities in Aroclor 1260 (0.07 percent) and is not detectable in other Aroclors (ATSDR 2000).

In this investigation, PCB congener 209 was detected in 15 of 80 (19%) of all blood samples and in 15 of 45 (33%) of blood samples with detectable concentrations of PCBs (table 5.2). When detected, PCB congener 209 accounted for 0.4 to 16.3 percent of the total PCBs (average of 2.7%). By comparison, in another published study, PCB 209, when detected, accounted for only 0.02 to 0.32 percent of the total blood PCBs (Burse, Groce, and Caudill 1994). Therefore, the relative percentage of PCB 209 in blood samples from this investigation is higher than that reported for other study populations.

Figure 5.1 shows that blood PCB 209 concentrations were highest in older residents of the community and were significantly correlated with the age of the participants ($r_s = 0.661$, $p < 0.02$). All of the people with detectable PCB 209 concentrations were adults (32 to 89 years old), except for one teenager (13 years old). Furthermore, PCB 209 was detected in eight of the nine adults who had blood total PCB concentrations in excess of 10 ppb, which was approximately the 90th percentile of the blood PCB concentrations in the participants of this study.

Highly chlorinated PCBs, such as PCB 209, are resistant to biological degradation. A biological half-life of 35 years for PCB 209 was estimated using the metabolic rate constants of Brown (Brown 1994) and assuming first-order elimination kinetics. Therefore, after PCB 209 is taken up by the body, it would persist for long periods of time. Furthermore, the relative proportion of PCB 209 to other congeners in

Table 5.2. Blood Serum PCBs: Congener-Specific Statistics.

PCB Congener	Number Detectable[a]	Percentage Detectable[b]	Mean Concentration (ppb)[c]	Total PCB Concentration (%)[d]
44	5	6 %	0.13	5.5
52	5	6 %	0.16	7.5
209	15	19 %	0.47	2.7

[a] Number of blood samples with a detectable concentration of congener (of 80 samples).

[b] Percentage of blood samples with a detectable concentration of congener (of 80 samples).

[c] Mean concentration of congener in blood samples with detectable levels of congener.

[d] Mean congener percentage of total PCBs in those samples with detectable levels of congener.

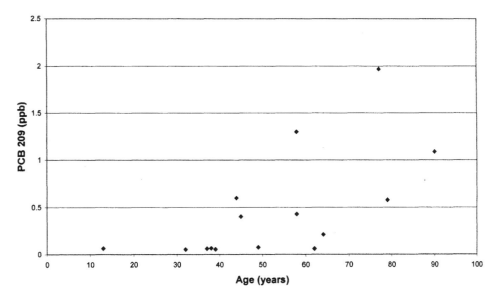

Figure 5.1.   Blood serum concentrations of PCB 209 (ppb) versus age (years).

blood could increase as lower chlorinated congeners are slowly eliminated from the body by preferential metabolism.

PCB congener 209 would also be expected to persist in the environment because of its resistance to biological, chemical, and physical degradation. In this investigation, PCB 209 was detected in 10 of 10 soil samples (14–650 ppb) and in 5 of 10 house-dust samples (0–65 ppb) collected from the homes of the participants with detectable blood concentrations of PCB 209. However, there was no significant correlation between the concentrations of PCB 209 in blood samples and the concentrations of PCB 209 in soil (r_s = –0.238) or house-dust (r_s = –0.178) samples from participants' homes.

PCB 209 has a long biological half-life, and it was detected primarily in blood samples from older people who had been long-term residents of the community. Furthermore, there was no correlation between PCB 209 concentrations in blood and the PCB 209 concentrations in soil or dust samples from the participants' homes. These results are consistent with the hypothesis that past exposures to PCB 209 could be a significant contributor to the participants' body burdens of PCB 209.

PCB Congeners 44 and 52

PCB congeners 44 and 52 are tetrachlorbiphenyls and are constituents of the most commonly used Aroclors. Lower chlorinated PCBs, such as PCB 44 and 52, which lack chlorine atoms on the meta and para positions of the biphenyl ring, are susceptible to biological degradation (Brown 1994). Using the metabolic rate constants of Brown (Brown 1994) and assuming first-order elimination kinetics, the biological half-lives of PCB 44 and 52 were estimated to be approxi-

mately 1.3 months. Therefore, after PCB 44 and 52 are taken up by the body, they would be rapidly eliminated, as compared to higher chlorinated PCBs. Thus the presence of these congeners in blood is evidence of recent exposure.

In this investigation, PCB congeners 44 and 52 were detected in five participants, or 6 percent of all blood samples, and in 11 percent of blood samples with detectable concentrations of PCBs (table 5.2). When concentrations of PCB congeners 44 and 52 were detected, they constituted a significant fraction of the total PCBs. Furthermore, other "episodic" PCBs (Hansen 2001), such as PCB 101, 110, 151, which are present only transiently in humans, were also detected in these same blood samples.

The ages of the participants with detectable concentrations of PCB 44 and 52 ranged from 13 to 62 years old. The concentration of total PCBs in blood samples with detectable concentrations of PCB 44 and 52 ranged from 0.7 to 4.6 ppb. PCB 44 and 52 were not detected in any of the participants who had high total PCB concentrations.

PCB 52 was detected in two of four soil samples (9–52 ppb) and in three of four house-dust samples (16-782 ppb) that were collected at the homes of participants with detectable blood concentrations of these congeners. The environmental samples were not analyzed for PCB 44. Lower chlorinated PCBs, such as 44 and 52, are more volatile than higher chlorinated biphenyls and could volatilize from the soil into the air. In June 2000, the EPA collected 24-hour ambient air samples at six locations in residential areas near the facility. Presumptive evidence for the presence of PCB 52 at a concentration of 2 ng/m^3 (nanograms per cubic meter) was reported in one of 12 air samples (USEPA 2000). The EPA did not analyze the air samples for PCB 44.

Consumption of fish is another potential source of exposure to lower chlorinated PCBs. Very few species of fish show cyp 2B1- and cyp 2B2-type enzyme induction that would be effective in metabolizing the lower chlorinated congeners (Gerstenberger, Gallinat, and Dellinger 1997; Hansen 1998). Consequently, fish often contain detectable concentrations of lower chlorinated PCBs (e.g., PCB 52 and 44) that are infrequently found in humans (Hansen 1998). In 1993, elevated PCB concentrations detected in fish prompted the Alabama Department of Public Health to issue a "no consumption fish advisory" for Choccoloco Creek, which receives drainage from the area around the plant. Nevertheless, many of the participants in this investigation reported that they ate fish from the local waterways in the past, and a few people may continue to do so. Recent fish consumption histories for the participants in this investigation were not available, so it is not known if recent fish consumption could account for the elevated PCB 44 and 52 levels seen in five of the study participants.

These results indicate that PCB 44 and 52 were detected in a few people who had relatively low blood concentrations of total PCBs. Although the presence of these congeners suggests recent exposure to PCBs, the source of the exposure is not known. PCB 44 and 52 were not detected in those participants with high total blood PCB concentrations. This finding supports the hypothesis that people with high total PCB concentrations have not had recent exposure to PCB mixtures containing these congeners.

CONCLUSIONS

In the general population, background exposures to trace levels of PCBs in food, such as fish, meat, and poultry, are the major source of PCB exposures (ATSDR 2000). The high concentrations of PCBs detected in blood samples from some residents of this community indicate that additional exposures have occurred. Elevated concentrations of PCBs were detected in fish from Choccolocco Creek (located near Anniston), which prompted the state to issue a fishing advisory in 1993. Additionally, there are reports that in the past, nearby residents who lived adjacent to the plant raised hogs that sometimes grazed on plant property. Tissues from one of these hogs were reported to be contaminated with PCBs at concentrations as high as 19,000 ppm (Grunwald 2002). Thus past consumption of locally raised animals and fish from nearby waterways may have been a significant source of PCB exposure.

The nine participants with blood PCB concentrations in excess of 10 ppb share many of the following characteristics: (1) they are older adults, aged 43 years and above; (2) they reported no known occupational exposure to PCBs; (3) they grew up in neighborhoods near the facility and lived there most of their lives; (4) they ate locally grown fruits and vegetables and locally raised chickens, eggs, and other animals; (5) they ate fish caught in local rivers and lakes; (6) the long-lived PCB congener 209 was detected in their blood and it constituted a higher percentage of the total PCBs than reported in other study populations; (7) the short-lived PCB congeners 44 and 52 were not detected in their blood; and (8) six of the nine reported that in the past they had eaten clay from the neighborhood. Geophagia, including clay eating, has been reported to be a cultural or social tradition in some southern societies (Vermeer and Frate 1979; Feldman 1986).

Once PCBs get inside the body, many of the congeners are resistant to metabolic degradation. The major PCB congeners detected in the participants of this investigation have long biological half-lives. Therefore, exposures to PCBs that occurred years, or even decades, ago could have contributed to the high body burdens of PCBs found in some participants who were long-term residents of the area. This hypothesis is supported by the strong correlations between blood PCB concentrations and age and length of residency in the area. Moreover, analyses of environmental data (soil and house dust) failed to show a correlation between current environmental concentrations of PCB and blood PCB concentrations.

Since 1995, the chemical manufacturing company has purchased, remediated, and restricted access to many of the off-site, PCB-contaminated properties. However, it is likely that prior to remedial activities, residents of the area had exposure to PCBs from eating locally raised animals, plants, and fish, and from contact with contaminated soil, sediment, surface water, and air. Exposures to PCBs in the past, when the facility was producing PCBs, likely exceeded current exposures and could be responsible for the elevated blood PCB concentrations seen in some older, long-term residents of the community.

DISCLAIMER

The findings and conclusions in this report are those of the authors and do not necessarily represent the views of the Centers for Disease Control and Prevention and the Agency for Toxic Substances and Disease Registry.

REFERENCES

Agency for Toxic Substances and Disease Registry (ATSDR). 2000. "Toxicological Profile for Polychlorinated Biphenyls (Update)." Atlanta: U.S. Department of Health and Human Services.

Barr, J., B., V. L. Maggio, D. B. Barr, W. E. Turner, A. Sjodin, C. D. Sandau, J. L. Pirkle, L. L. Needham, and D. G. Patterson Jr. 2003. "New High-Resolution Mass Spectrometric Approach for the Measurement of Polychlorinated Biphenyls and Organochlorine Pesticides in Human Serum." *J. Chromatography B.* 794:137–48.

Brown, J. F. 1994. "Determination of PCB Metabolic, Excretion, and Accumulation Rates for Use as Indicators of Biological Response and Relative Risk." *Environ. Sci. Technol.* 28: 2295–2305.

Burse, V. W., D. F. Groce, S. P. Caudill, M. P. Korver, D. L. Phillips, P. C. McClure, C. R. Lapeza, S. L. Head, D. T. Miller, D. J. Buckley, J. Nessif, R. J. Timperi, and P. M. George. 1994. "Determination of Polychlorinated Biphenyl Levels in the Serum of Residents and in the Homogenates of Seafood from the New Bedford, Massachusetts Area: A Comparison of Exposure Sources through Pattern Recognition Techniques." *Sci. Total Environ.* 144:153–77.

Calabrese, E. J., R. Barnes, E. J. Stanek, H. Pastides, C. E. Gilbert, P. Venman, X. Wang, A. Lasztity, and P. T. Kostecki. 1989. "How Much Soil Do Young Children Ingest: An Epidemiologic Study." *Regul Toxicol. and Pharmacol.* 10:123–37.

Calabrese, E. J., E. J. Stanek, C. E. Gilbert, and R. M. Barnes. 1990. "Preliminary Adult Soil Ingestion Estimates: Results of a Pilot Study." *Regul Toxicol. and Pharmacol.* 12:88–95.

Feldman, M. D. 1986. "Pica: Current Perspectives." *Psychomatics* 27 (7): 519–23.

Gerstenberger, S. L., M. P. Gallinat, and J. A. Dellinger. 1997. "Polychlorinated Biphenyl Congeners and Selected Organochlorines in Lake Superior Fish, USA." *Environ Toxicol. Chem.* 16 (11): 2222–28.

Gerstenberger, S. L, D. R. Tavris, L. K. Hansen, J. Pratt-Shelly, and J. A. Dellinger. 1997. "Concentrations of Blood and Hair Mercury and Serum PCBs in an Ojibwa Population That Consumes Great Lakes Region Fish." *Clin. Toxicol.* 35:377–86.

Grunwald, M. 2002. "In Dirt, Water, and Hogs, Town Got Its Fill of PCBs." *Washington Post,* January 1, p. A17.

Hansen, L. G. 1998. "Stepping Backward to Improve Assessment of PCB Congener Toxicities." *Environ. Health Perspect.* 106: 171–89.

———. 2001. "Identification of Steady State and Episodic PCB Congeners from Multiple Pathway Exposures." In *PCBs: Recent Advances in Environmental Toxicology and Health Effects,* edited by L. W. Robertson and L. G. Hansen, 47–56. Lexington: University Press of Kentucky.

Miller, D. T., S. K. Condon, S. Kutzner, D. L. Phillips, E. Kreuger, R. Timperi, V. W. Burse, J. Cutler, and D. M. Gute. 1991. "Human Exposure to Polychlorinated Biphenyls in Greater New Bedford, Massachusetts: A Prevalence Study." *Arch. Environ. Contam. Toxicol.* 20:410–16.

Orloff, K. G., S. Dearwent, S. Metcalf, S. Kathman, and W. Turner. 2001. "Human Exposure to Polychlorinated Biphenyls in a Residential Community." *Arch. Environ. Contam. Toxicol.* 44: 125–31.

Paustenbach, D. J., B. L. Finley, and T. F. Long. 1997. "The Critical Role of House Dust in Understanding the Hazards Posed by Contaminated Soils." *Int. J. Toxicol.* 16:339–62.

Stanek, E. J., and E. J. Calabrese. 1995. "Daily Estimates of Soil Ingestion in Children." *Environ. Health Perspect.* 103 (3): 276–85.

Turner, W., E. DiPietro, C. Lapeza, V. Green, J. Gill, and D. G. Patterson Jr. 1997. "A Fast Universal Automated Cleanup System for the Isotope-Dilution High-Resolution Mass Spectrometric Analysis of PCDDs, PCDFs, Coplanar PCBs, PCB Congeners, and Persistent Pesticides from the Same Serum Sample." *Organohalogen Compounds* 31:26–31.

U.S. Environmental Protection Agency, Region IV (USEPA). 2000. "Science and Ecosystem Support Division." Polychlorinated Biphenyl Ambient Air Study. Anniston, Ala. June.

Vermeer, D. E., and D. A. Frate. 1979. "Geophagia in Rural Mississippi: Environmental and Cultural Contexts and Nutritional Implications." *Am. J. Clin. Nutr.* 32: 2129–35.

6

A Simple Case for Subdividing Populations Exposed to PCBs by Homologue Profile

Tomás Martín-Jiménez, *University of Illinois at Urbana-Champaign*
Larry G. Hansen, *University of Illinois at Urbana-Champaign*

We examined serum PCB specific congener data gathered for litigation and released as public record, comparing the homologue proportions in 29 residents of Anniston, Alabama, with 7 residents of southwest Chicago, Illinois. As expected, the average profiles differ substantially, but within the groups there are major subdivisions with even more striking differences. It is possible to account for the differences based on unique exposures if the subpopulations are recognized, but an inaccurate picture emerges if the entire populations are averaged together. Further patterns emerge once the dissimilar residue profiles are separated and this may have strong implications for classifying PCB exposures in epidemiology studies.

INTRODUCTION

Human serum PCB data are commonly averaged to compare "exposed" versus "reference" populations. This may be satisfactory for accidental and/or occupational exposures which overwhelm environmental plus food chain PCB exposures. On the other hand, persons living in or near PCB-contaminated environments are potentially exposed to multiple significant sources by multiple routes. Each source in each PCB-contaminated environment will have a different profile of congeners depending on the focal local source, history of weathering and secondary influx, and the properties of the exposure media themselves. Averaging merely erases the useful information in individual homologue and congener patterns (Hansen 2001).

It is therefore useful to attempt to differentiate persons and subgroups, according to some logical residue profile criterion, within larger populations which have been exposed to elevated levels of PCBs. Detection of unique subgroups will help to determine probable exposure sources, probable maximum body burdens, expected health effects, and possible medical intervention. However, by the time exposure has been verified, human samples are often taken long after the highest exposures. Nevertheless, taking the variations in toxicokinetics of individual congeners into account, even these delayed measurements can be reasonably interpreted on an individual basis (see chapter 7, this volume).

Some information can be gleaned from PCB congener profiles, providing indications of residue age, source(s) of exposure, and recent/current exposure. Relative proportions of readily metabolized ("episodic"), moderately persistent, and highly persistent congeners can also be used to estimate historical PCB burdens from more recent sample analyses. Such estimations were used to successfully characterize exposures of 11 residents of Anniston, Alabama (Hansen, DeCaprio, and Nisbet 2003). This chapter deals with homologue profiles derived from detailed serum PCB analyses for 29 additional residents of Anniston. The objective was to categorize them according to the most similar homologue patterns to facilitate toxicokinetic extrapolations. In the process, certain relationships were revealed which will be of use in characterizing and subdividing other PCB-exposed populations.

METHODS

Detailed serum PCB analyses for 29 residents of Anniston, Alabama, and 7 residents of southwest Chicago, Illinois, were available from interpretations developed by request for litigation. In both cases (Anniston and Chicago) initial serum

PCB analyses were considered unsuitable, so a second set of samples were sent to Axys Laboratories (Sidney, British Columbia) with intensive chain-of-custody monitoring. The following analysis of the comparative data was possible because both sets of samples were analyzed by the same method in the same ("gold standard") laboratory, even though at different times. The analytical results and quality control/quality assurance data were thoroughly peer reviewed by plaintiff and defense expert witnesses. All of these experts were knowledgeable regarding PCBs and accepted the analyses as reliable. Interpretations, however, differed.

All Anniston residents sampled were greater than 30 years of age, having lived in Anniston at least during the early 1970s. The Chicago residents varied in age from <10 years to >50 years. Because recent settlement disbursements were still under negotiation, we did not attempt to obtain permission from individuals to utilize information which, although public record, might reveal identities. We are only using data analyzed on request, the data and analyses being part of formal public records (reports, exhibits, depositions); therefore, we have not incorporated demographic data into the evaluations, relying only on PCB homologue and congener relationships to characterize the individuals and their exposures.

RESULTS AND DISCUSSION

Comparing mean homologue percentages for the 29 Anniston residents with the 7 Chicago residents, it is possible to detect what appear to be subtle differences in lower chlorinated and higher chlorinated homologues between the two populations (fig. 6.1).

In fact, the proportions of tri-CBs and tetra-CBs in the Chicago residents are significantly greater ($p < 0.01$) than those of the Anniston residents. Conversely, the proportions of octa-CBs ($p = 0.033$) and, especially, nona-CBs and deca-CBs ($p < 0.001$) are significantly greater in the Anniston residents. Hexa-CBs, especially CBs 138 and 153, clearly dominate in most human PCB residue profiles, and they are closely matched by hepta-CBs, especially CB 180 (Gladen, Doucet, and Hansen 2003; Hansen 1999; Hansen 2001). These 3 CBs (138, 153, 180) are almost always highly correlated in human residues so provide little information of use in differentiating residue profiles; on the other hand, their dominant presence encourages neglect of major differences in homologue groups with fewer or greater numbers of chlorine substituents. The most useful information can be found in the relative proportions of these more extreme, but less dominant, groups.

For example, it can be deduced that the Anniston residues are older ("aged" longer so that the lower chlorinated homologues are depleted). The hexa-CBs and hepta-CBs should be expected to remain rather constant because they are generally more persistent and make up a larger proportion of most PCB exposure sources. Thus, reducing the proportions of lower chlorinated homologues has the effect of slightly increasing the proportions of homologues with greater numbers of chlorine substituents. Nevertheless, it appears that the Anniston residents have unusually high proportions of deca-CB (CB 209) (Gladen, Doucet, and Hansen 2003; Hansen 1999; Hansen 2001). The individual data revealed a large range of values (0.5 to 10%) for deca-CB; therefore, the 9 individuals with values greater than 3 percent were averaged and compared to the average of the

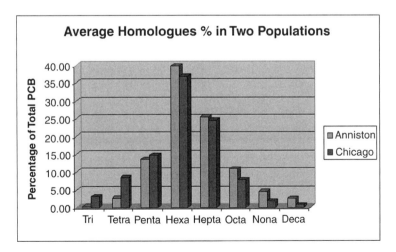

Figure 6.1. Comparison of average homologue proportions between Anniston and Chicago residents. In general, Anniston residents exhibit a heavier congener profile.

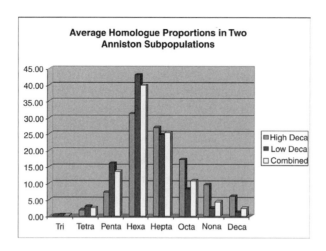

Figure 6.2.   Comparison of average homologue proportions for two subpopulations in Anniston. The group with high proportion of CB 209 have a heavier homologue profile.

20 remaining individuals with less than 3 percent deca-CB (fig. 6.2).

This clearly shows that the two subpopulations of Anniston residents are quite different. Tri-CBs and tetra-CBs were low, and hepta-CBs were not different. Those with high proportions of CB 209 have significantly lower proportions of penta-CBs and hexa-CBs, while octa-CBs, nona-CBs, and deca-CBs were significantly higher (all p < 0.001). The profile of the subpopulation with heavier homologues could indicate that the residue is more aged with much less recent exposure; however, the extremely high proportion of CB 209 makes it more likely that these individuals were exposed to the unique "montar" mixture (Hermanson and Johnson 2007) or some other more highly chlorinated PCB such as those used in foundries (Erickson 1997). By averaging the data simply because the persons lived in the same general area, this important information is lost.

The average homologue profiles indicate trends. However, by looking at the individual data and arranging by some decrement, more specific relationships can be found. Mono-CBs and di-CBs are labile and of low proportions, and analytical results are not as dependable. Tri-CBs are often dominated by CB 28, which, as with CB 74, is relatively persistent and may be a product of slow dechlorination (Brown 1994; Cho et al. 2003). Tetra-CBs may be highly variable, and the group contains persistent (e.g., CBs 47, 66, 74) as well as highly labile congeners. Penta-CBs appear to be the pivotal homologue group for separating recent exposures from aged residues. Therefore the individual data were arranged in order of increasing proportions of penta-CBs (2.3 to 23.1%). This arrangement helped to detect the widely divergent proportions of the often dominant penta-CB 118 (0.6 to 12.4%) and the ratio of CB 118 to CB 156

(118:156 ratio ranged from 0.15 to 6.8). This information, the low ratios of the metabolizable CB 118 to the persistent CB 156, is helpful in detecting induction of CYP1A-like enzymes which metabolize mono-*ortho* PCBs with chlorines in *meta* and *para* positions (Brown 1994; Brown and Lawton 2001; Masuda 2001) and was used to adjust estimated half-lives of elimination for specific individuals in chapter 7 of this volume.

REFERENCES

Brown, J. F., Jr. 1994. "Determination of PCB Metabolic, Excretion and Accumulation Rates for Use as Indicators of Biological Responses and Relative Risk." *Environ. Sci. Technol.* 28:2295–2305.

Brown, J. F. J., and R. W. Lawton. 2001. "Factors Controlling the Distribution and Levels of PCBs after Occupational Exposure." In *Recent Advances in the Environmental Toxicology and Health Effects of PCBs,* edited by L. W. Robertson and L. G. Hansen. Lexington: University of Kentucky Press.

Cho, Y. C., R. C. Sokol, R. C. Frohnhoefer, and G. Y. Rhee. 2003. "Reductive Dechlorination of Polychlorinated Biphenyls: Threshold Concentration and Dechlorination Kinetics of Individual Congeners in Aroclor 1248." *Environ. Sci. Technol.* 37:5651–56.

Erickson, M. D. 1997. *Analytical Chemistry of PCBs.* 2nd ed. Boca Raton, Fla.: Lewis Publishers.

Gladen, B. C., J. Doucet, and L. G. Hansen. 2003. "Assessing Human Polychlorinated Biphenyl Contamination for Epidemiologic Studies: Lessons from Patterns of Congener Concentrations in Canadians in 1992." *Environ. Health Perspect.* 111:437–43.

Hansen, L. G. 1999. *The Ortho Side of PCBs: Occurrence and Disposition.* Boston: Kluwer Academic.

———. 2001. "Identification of Steady State and Episodic PCB Congeners from Multiple Pathway Exposures." In *Recent Advances in the Environmental Toxicology and Health Effects of PCBs,* edited by L. W. Robertson and L. G. Hansen, 47–56. Lexington: University of Kentucky Press.

Hansen, L., A. DeCaprio, and I. Nisbet. 2003. "PCB Congener Comparisons Reveal Exposure Histories for Residents of Anniston, Alabama, USA." *Fresenius Environ. Bull.* 12:181–89.

Hermanson, M. H., and G. W. Johnson. 2007. "Polychlorinated Biphenyls in Tree Bark from Anniston, Alabama." *Chemosphere* 68:191–98.

Masuda, Y. 2001. "Fate of PCDF/PCB Congeners and Change of Clinical Symptoms in Patients with Yusho PCB Poisoning for 30 Years." *Chemosphere* 43:925–930.

7

Toxicokinetic Extrapolation of PCB Exposure in Anniston, Alabama

Tomás Martín-Jiménez, *University of Illinois at Urbana-Champaign*
Larry G. Hansen, *University of Illinois at Urbana-Champaign*

Human PCB residue data are often limited to a single measurement from a sample taken years after the most intense exposure. This may not accurately reflect actual exposures because of the distinct profiles of PCB congeners associated with different exposure sources and the dramatically different metabolism and elimination kinetics of individual PCBs. Detailed PCB congener analyses were available for a number of residents of Anniston, Alabama. Our objective was to determine if exposure histories could be reconstructed based on PCB homologue and congener profiles. The level of atmospheric contamination had declined from about 300,000 ng/m^3 near the manufacturing plant in 1969 to 20,000 ng/m^3 in 1970 and, further, to an average of 27 ng/m^3 in the late 1990s. Since vapor phase PCBs generally accumulate in plant foliage and deposit on surfaces, additional routes of exposure would have also been declining during this period, although not as rapidly.

Homologue profiles were first used for 29 residents of Anniston to determine their patterns, variability, degree of aging of the residues, and evidence of recent exposures. Suspected recent exposures were confirmed or denied by the proportions of specific transient ("episodic") PCBs in the residue profile. Using these methods, a resident with 18 ppb serum PCB in 2002 was estimated to have had levels between 1,400 and 2,200 ppb in the early 1970s because of the aged residue. On the other hand, a resident with 60 ppb serum PCB in 2002 was estimated to have had levels between 1,100 and 1,400 ppb in the early 1970s; this lower estimate was due to the use of longer apparent half-lives because of evidence of past exposure to more highly chlorinated PCBs, plus some level of current exposure.

INTRODUCTION

For decades, the health of humans and animals has been constantly threatened by the production and release of polychlorinated biphenyls (PCBs) into the environment. PCBs were used for industrial purposes from the mid-1930s until the late 1970s (1977 in the United States). Although their production ended almost three decades ago, in-service uses continued, and most of the remaining world production of PCBs is still in the environment. PCBs are a ubiquitous group of synthetic organic chemicals that contain 209 congeners with varying physicochemical features. Toxicokinetics vary widely among congeners and account for the variety of disposition profiles that can be observed for total PCBs depending on the particular congener mixture (Hansen 1999; Safe 1994). For example, lightly chlorinated congeners tend to be metabolized and eliminated from the body at much higher rates than highly chlorinated congeners. Therefore, toxicokinetic differences among congeners have to be accounted for in order to estimate past levels of exposure in individuals based on more recent blood PCB residue measurements. These estimations are often necessary when establishing public health policies, during medical surveillance or to solve liability litigations between affected individuals and PCB manufacturing companies.

One such case is that of the residents of Anniston, Alabama, who were highly exposed to PCBs released from a local manufacturing plant owned by Monsanto Chemical Company from the 1930s into the 1970s. Although the presence of environmental residues had been reported since the late 1960s, the extent of the contamination did not become well known until 1993 (ADPH 1996). Rates of discharge were greatest before 1969, decreased during the 1970s and 1980s after cessation of manufacture in 1971, and decreased even more during the 1990s after remediation steps were taken. The total production at this location was estimated at about 400,000 metric tons with emissions into the atmosphere from 1953 to 1971 of at least 20.5 metric tons (MT) (Hermanson and Johnson 2007). Air samples collected in Anniston in 1997–98 showed a 13-month average concentration of 27 ng/m³ and a maximum concentration of 80 ng/m³. The air concentration of total PCBs decreased exponentially as the distance from the plant increased. Interestingly, certain specific areas in Anniston showed a unique, heavier congener pattern that is not characteristic of the Aroclors produced in Anniston (Hermanson and Johnson 2007). This pattern, which contributed to congener profiles observed elsewhere in Anniston, might be related to emissions of high-temperature distillation residues known as Montars (Hermanson and Johnson 2007) or to heavier than usual Aroclors used in foundries (e.g., Aroclor 1268) (Erickson 1997).

Anniston residents were contaminated by a variety of means including eating fish from contaminated creeks and rivers, eating meat and vegetables from locally raised livestock and local gardens, respectively, swimming in ponds and creeks, breathing contaminated air and dust, and so on. Ingestion of soil (geophagia) contaminated with PCBs also contributed to the total exposure, in particular in pregnant women and children. The latter would have also been exposed in utero, by drinking breast milk, and by playing in soil and creeks (Hansen, DeCaprio, and Nisbet 2003; Orloff et al. 2003; chapter 5, this volume). In spite of the high levels of environmental contamination, no adequate information on historical individual or population exposure is available, other than that provided by measured levels of PCB residues in blood or fat many years after the onset of exposure. This information is insufficient for a detailed analysis of exposure and health issues. To make the information more useful requires the use of toxicokinetic techniques and reasonable assumptions in order to estimate what the blood concentration and PCB congener profile may have been in individuals at the peak of environmental contamination. We have two objectives: first, to discuss the toxicokinetic principles involved in back-extrapolating from current to past concentrations of PCB residues in blood, and second, to reconstruct exposure histories in selected residents from Anniston, Alabama, based on their current PCB homologue and congener profiles. These back calculations were based on human data from other exposures and are similar to more formal models for 2,3,7,8-tetrachlorodibenzo-p-dioxin (TCDD) (Aylward et al. 2005); however, this chapter presents a suggested approach

for PCBs and not a validated model. With adequate analytical data, such an approach with PCBs might be preferable to the models used for single chemicals because the multiple congeners each have unique toxicokinetic properties and the PCB congener relationships alone can reveal much about the individuals' physiology as well as exposure history (Hansen 1999; Hansen, DeCaprio, and Nisbet 2003; Orloff et al. 2003; chapter 5, this volume).

MATERIALS AND METHODS

Study Group

In this preliminary study, we analyzed PCB homologue profile data from 3 adults selected from a group of 29 residents from Anniston, Alabama, who were plaintiffs in a class action civil litigation against Monsanto Chemical Company. Serological data were obtained in 2002 and included a single-point total PCB concentration measurement per individual. Demographical information collected from these individuals was not used at this stage in order to protect the identities of those whose settlements were still pending. Demographical information can be very useful to increase the accuracy of the toxicokinetic parameter estimates (e.g., half-life). Indeed, it is well known that factors such as pregnancy, obesity, age, distance from the plant, and so on can affect PCB congener disposition, and a model that accounts for those variables is likely to yield more accurate and precise predictions. Future studies will make use of this information and will include all the individuals in our dataset. The three adult female subjects considered in our analysis:

Subject A: The total PCB (wet) measured was 60.1 ng/ml serum. Subject A exhibited a larger than usual level of congener 209 (6.07%). The ratio of congeners 118/156 was 4.31. The sum of trichlorobiphenyl and tetrachlorobiphenyl homologues was 3.4 percent.

Subject B: The total PCB (wet) measured was 17.7 ng/ml serum. The level of congener 209 was 0.5%. The ratio of congeners 118/156 was 6.78. The sum of trichlorobiphenyl and tetrachlorobiphenyl homologues was 3.4 percent. Subject B exhibited a lower proportion of heavy congener homologues compared to A or C.

Subject C: The total PCB (wet) measured was 43.9 ng/ml serum. Subject C exhibited a larger than usual level of congener 209 (4.6%). The ratio of congeners 118/156 was 1.46. The sum of trichlorobiphenyl and tetrachlorobiphenyl homologues was 1.3 percent.

The three subjects had total PCB concentrations in serum higher than those reported in the 1990s in groups from the

U.S. population with modest exposure from a variety of sources (ATSDR 2000a, 2000b; Orloff et al. 2003).

Blood Sampling and Chemical Assay

Blood samples had been taken from some of the individuals in our study group during 1998 for limited PCB analysis. In 2002, blood samples were assayed for all 209 PCB congeners by a contract laboratory (Axys Analytical Services, Ltd., Sidney, British Columbia, Canada). These data, along with the quality control and quality assurance data, were provided to expert witnesses for plaintiffs and defense in *Tolbert et al. vs. Monsanto et al.* (U.S. District Court, Northern District of Alabama, Civil Action No. CV-01-C-1407-W). PCB experts for both plaintiffs and defense were commissioned to critically examine not only the data but also the quality-control/quality-assurance data and procedures as well. The PCB experts from both sides agreed that the data were reliable, although opinions on the significance were opposed. Thus, the data are public record, but care has been taken to protect individual identities since disbursement of the settlement was still under negotiation.

Data Analysis

Data were analyzed individually after considering the limitations from having one single measurement per individual and the toxicokinetic principles that could be applied to overcome some of those limitations.

Toxicokinetic principles. Without having historical residue data, the problem at hand is one of extrapolating back from current residues to a previous minimum/maximum range that could account for the current body burden. Before reviewing the toxicokinetic principles and the criteria followed, we must understand the need to allow for wide intervals for our concentration predictions at this stage. Single concentration measurements in blood years after peak exposure do not permit determination of the actual toxicokinetic features of the chemical of interest in individuals. Therefore we need to rely on basic principles, safe assumptions about exposure history, and the "fingerprint" provided by the profile of PCB congeners in the measured blood concentration. This fingerprint is the most valuable piece of collective data because quantitative relationships among PCB congeners are known so that current proportions can reveal the evolution of the fingerprint.

The problem that we are trying to solve involves three elements of the concentration-time disposition curve (fig. 7.1). The first of these elements, *current concentration,* is the single blood concentration measured in individuals at the present time or a series of years after the peak exposure or concentration may have taken place. The second element is referred to in figure 7.1 as *slope;* it allows tracing back for an

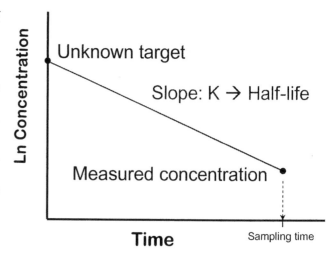

Figure 7.1.   Extrapolating from current concentration back to past exposure.

adequate period from the measured concentration to the unknown target concentration. This period spans from the moment when peak concentration occurred (or more accurately the end of pseudo-steady-state concentration) to the current time. The *time from peak to current concentration* constitutes, therefore, the third element in our paradigm. In theory, since we know the measured concentration years after exposure, we only need to know the value of the slope and the time from peak to current concentration in order to predict what the initial concentration may have been in a particular individual. Presented in this way, the task seems relatively easy; unfortunately, reality is much more complex and does not allow us to take a direct road back in time. There are several problems associated with each of the three elements that we need to understand before a meaningful back-extrapolation can be attempted. For ease of presentation, we will discuss these elements in the following sequence: slope, measured concentration, and time from peak to current concentration.

Slope. Toxicokinetic studies in several species have revealed the different biotransformation processes affecting PCB congeners as well as quantified the extent of absorption, distribution, and elimination for individual congeners or groups of congeners (Brown 1994; Brown and Lawton 2001; Brown et al. 1989; Hansen 1999; Hansen and Welborn 1977; Masuda 2001; Shirai and Kissel 1996). The bioavailability, extent of metabolism, adipose-plasma distribution ratios, and excretion patterns of selected congeners by different routes, as well as the extent of metabolism, adipose-plasma distribution ratios, and excretion patterns for total PCB, have also been studied. There is much less information, however, on how all these different aspects of PCB congener disposition integrate into the concentration-time profile of

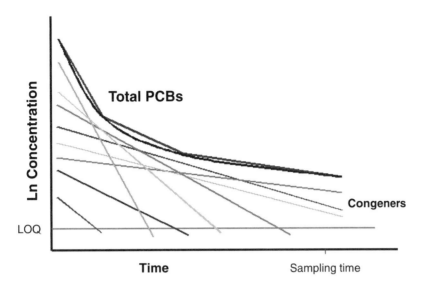

Figure 7.2.   The slope of total PCB decay is not constant at any moment.

total PCBs for years after exposure. The first consideration pertains to the assumption of first-order toxicokinetics, which establishes that the rate of decay of plasma concentration is independent of the magnitude of exposure. This may or may not be true in each case. It is well recognized that saturation or, especially, induction of enzymes responsible for enzymatic metabolism may transiently affect the clearance and half-life of PCBs (Brown and Lawton 2001; Masuda 2001). It is also known that chlorinated aromatics are excreted directly from the central compartment into the intestinal lumen, and this rate is greater at higher burdens and reduced exposures because of favored equilibrium (Aylward et al. 2005; Bergman, Larsen, and Bakke 1982; Saghir, Koritz, and Hansen 1999; Sundlof et al. 1982). Concentration-dependent elimination of PCBs is very rapid immediately after removal of a major source of exposure; in occupationally exposed individuals serum PCB concentrations are reduced by 50 percent 0.25–1.0 years after PCB use is discontinued (NIOSH 1977).

A second consideration, at least for individual congeners, is the basic structure of the blood concentration-time profile. Although figure 7.1 depicts PCB decay in blood as a straight line, in reality, the concentration-time profile can be better represented by a multiphase curve with typically two or three segments. To make matters more complex, when we consider the disposition of total PCBs, even after logarithmic transformation of the concentration axis, the concentration-time profile can be represented neither by a single straight line nor by a multisegment line (multicompartmental profile). As presented in figure 7.2, the decay of total PCBs in blood does not conform to a straight line, but to a curve.

In other words, we cannot properly talk about a half-life for total PCBs because the slope of the disposition curve is constantly changing on account of the large number of chemicals (up to 209 congeners) disappearing from blood at different rates. Not even the curvature is constant, but changes depending on which congeners are left in plasma at any given moment. The decay will be faster early after peak concentration (or pseudo-steady-state concentration), as compared to several years after the peak, because the short-lived congeners disappear more quickly from blood than the more persistent congeners. Only when a few congeners with very similar half-lives are left in the body, many years after exposure, will the decay profile approximate a straight line. Individual PCB congeners exhibit multiphasic elimination kinetics; nevertheless, most of the curvilinear structure of the total PCB decay in blood is caused by its composite nature. At every moment, the elimination rate constant is a weighted average of the elimination rate constants of all the congeners present in the mixture. Since the rates of decay vary for each individual congener, there will be different relative proportions of congeners in blood at different times and, therefore, the elimination rate "constant" will change continuously. Although this curved line could be simulated if the compartmental kinetic features of each congener in the mixture were to be known, the best we can do at this stage is simplifying the curve by using different "pseudo-half-lives" for different periods after peak exposure (figs. 7.2 and 7.3). When traveling back in time, these half-lives will be shorter the closer we get to the peak exposure or pseudo-steady-state phase.

Measured concentration. In addition to analytical uncertainty, measured concentration can be affected both by the

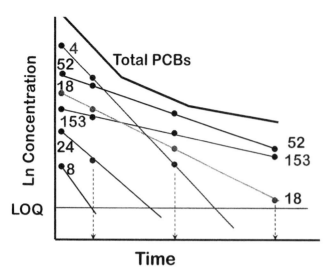

Figure 7.3.   Comparison of (arbitrary) PCB congener half-lives during selected time periods and the effect on average total PCB half-life.

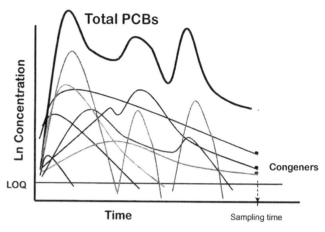

Figure 7.4.   Blood levels of PCBs result from multiple exposures and routes.

rate of decay of congeners from an old exposure and by the relatively sudden effect of recent exposure. If there is evidence of recent exposure, corrective measures have to be adopted to minimize the ensuing bias in our predictions. Figure 7.1 presents the disposition of PCB as a continuum resulting from a single exposure event and a constant first-order elimination process. In reality, as depicted in figure 7.4, multiple exposure events may take place after the first peak concentration is achieved. Individuals can be exposed many times by different routes (transdermal, inhalation, and oral, which could be highly variable depending on the source ingested); therefore, the measured concentration reflects not only the initial exposure but also any ensuing exposure, whether recent or historical.

As depicted in figure 7.5, if we were to trace back from the measured concentration to the initial concentration using the slope, even if the latter were to be 100 percent accurate, we might be overestimating the initial exposure due to more recent (one or many) exposure events. To compensate for the potential effect of this source of bias, if there is evidence of recent exposure, two types of adjustments can be made. First, we can modify the half-life during the final segment of our "multilinearized" PCB decay profile. The selected half-life should be longer than initially estimated and applied for an adequate amount of time. Second, we can adjust the measured concentration and estimate what the latter may have been if no recent exposure had taken place. Assuming a constant volume of distribution of the chemicals, the ratio of half-lives, with and without recent exposure, would approximately parallel the ratio of clearances, which in turn would approximately represent the ratio of concentrations. As dis-

cussed in the next section, the current half-life can be estimated from the congener profile of the total PCB measured, and the half-life that would be present, if no recent exposure had occurred, can be estimated from the original congener (or homologue) profile and the half-lives of the individual congeners (or homologues).

Time from exposure. A very delicate component of the estimation procedure and one that needs to be reasonably well estimated in order to minimize prediction bias is time from exposure. Figure 7.6 represents two hypothetical cases of the same total exposure but at different exposure rates. In A, the exposure takes place during 20 time units at exposure levels that decrease with time while in B the exposure takes place along 40 time units at exposure levels half those of A. As we can appreciate, if we were to extrapolate in both cases from the concentrations obtained after 80 time units back to the concentrations at time 0, we would obtain very different predicted concentrations. For case A, we should extrapolate back during 60 time units while in the case of B we should extrapolate back 40 time units. Furthermore, in both cases, although more clearly visible in B, there is a pseudo-steady-state level that persists for a undefined amount of time, which results from continuous or regularly spaced exposure at decreasing levels. This may well reflect a situation in which a very high initial contamination persisted for a certain period of time at decreasing levels, so that, during that period, a fraction of the elimination rate is offset by a decreasing but significant entry of PCBs into the system. In order to account for this, a new shallower slope should be considered during the period corresponding to the pseudo-steady state. Interestingly, this may require the application of a shallower segment (longer half-life) to trace back from the pseudo-peak at the end of the pseudo-steady-state phase to the real peak. In summary, we would travel back from the

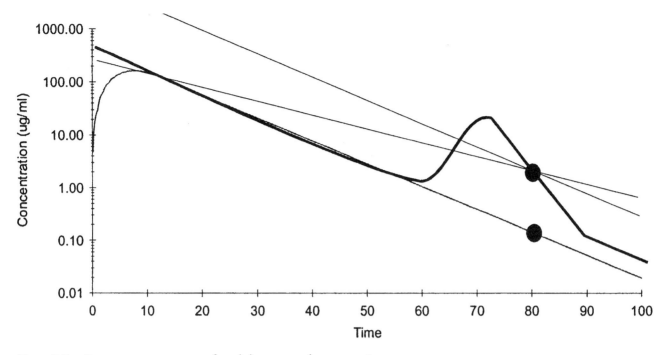

Figure 7.5. Recent exposure may confound the measured concentration.

measured concentration by using several segments of increasing steepness, and after reaching the pseudo-peak we would use a shallower segment.

Toxicokinetic extrapolation. For the purpose of this preliminary evaluation and for clarity of presentation, we decided to use homologue groups rather than individual congeners for the extrapolations. We understand that this hampers the precision of our estimations and predictions, but these will be refined in future studies, which will include both individual congener and demographic data. We have chosen not to use demographic data at this stage in order to protect the identity of the affected individuals.

The most variable and uncertain parameters to estimate are the average half-lives of the PCB homologue groups. Very broad values of elimination half-lives have been estimated for TCDD, PCB congeners, and total PCB in humans; however, when sources of variability and nonlinear toxicokinetics are considered, the variability is reduced (Aylward et al. 2005; Shirai and Kissel 1996). Some variability, of course, is due to inter-individual differences. Much of the variability is due to the ongoing exposures artifact, which has a larger impact as the body burden is lowered (Shirai and Kissel 1996). Thus the half-life of PCB 153, for example, may be between 4 and 5 years in Yusho patients for the

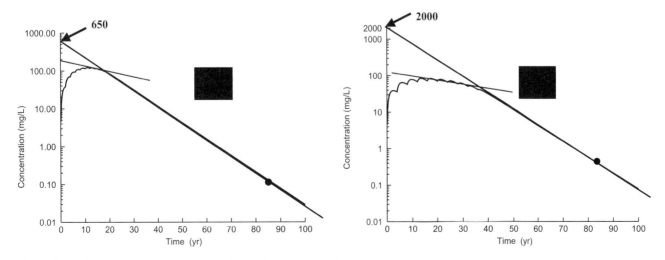

Figure 7.6. The time to pseudo-peak is affected by the rate and period of exposure.

Table 7.1. Half-Lives for PCB Homologue Groups as Derived from Dominant Congeners in the Profiles and Literature Estimations of Congener Half-Lives.

Homologue Group	Dominant Congeners		Estimated Half-Life (Years)		
	Episodic	Persistent	Initial	Induced	Delayed
Mono-CBs	All	-	0.2	-	-
Di-CBs	All	15	0.2	-	-
Tri-CBs	18, 31	28	0.5	0.25	1.0
Tetra-CBs	44, 52, 70	47, 66, 74	1.0	0.75	1.5
Penta-CBs	95, 101, 110	99, (105,118)	2.0	0.75	3.0
Hexa-CBs	132, 149, 151	138, 153, 156	4.0	3.5	8.0
Hepta-CBs	185	170, 180, 187	5.0	5.0	8.0
Octa-CBs	-	194, 203	6.0	6.0	9.0
Nona-CBs	-	206	7.0	7.0	9.0
Deca-CB	-	209	9.0	9.0	10.0

first 15 years, and then the apparent half-life might lengthen to around 15 years as the body burden declines (Masuda 2001). Considering multiple routes of parent PCB elimination, such as slow excretion in addition to more rapid metabolism and excretion, actual half-lives in excess of 8–10 years are unlikely. For most of the persistent PCBs, half-lives of 2–6 years are most likely (Hansen 1999; Shirai and Kissel 1996), while some labile congeners may have half-lives of less than 30 days (Brown and Lawton 2001; Hansen 1999; Shirai and Kissel 1996). Congener and average homologue half-lives were estimated with guidance from the literature (Brown 1987; Brown and Lawton 2001; Chen et al. 1982; Hansen 1999; Hansen and Welborn 1977; Masuda 2001; Masuda, Kagawa, and Kuratsune 1974; Shirai and Kissel 1996).

Estimations of homologue half-lives, more appropriate for initial extrapolations, must take into account the dominant congeners within the homologue group of the residue profile of concern. These estimates will be adjusted according to the age of the residue (see above), evidence of current or recent exposures (such as the presence of labile or "episodic" congeners), and evidence of enzyme induction, especially of CYPs 1A. Regardless of the agent inducing CYPs 1A, this induction has a profound influence on PCB congener compositions. In the Yusho and Yucheng patients, co-exposure to PCDFs (polychlorinated dibenzo-furans) was presumed to have caused significant enzyme induction, resulting in a "Type A" profile (Brown and Lawton 2001; Brown et al. 1989; Chen et al. 1982; Masuda 2001; Masuda, Kagawa, and Kuratsune 1974). This is most readily characterized by the ratio of PCB 118 (readily metabolized in induced individuals) to PCB 156 (PCB 156 = PCB 118 + 1 chlorine and is refractory to metabolism). The usual ratio, 118:156, is

between 4:1 and 10:1, depending on the exposure source and the age of the residue. In this dataset, PCB 156 was not resolved from the trace homologue, PCB 157; thus our ratio includes the minor contribution of PCB 157, which is even more refractory to metabolism. Ratios below about 1:2.5 are indicative of CYP 1A induction and will accelerate the elimination of PCBs.

Taking these factors into account, the following half-lives were established for each homologue group under varying conditions. MonoCBs and DiCBs are considered labile as a group. After assigning average values of half-life for each homologue group estimated from values for individual congeners published in the literature (table 7.1), we weighted each homologue half-life by the proportion of that homologue present in the blood concentration measurement. The half-life of the total residue at blood sampling was estimated by averaging the weighted homologue half-lives. This average value was compared with the half-life that would be present if no recent exposure had taken place. The latter was estimated by calculating the relative disappearance of homologues from the initial profile, based on the corresponding homologue half-lives from table 7.1 (initial, induced, or delayed). The time to back extrapolate was estimated at 30 years (2002 to 1972) based on historical environmental exposure information. For the first 15 years, we used the half-lives from column "initial" on table 7.1. If there was evidence of early induction (118/156 ratio = <2.0), we used the "induction" column for that period. For the second 15 years, we used the half-lives from the column "delayed" (table 7.1). The initial homologue profile was estimated in two different ways. First, if the individual did not exhibit a high level of congener 209 in blood (see chapter 5, this volume, as well), we estimated the initial homologue profile by obtaining the weighted average of the homologue profiles according to the following production breakdown of recorded PCB production in the Anniston plant (5): Aroclor 1242 (67% of total), 1254 (16%), 1260 (7%), 1248 (4%), 1262 (3%), 1221 (2%), 1268 (1%). This led to the following homologue breakdown: 1 (0.54%), 2 (10.58%), 3 (32.21%), 4 (16.73%), 5 (24.94%), 6 (9.17%), 7 (4.82%), 8 (1.01%), 9 (0.1%), and 10 (0.01%).

Second, if the individual did exhibit a high level of congener 209, the initial homologue profile was assumed to be much heavier (high proportion of highly chlorinated congeners) and was estimated to be even heavier than that of Aroclor 1268. This approach is rather conservative, since the half-lives used to back-extrapolate would be much larger. If the ratio of the current estimated half-life to the half-life that would be present if no exposure had occurred was less than one, the measured concentration (starting point of our back-extrapolation) was multiplied by this ratio in order to compensate for the confounding effect of recent exposure. Recent exposure was suspected if the levels of trichlorobiphenyl and tetrachlorobiphenyl homologues in the mixture were higher than those expected in aged residues such as these three.

RESULTS

Historical environmental contamination data indicate that PCBs were manufactured in Anniston, Alabama, until 1971, with a minimum of plant emissions to the atmosphere from 1953 to 1971 of 20.5 MT (Hermanson and Johnson 2007). The atmospheric concentration decreased sharply just one year before (emission controls) and, again, one year after cessation of the exposure. It is likely that the decrease in the actual level of contamination was not so drastic due to accumulation and circulation among the atmosphere and diverse sink compartments (plants, air, soil, water, etc.). Therefore, for our calculations, we are assuming that contaminated individuals reached two pseudo-steady-state levels, one by 1969–71 and a lower level by 1974–75; after that there was a net decline in blood levels, with transient increases due to localized contamination and exposure.

Subject A: As shown in table 7.2, the total concentration (PCB wet) measured on this individual was 60.1 ng/ml serum, which is considered very high (ATSDR 2000b; Orloff et al. 2003; chapter 5, this volume). The congener ratio 118/156+157 was 4.31, which likely indicates that some minor induction was present some time earlier, but there is little justification to use a shorter half-life. The proportion of tetra-CBs in this individual was 3.17, which may indicate some level of recent exposure. Therefore, we modified the measured concentration by the ratio of the average current estimated average half-life (7.3 years) and the average expected half-life (8.65 years), which resulted in a concentration of 51 ng/ml instead of 60.1 as the starting point. As discussed earlier, assuming linear toxicokinetics, this ratio roughly represents the ratio of total clearance, which in turn represents the ratio of concentrations. The proportion of congener 209 was 6.07. The higher than usual level of congener 209 in this individual might be an indication of a very old residue as well as the possibility (Hermanson and Johnson 2007; chapter 5, this volume) of this individual having been exposed to a congener mixture containing a high proportion of highly chlorinated homologs. Half-lives for the different segments were calculated from the weighted homologue average half-lives at different times after original exposure and were 8.61 years for the period 2002–1987 and 5.62 years for the period 1987–72. Starting at the measured concentration in 2002 and doubling the concentration every half-life back in time, we obtained a range of 1,085–1,388 ng/ml between 1972 and 1970. This was within the range of total PCBs found in capacitor workers in 1976 (mean = 1,500 ng/ml) (Lawton et al. 1985b) when corrected for a decimal error (Lawton et al. 1985a). It was, however, less than 10 percent of the maximum level observed (21,000 ppb) (Lawton et al. 1985a; Lawton et al. 1985b).

Subject B: The total concentration (PCB wet) measured on this individual was 17.7 ng/ml serum, moderately higher than fish consumers (ATSDR 2000b; Orloff et al. 2003). The congener ratio 118/156 was 6.78, which likely indicates no induction, so there is no justification to use a shorter half-life. The proportion of tetra-CBs in this individual was 3.2, which may indicate some level of recent exposure,

Table 7.2. Predicted Pseudo-Steady-State Concentration for Three Residents from Anniston, Alabama.

Subject	Concentration (ng/ml)	118/156+157 Ratio	Congener 209	Predicted Serum Concentration (1972–70)
A	60.1	4.31	6.07%	1,085–1,388 ng/ml
B	17.7	6.78	0.5%	1,424–2,150 ng/ml
C	43.9	1.46	4.6%	937–1,200 ng/ml

younger residues, or a much less heavy congener pattern than in the case of subject A. The presence of significant amounts of episodic penta-CBs (95 + 101 + 110 = 1.1%) and hexa-CBs (132 + 149 + 151 = 2.1%) favors recent exposure. However, the estimated current and expected half-lives were very close and no adjustment of the measured concentration was applied. The proportion of congener 209 was 0.5, which indicated the possibility of an initial profile with higher proportion of less chlorinated congeners, as compared to subject A. Therefore, the initial homologue profile was obtained from the weighted proportions of the different Aroclors, as described in the previous section. Half-lives for the different segments were calculated from the weighted homologue averages at different times after original exposure, and were 7.94 years for the period 2002–1987 and 3.38 years for the period 1987–72. Starting at the measured concentration in 2002 and doubling the concentration every half-life back in time, we obtained a range of 1,424–2,147 ng/ml between 1972 and 1970. This was slightly higher than the previous subject, but still well within the range of serum levels for capacitor workers.

Subject C: As shown in table 7.2, the total concentration (PCB wet) measured on this individual was 43.9 ng/ml serum, which is considered very high (ATSDR 2000b; Orloff et al. 2003). The congener ratio 118/156 was 1.46, which likely indicates that metabolic induction was present at some earlier time, so a shorter half-life (table 7.1, induced) was used for the first 15-year period. The proportion of tetra-CBs in this individual was 1.3, and penta-CBs were

only 7.2 percent, which strongly indicates very low recent exposure and also supports some degree of CYP induction. Therefore, we did not modify the measured concentration by the ratio of the current and expected average half-lives. The proportion of congener 209 was 4.6. The higher than usual level of congener 209 in this individual might be an indication of a very old residue as well as the possibility of this individual having being exposed to a congener mixture containing high levels of highly chlorinated homologs (Montars) or other sources, such as Aroclor 1268, used in foundries. Half-lives for the different segments were calculated from the weighted average homologue half-lives at different times after original exposure and were 8.63 years for the period 2002–1987 and 5.6 years for the period 1987–72. Starting at the measured concentration in 2002 and doubling the concentration every half-life back in time, we obtained a range of 937 to 1,200 ng/ml between 1972 and 1970.

DISCUSSION

Prediction of initial concentrations of total PCB in blood at the time of peak concentration or the end of pseudo–steady state was possible by applying basic toxicokinetic principles to the PCB congener profiles observed at the present time. Assumptions of dose linearity, correlation between current proportions of congener 209 and level of chlorination of the original residue, correlation between congeners 118/156 ratio and level of initial enzymatic induction, correlation between

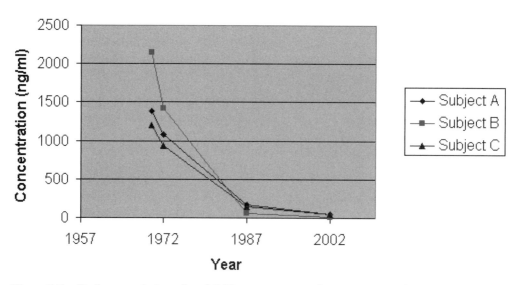

Figure 7.7.   Back-extrapolation of total PCB concentrations for Anniston residents.

proportions of homologues with low chlorination and magnitude of recent exposure, and constancy of the volume of distribution of PCBs were adopted in order to obtain valid predictions. As presented in figure 7.7, the measured total PCB concentration in 2002 was not correlated with the predicted concentration between 1970 and 1972 when all the individuals were considered. The predicted target concentration for subject B, which exhibited a lower current PCB concentration, was indeed higher than that of the other two subjects. Evidence of an original heavier congener profile for subjects A and C justifies this difference. A heavier original profile led us to estimate longer half-lives, resulting in lower initial concentration predictions for A and C. Using this conservative approach is more appropriate when the results of the data analysis may have legal implications. If the estimated original congener profile for subjects A and C had been the same as that for subject B, the predicted target concentration range between 1972 and 1970 would have been 4,400–6,600 ng/ml for subject A and 2,900–4,300 ng/ml for subject C. In any case, the exposure-time concentrations estimated in this study, using the most conservative approach, were well above the normal limits (ATSDR 2000b; Orloff et al. 2003) but within average ranges for capacitor workers at the time (Lawton et al. 1985a; Lawton et al. 1985b). The highest level reported for the capacitor workers in 1976–79 was above 21,000 ppb; in a serum sample drawn from an Anniston resident in 1996, the total PCB level was 2,112 ppb (ATSDR 2000a). This clearly indicates that the Anniston residents considered in this study were exposed 30 years previously to very high levels of environmental PCBs. The exposure (essentially 24 hours/day x 7 days/week) was without benefit of protective clothing or ventilation because the residents, unlike the capacitor workers, were unaware of the presence of PCBs.

The current method of approach results in values consistent with numerous laboratory and epidemiological studies illustrating the accuracy of the estimations. With the high-quality and detailed data available for these subjects, much information regarding individual personal characteristics was derived from the known relationships among PCB congeners, something that is much more difficult with single compounds. To improve precision, one could construct a model that would need further refining and validation with experimental data in a long-lived species such as swine, retrospective data in humans, and more detailed analysis of trend studies. Future studies that include PCB congener profiles, rather than homologues, could be extended to a much larger population if the analytical data were adequate, and could be extended considering their demographic information as well. Furthermore, multivariate statistics techniques can be applied that allow incorporation of inter-individual variability through the use of Monte Carlo simulations and sensitivity analysis.

REFERENCES

Agency for Toxic Substances and Disease Registry (ATSDR). 2000a. "Health Consultation. Anniston, Calhoun County, Alabama. Evaluation of Soil, Blood and Air Data from Anniston, Alabama. Atlanta." Report CERCLIS No. ALD004019048.

———. 2000b. "Toxicological Profile for Polychlorinated Biphenyls." Update. Atlanta: Department of Health and Human Services.

Alabama Department of Public Health (ADPH). 1996. "Cobbtown/Sweet Valley Community PCB Exposure Investigation. Anniston, Calhoun County, Alabama." Birmingham: Alabama Department of Public Health for ATSDR.

Aylward, L. L., R. C. Brunet, G. Carrier, S. M. Hays, C. A. Cushing, L. L. Needham, D. G. Patterson Jr., P. M. Gerthoux, P. Brambilla, and P. Mocarelli. 2005. "Concentration-Dependent TCDD Elimination Kinetics in Humans: Toxicokinetic Modeling for Moderately to Highly Exposed Adults from Seveso, Italy, and Vienna, Austria, and Impacts on Dose Estimates for the NIOSH Cohort." *J. Exposure Analysis and Environ. Epidemiol.* 15:51–65.

Bergman, A., G. Larsen, and J. E. Bakke. 1982. "Biliary Secretion, Retention and Excretion of Five ^{14}C-Labeled Polychlorinated Biphenyls in the Rat." *Chemosphere* 11 (3): 249–53.

Brown, D. P. 1987. "Mortality of Workers Exposed to Polychlorinated Biphenyls—an Update." *Arch. Environ. Health* 42 (6): 333–39.

Brown, J. F., Jr. 1994. "Determination of PCB Metabolic, Excretion and Accumulation Rates for Use as Indicators of Biological Responses and Relative Risk." *Environ. Sci. Technol.* 28: 2295–2305.

Brown, J. F., Jr., and R. W. Lawton. 2001. "Factors Controlling the Distribution and Levels of PCBs after Occupational Exposure." In *Recent Advances in the Environmental Toxicology and Health Effects of PCBs,* edited by L. W. Robertson and L. G. Hansen, 103–9. Lexington: University of Kentucky Press.

Brown, J. F., Jr., R. W. Lawton, M. R. Ross, J. Feingold, R. E. Wagner, and S. B. Hamilton. 1989. "Persistence of PCB Congeners in Capacitor Workers and Yusho Patients." *Chemosphere* 19:829–34.

Chen, P. H., M. L. Luo, C. K. Wong, and C. J. Chen. 1982. "Comparative Rates of Elimination of Some Individual Polychlorinated Biphenyls from the Blood of PCB-Poisoned Patients in Taiwan." *Food Chem. Toxicol.* 20 (4): 417–25.

Erickson, M. D. 1997. *Analytical Chemistry of PCBs.* 2nd ed. Boca Raton, Fla.: Lewis Publishers.

Hansen, L., A. DeCaprio, and I. Nisbet. 2003. "PCB Congener Comparisons Reveal Exposure Histories for Residents of Anniston, Alabama, USA." *Fresenius Environ. Bull.* 12 (2): 181–89.

Hansen, L. G. 1999. *The Ortho Side of PCBs: Occurrence and Disposition.* Boston: Kluwer Academic.

Hansen, L. G., and M. E. Welborn. 1977. "Distribution, Dilution and Elimination of Polychlorinated Biphenyl Analogs in Growing Swine." *J. Pharm. Sci.* 66 (4): 497–501.

Hermanson, M. H., and G. Johnson. 2007. "Polychlorinated Biphenyls in Tree Bark near a Former Manufacturing Plant in Anniston, Alabama." *Chemosphere* 68:191–98.

Lawton, R. W., J. F. Brown Jr., M. R. Ross, and J. Feingold. 1985a. "Comparability and Precision of Serum PCB Measurements." *Arch. Environ. Health* 40 (1): 29–37.

Lawton, R. W., M. R. Ross, J. Feingold, and J. F. Brown Jr. 1985b. "Effects of PCB Exposure on Biochemical and Hematological Findings in Capacitor Workers." *Environ. Health Perspect.* 60:165–84.

Masuda, Y. 2001. "Fate of PCDF/PCB Congeners and Change of Clinical Symptoms in Patients with Yusho PCB Poisoning for 30 Years." *Chemosphere* 43 (4–7): 925–30.

Masuda, Y., R. Kagawa, and M. Kuratsune. 1974. "Comparison of Polychlorinated Biphenyls in Yusho Patients and Ordinary Persons." *Bull. Environ. Contam. Toxicol.* 11:213–16.

National Institutes of Occupational Safety and Health (NIOSH). 1977. "Occupational Exposure to Polychlorinated Biphenyls (PCBs)." Department of Health, Education and Welfare, NIOSH, Publication No. 77–225.

Orloff, K. G., S. Dearwent, S. Metcalf, S. Kathman, and W. Turner. 2003. "Human Exposure to Polychlorinated Biphenyls in a Residential Community." *Arch. Environ. Contam. Toxicol.* 44 (1): 125–31.

Safe, S. H. 1994. "Polychlorinated Biphenyls (PCBs): Environmental Impact, Biochemical and Toxic Responses, and Implications for Risk Assessment." *Crit. Rev. Toxicol.* 24 (2): 87–149.

Saghir, S. A., G. D. Koritz, and L. G. Hansen. 1999. "Short-Term Distribution, Metabolism, and Excretion of 2,2',5-Tri-, 2,2',4,4'-Tetra-, and 3,3',4,4'-Tetrachlorobiphenyls in Prepubertal Rats." *Arch. Environ. Contam. Toxicol.* 36 (2): 213–20.

Shirai, J. H., and J. C. Kissel. 1996. "Uncertainty in Estimated Half-Lives of PCBs in Humans: Impact on Exposure Assessment." *Sci. Total. Environ.* 187 (3): 199–210.

Sundlof, S. F., L. G. Hansen, G. D. Koritz, and S. M. Sundlof. 1982. "The Pharmacokinetics of Hexachlorobenzene in Male Beagles: Distribution, Excretion, and Pharmacokinetic Model." *Drug Metab. Dispos.* 10 (4): 371–81.

8

PCB Sources and Human Exposure in the Slovak Republic

Anton Kočan, *Slovak Medical University*

Ján Petrík, *Slovak Medical University*

Beata Drobná, *Slovak Medical University*

Jana Chovancová, *Slovak Medical University*

Stanislav Jursa, *Slovak Medical University*

Branko Balla, *Slovak Medical University*

Tomáš Trnovec, *Slovak Medical University*

INTRODUCTION

Major sources of environmental contamination with PCBs in the Slovak Republic are (1) PCB production in eastern Slovakia during the period 1959–84 and (2) PCB use in electrical and heat exchanging equipment and as a paint additive. Consequently, PCBs are present in environmental compartments such as air, surface water, and soil-contaminated crops and in human foods such as meats, milk, and eggs. A higher PCB content in water sediment, surface water, soil, and air has resulted in high exposure of wild fish and food products coming from cattle, swine, and poultry kept in the polluted area and fed with locally produced feed. (Kočan et al. 2001; Petrík et al. 2001b; Chovancová et al. 2003; Drobná et al. 2003).

RESULTS

PCB Production

The Chemko chemical factory situated in eastern Slovakia at Strazske town in the Michalovce District produced PCBs for 25 years, starting in 1959 and closing production in January 1984. In total, about 21,482 t (tonnes, or metric tons) of PCB formulations called Delor (Delotherm, Hydelor) were produced (Kočan et al. 2001). A trend in the production is shown in figure 8.1. The producer has declared that 9,869 t (46%) was exported in particular to the former German Democratic Republic. The rest, that is, 11,613 t, was used inside former Czechoslovakia chiefly as heat exchanger fluids, capacitor and transformer dielectric fluids, and paint additives.

Moreover, it is estimated that about 1,600 t of PCB wastes (mainly distillation residues containing highly chlo-rinated biphenyls and terphenyls, as well as PCDFs) were generated during the production. Most of these wastes are still stored waiting for safe and definitive disposal. Especially during the first half of the production, large amounts of PCBs were released into the environment (mainly into watercourses and soil) due to poor technological measures. In addition, former ignorance about the environmental persistency of PCBs and their harmful effects on living organisms caused that almost no attention was paid to PCB releases at that date. A part of the Chemko PCB waste entered the environment mainly via the factory effluent canal causing the contamination of the Laborec River and Zemplinska Sirava Lake. Another part could lie at Chemko landfill sites in a form of contaminated soil mixed with miscellaneous production waste. It should be noted that during the production a part of the distillation residues was mixed with heavy oil and used as fuel in a factory heating plant. About 115 t of unsold and used PCBs stored at Chemko were burned in the 1970 and 1980s in Slovakia's cement kilns.

The Chemko PCB production was based on the direct catalytic ($FeCl_3$) chlorination of pure biphenyl also manufactured in Chemko by the high-temperature pyrolysis of benzene. When the expectable stage of chlorination was achieved a reaction mixture was neutralized with sodium hydroxide and vacuum distilled. Lower chlorinated compounds were distilled out of the mixture using steam-jet air pumps. Those compounds condensed in vacuum-pump water and finally reached a Chemko effluent canal. Distillation residues containing highly chlorinated biphenyls and terphenyls, and inorganic soils were filled into drums and stored. Dioxin-like compounds such as polychlorinated dibenzofurans (PCDFs) have not been analyzed in the distillation

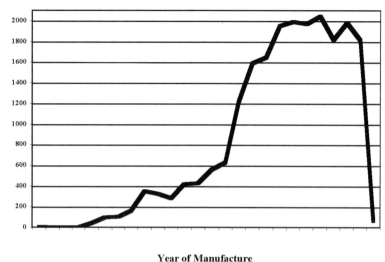

Year of Manufacture

Figure 8.1. Amounts of PCB formulations manufactured by Chemko Strazske in 1959–84.

residues and any other wastes. Because the residues were heated up to high temperature (sometimes about 500 degrees C) due to neutralization reactions, one can expect a high content of PCDFs and other dioxin-like compounds. Leaks in production installations caused by corrosion and their maintenance, adsorbents from PCB refinement, and pouring away test samples from the production were additional sources of environmental pollution.

There was an attempt to manufacture polychlorinated terphenyls in Chemko. About 35 t were produced of which about 30 t were used at glue production for the furniture industry. Production devices as well as those for the PCB production corroded to a large extent resulting in leakage of reaction mixtures.

PCB Use and Waste around Slovakia

PCB formulations are still used in Slovakia in some closed systems. According to inventory estimation done in 1997 PCBs have been used in power capacitors only, which are located mainly at transformer and switchgear units. The capacitors can contain from 1.4 to 20 kg PCBs. Since operators had not often known that the capacitors contained PCBs these devices have not been disposed of in a proper way. As their casing is fragile, PCBs from damaged devices can easily enter the environment.

There is insufficient information on the present use of PCBs in heat exchanging systems, which were an important source of environmental contamination with PCBs in the past. A lot of heat exchangers around Slovakia used PCB fluids (for example, facilities preparing asphalted gravel for road construction). Because of leakage, hundreds of tons of PCBs were released into soil under the exchangers. Several

years ago some companies managing those facilities collected used PCBs and ensured their disposal abroad by incineration. It seems that the fillings of transformers operating in Slovakia have already been refilled with non-PCB-containing fluids.

Because PCBs were also used as a paint additive (their content was up to 21%), old paint coatings can release these pollutants causing environmental contamination.

Besides Chemko having more than 1,000 t of PCB waste, tens of tons could still be present in heat exchangers at the former intensive users of PCBs. Up to 1,000 t of PCBs could be encased in power capacitors that are still in use or disabled and stored at their users. A graphical illustration of the 1997 inventory estimation is presented in figure 8.2.

Environmental Pollution with PCBs in the Polluted and Background Areas

The Michalovce District (polluted area, Chemko factory) and the Stropkov District (control and background area) highlighted on a schematic map of Slovakia (fig. 8.3) have been chosen for the project as well as for a pilot project whose results are presented in this review (Kočan et al. 2001; Petrík et al. 2001b; Chovancová et al. 2003; Chovancová, Kočan, and Petrík 2004).

PCBs, which were produced for 24 years in the Chemko chemical factory, have caused increased contamination of the district environment. The Stropkov District, about 60 km upwind and upstream from the Michalovce District, has no extraordinary sources of PCB pollution. Both districts, however, are characterized by a dominant river (Laborec in the Michalovce District and Ondava in the Stropkov District) and a large lake (Zemplinska Sirava vs. Domasa; both lakes

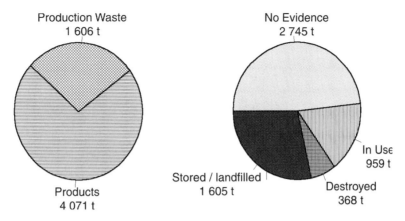

Figure 8.2.   Pie chart of the outcomes of Slovakia's PCB inventory estimation in 1997 (Kočan et al. 1998).

were artificially built for recreational, industrial and/or irrigation purposes) fed by Laborec waters and Ondava ones, respectively (fig. 8.4). Because of the geographical closeness of the districts, human populations living there are alike as regards nutrition and social habits. A map of the most polluted part of the Michalovce District showing the Chemko factory, its effluent canal, Laborec River, and Zemplinska Sirava Lake (with its filling and discharging canals) is in figure 8.5.

Data obtained within research projects aimed at PCB analysis in human blood, adipose tissue, and milk collected in several of Slovakia's districts have shown that PCB levels were substantially higher in the Michalovce District than in the rest of Slovakia (Petrík et al. 1991, 2001a; Kočan et al.

1994a, 1994b, 1995, 1996). Therefore, in 1997–98 a pilot research project, the Burden of the Environment and Human Population in an Area Contaminated with PCBs, already focused on this district was realized. It defined a state of (1) environmental pollution with PCBs involving ambient air, water sediment, soil, and wildlife (fish and game animals) in the Michalovce and Stropkov Districts selected as a control and background area, (2) food contamination, (3) human exposure, and (4) human health effects, in particular thyroid ones.

Ambient Air

Six 24-hour high-volume ambient air samples were collected in September 1997 at the distance of 1 to 30 km from

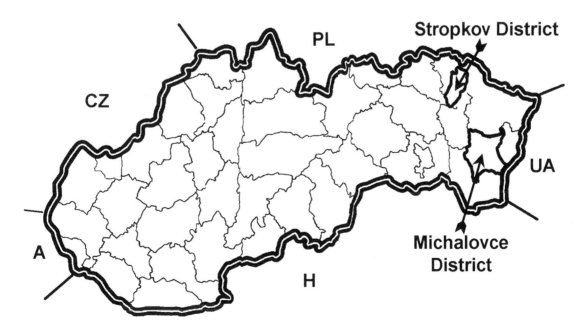

Figure 8.3.   The position of the polluted area (Michalovce District) and control area (Stropkov District) with background PCB pollution within the Slovak Republic.

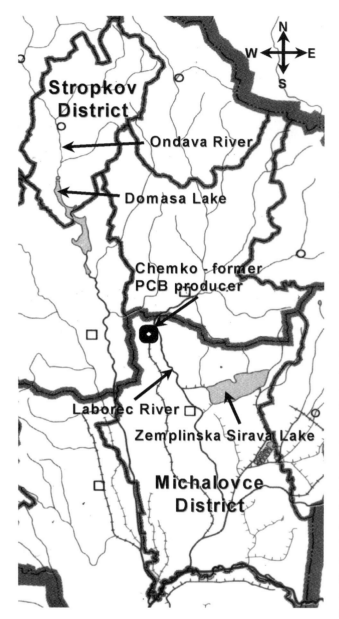

Figure 8.4.   More detailed map of the polluted and background areas with rivers and lake. The population of the Michalovce District is 108,000; the Stropkov District population is 20,500.

the Chemko factory (Michalovce District). Six ambient air samples were taken in the control area (Stropkov District) as well. All the samples were collected in residential areas (Kočan et al. 2001).

PCB concentrations (the sum of all congeners determined) in ambient air are presented in figure 8.6. Several times higher values were found close to the Chemko factory and its landfill and storage sites in Strazske (sampling sites Vola, Strazske). Somewhat increased levels were observed in towns (Michalovce, Stropkov); this might be influenced by more intense traffic in the towns causing the higher air occurrence of PCB-containing dust particles. PCB concen-

trations at the remote sites of the Michalovce District were similar to those in the control area. As expected, the air samples contained lower chlorinated congeners rather than higher chlorinated ones (Kočan et al. 1998).

Soil

Thirty-one samples were taken from an upper soil layer (20 cm) near the disposal sites of the PCB manufacturer (Chemko), agricultural fields in Michalovce and Stropkov, and at several plants around Slovakia preparing gravel coated with asphalt and using PCBs in their heat exchangers. Each sample was created by averaging 5 to 10 soil plugs taken by a steel tube 2 cm in diameter lined with a disposable PTFE sheet hammered into the soil at 10-m span following the pattern of "dice five" (Kočan et al. 2001).

Soil samples taken close to the waste disposal sites of Chemko and plants preparing asphalted gravel contained PCB levels as illustrated in figure 8.7. As can be seen in figure 8.8, PCB levels in the soil samples taken from the agricultural fields near some towns and villages were considerably lower (on average 0.008 mg/kg) than those taken from the neighborhood of disposal sites and plants mixing asphalt and gravel (compare with fig. 8.7).

Plants situated close to quarries, that is, in mountains on rocky ground, prepared gravel coated with asphalt to be used for road construction. Because of the inherent looseness of heat-exchange systems filled with PCBs, hundreds of tonnes of PCBs leaked from those systems. PCB concentrations peaked at 53,000 mg/kg in a soil sample taken under one of these heat exchangers. However, high PCB levels were also observed in agricultural (35 and 38 mg/kg) or forest soil (3.9 and 7.5 mg/kg) taken near those plants. As soil from the vicinity of a major Chemko landfill site which could be contaminated only by air-transported particles contained increased PCB levels (0.4 to 5.8 mg/kg), it is likely that this dump contains large quantities of PCBs. Similarly, PCB waste might be present at the municipal dump of Michalovce (the district town) since increased PCB concentrations were found in a sample collected close to the dump (0.17 mg/kg).

Surface Water and Water Sediment

Thirty-three 5-L water samples were taken from the open effluent canal of the PCB manufacturer, the Laborec River (the canal is emptying into it), Zemplinska Sirava Lake (a large water reservoir serving for flood-control, irrigation, and recreational purposes supplied from the Laborec and several Vihorlat mountain streams), Ondava River, and Domasa Lake (the latter two situated in the control area). Where possible, sediment samples (N = 30) were collected at the same sites using an Ekman sampling grab.

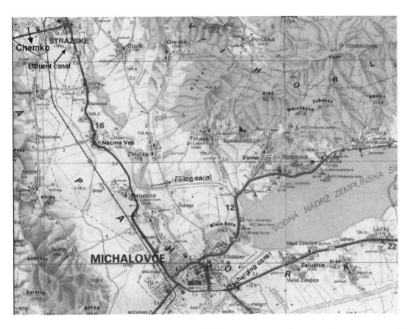

Figure 8.5. Map of a part of the Michalovce District showing the position of the Chemko factory, its effluent canal emptying into the Laborec River, and Zemplinska Sirava filling and discharging canals. The district capital is Michalovce town, with a population of 41,000.

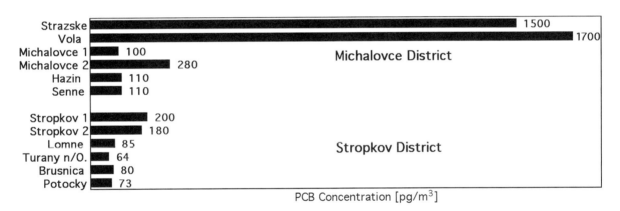

Figure 8.6. PCB concentrations (the sum of all congeners determined) in ambient air samples collected in the residential areas of the Michalovce and Stropkov Districts.

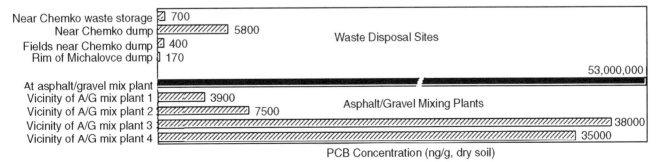

Figure 8.7. PCB levels (the sum of all congeners determined) in soil samples collected in the vicinity of asphalt/gravel mixing plants and the waste disposal sites of the Chemko factory.

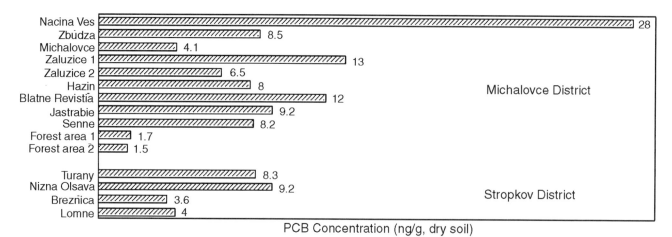

Figure 8.8.   PCB levels (the sum of all congeners determined) in soil samples collected at the agricultural fields of the Michalovce and Stropkov Districts.

Figure 8.9 shows PCB concentrations determined in the sediment samples. As expected, the highest levels were found in a muddy part of an effluent canal flowing from the Chemko factory (one of the samples contained 5,000 mg/kg). As the effluent canal between the factory and its merging with a municipal sewage canal contained sandy sediment (the bed of the canal is concreted over) having low adsorbing effects for PCBs, a lower value (48 mg/kg) was found. No doubt the polluted effluent canal emptying into the Laborec River has caused its contamination (3.9 mg/kg). Zemplinska Sirava Lake (33.5 km² in surface area), which is partly filled from the Laborec, contained several hundred times higher PCB levels compared with a similar lake in the control area (1.7 to 3.1 mg/kg vs. 0.007 to 0.01 mg/kg).

Based on these findings, one can estimate that at least tons of PCBs are still adsorbed in the sediments of the efflu-

ent canal, Zemplinska Sirava, and Laborec. As can be seen in figure 8.10, the sediment has caused the contamination of surface water, although in this case the difference between the polluted and control area is not so substantial.

The levels of PCDD/Fs, non-ortho, and mono-ortho PCBs in sediments originated from Michalovce District were significantly higher than those from the background area (table 8.1).

Not surprisingly, the highest values of dioxin-like PCBs and PCDFs were determined in sediments from the industrial-effluent canal from the Chemko factory. Increased values also were found in sediments from the Laborec River and the water reservoir of Zemplinska Sirava. Low concentrations of PCDDs (polychlorinated dibenzo-p-dioxins) were measured in the Laborec and Ondava Rivers and in Domasa Lake. Moderate PCDD contamination was observed in Zemplin-

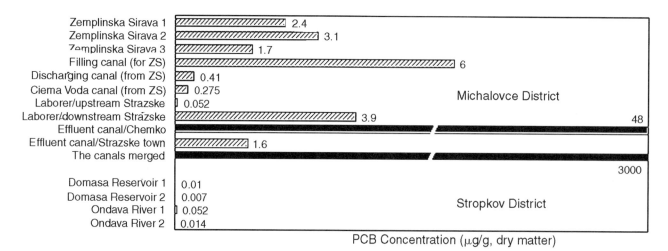

Figure 8.9.   PCB levels (the sum of all congeners determined) in bottom sediment samples taken from some watercourses in the Michalovce and Stropkov Districts.

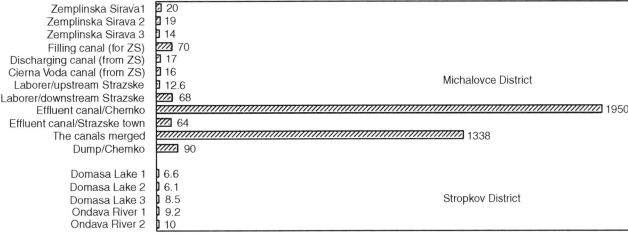

Figure 8.10.   PCB levels (the sum of all congeners) in surface water samples (solid particles were filtered off) taken from some watercourses in the Michalovce and Stropkov Districts.

ska Sirava. The PCDD:PCDF ratio reflects the source of possible contamination. It was found that the TEQs of PCDD/F values of the samples collected in the vicinity of the Chemko effluent canal derive mainly from PCDFs. The other samples with low PCDD/F TEQ concentrations show approximately similar contributions of PCDDs and PCDFs (Chovancová, Kočan, and Petrík 2004; Petrík, Chovancová, and Kočan 2003). PCB PCDD/Fs levels from some other selected water courses in Slovakia are also reported in the table 8.1.

Table 8.1. Levels of PCBs, PCDD/Fs, and Dioxin-Like PCBs in Sediments, Slovak Republic.

	PCDD/Fs	0-Ortho-PCBs	1-Ortho PCBs	PCBs
		(WHO-TEQ, $pg/g_{D.W.}$)		$(ng/g_{D.W.})$
Ondava River (above Bukocel)	0.27	0.17	0.032	19
Ondava River (below Bukocel)	0.53	0.53	0.11	27
Ondava River (below Kijov. Creek)	0.52	0.81	0.16	40
Effluent canal (below Chemko)	56	233	113	108,611
Effluent canal (below WWT plant)	341	1963	1236	733,210
Effluent canal (road Strazske-Michalovce)	586	1992	1039	566,726
Laborec (Vola)	0.23	11	6	1,377
Laborec (Nacina Ves)	0.098	6.8	3.6	967
Laborec (Petrovce)	261	712	431	98,445
Zemplinska Sirava (Biela Hora)	20	53	20	7,308
Zemplinska Sirava (Medvedia Hora)	21	74	22	5,511
Zemplinska Sirava (Kusin)	13	27	7.6	2,277
Domaša (dam)	0.1	0.2	0.039	36
Domaša (Holcikovce)	0.68	0.16	0.029	16
Domaša (Nova Kelca)	0.52	0.16	0.038	11
Nitra River (above NChZ Co.)	1.6	0.26	0.05	22
Nitra River (below NChZ Co.)	0.66	0.41	0.11	133
Nitra River (Chalmova)	1.1	0.57	0.11	74
Vah River (Liskova)	0.68	1.1	0.64	116
Vah River (Cernova)	0.73	1	1.3	204
Vah River (Krpeaany Reservoir)	1	0.46	0.23	45

Figure 8.11.   PCB levels (the sum of all congeners) in lipids from fish caught in some watercourses in the Michalovce and Stropkov Districts.

Wildlife

Sixty-nine fishes caught in the polluted area (the Laborec River and Zemplinska Sirava Lake) and 48 fishes caught in the control area (the Ondava River and Domasa Lake) were pooled, taking into consideration genera and catching sites to get 20/11 samples related to the Michalovce/Stropkov waters. Meat samples from 9 game animals, such as deer, roe deer, wild boar, and duck, killed in the Michalovce District, 9 in the adjacent Sobrance District (the samples came from the Vihorlat Hills that extend over both districts), and 16 in the Stropkov District were used for analysis.

Like the sediment, fish from the Michalovce waters contained much higher PCB levels than those from the Stropkov ones (fig. 8.11). The maximum lipid adjusted value peaked even at 900 mg/kg. It was confirmed that predatory fishes are more exposed than fishes feeding on plankton or benthic food. The environmental pollution of the Michalovce District with PCBs is reflected in exposure of forest and field wildlife as well (fig. 8.12).

Food Contamination

Six hundred fifty-nine food samples bought in shops and markets in the Michalovce and Stropkov Districts were pooled to 54 samples from Michalovce and 49 from Stropkov (e.g., everyday milk samples from a dairy collected within one month were pooled). Additionally, 108 food samples from the Michalovce District and 43 from the Stropkov District originating from animals kept on small family farms located at family houses in villages were collected. These farm-raised samples were not pooled. All the food samples were of animal origin, such as cow's milk, butter, beef, pork, salamis, frankfurters, chicken, lard, and eggs (Drobná et al. 2003).

As can be seen in figure 8.13, PCB levels in the farm-raised (homemade) food samples, especially chicken eggs, were, on average, higher than those kept on big farms, where animals are fed with feedstuff mixtures often imported from other parts of Slovakia and possibly less contaminated with PCBs. Chicken eggs from small family farms in the Michalovce District especially contained very high PCB amounts. This could be a result of any of the following:

1. Feed for animals raised at family houses contains a higher portion of plants, such as grass, beet, potatoes, vegetable, and so on, than feed used within large-scale breeding.
2. Animals, in particular, chicken and other livestock, kept on the small farms close to family houses used to have a run and thus could reach contaminated water, soil, or waste more easily than animals kept on closed stables typical of large-scale breeding.
3. Animals raised at family houses live longer than animals from large-scale breeding, and it is known that the POP levels increase with age.

The total concentrations of PCDD/Fs, non-ortho, and mono-ortho PCBs expressed as WHO-TEQs in the retail food-stuffs analyzed from both districts ranged from 0.21 pg/g fat in pork to 4.26 pg/g fat in beef. TEQ PCDD/F levels of all retail food samples analyzed did not exceed the EU legisla-

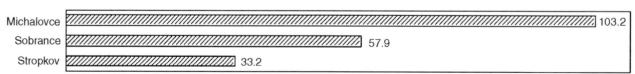

Figure 8.12.   PCB levels (the sum of congeners 28, 52, 101, 118, 138, 153, 156, 170, and 180) in lipids from game animals shot in forests and fields in the Michalovce, Sobrance, and Stropkov Districts.

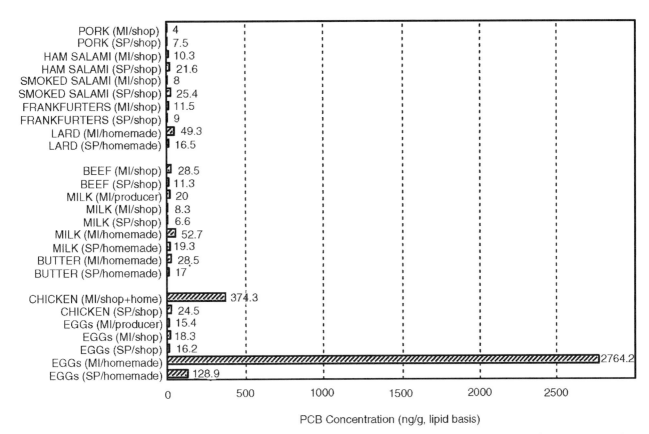

Figure 8.13.   PCB (the sum of congeners 28, 52, 101, 118, 138, 153, 156, 170, and 180), HCB, and p,p′-DDE levels in foods of animal origin collected in the Michalovce and Stropkov Districts.

tion limits. Extremely high levels of PCDFs (30.0 pg TEQ/g fat) and dioxin-like PCBs (225.4 pg TEQ/g fat) were found in chicken eggs from the small private farms located in the Michalovce District (Chovancová et al. 2003; Chovancová, Kočan, and Petrík 2004).

Human Exposure

In October 1998 blood samples from 215 people (107 males and 108 females) living long term (more than 10 years) in the Michalovce District and from 205 people (101 males and 104 females) living long term in the Stropkov District were taken. Among those 215 blood donors from the Michalovce District there were 11 fishermen catching fish in the Laborec River and Zemplinska Sirava Lake. In addition, blood was taken from 38 people (27 males and 11 females) working in former PCB production in Chemko Strazske (Pavuk et al. 2003).

Higher PCB content in some foods available in the polluted Michalovce District has evidently caused increased exposure of the human population of that district (fig. 8.14). Average PCB concentration of 4.2 ppm (the sum of PCB-28, 52, 101, 138, 153, 118, 156, and 170) in human blood lipids taken from the human general population living long term in the Michalovce District was 3.5 times higher than that in the Stropkov District (1.2 ppm). People eating foods coming from animals kept on small family farms at their houses contained significantly higher PCB levels in their blood serum lipids.

While an average 8.6 ppm was found in the blood lipids of the former Chemko workers, fishermen's blood lipids contained 11.2 ppm on average (the maximum value was 60 ppm). PCB levels in men compared to women were somewhat higher in both the Michalovce and Stropkov Districts.

The European Union's 5th Framework Programme, the PCBRISK project (No. QLK4-CT-2000-00488), was aimed at evaluating the PCB (PCB-28, 52, 101, 123[+149], 118, 114, 153, 105, 138[+163], 167, 156[+171], 157, 180, 170, and 189) and selected organochlorine pesticide (a-HCH, b-HCH, g-HCH, HCB, pp′-DDT, and pp′-DDE) exposure of adults (equally men and women) and children (equally boys and girls, eight to nine years old) living in the towns and villages of the Michalovce District adjacent to the Laborec River, downstream from the Chemko plant where higher PCB human levels could be expected, and in the towns and villages of the Svidnik and Stropkov Districts, several tens of

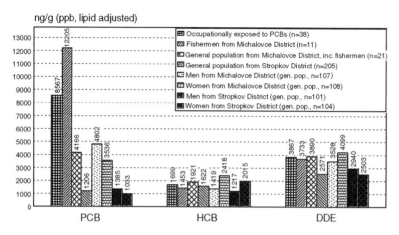

Figure 8.14.   PCB (the sum of congeners 28, 52, 101, 118, 138, 153, 156, 170, and 180), HCB, and p,p′-DDE levels in blood serum lipids from various human population groups living in the Michalovce and Stropkov Districts.

kilometers upstream from the Laborec, where lower PCB exposure was expected. The exposure of subgroups of adults to dioxins, furans, and dioxin-like PCBs was evaluated as well (Kočan et al. 2004).

Out of 2,049 blood samples taken from adults (1,011 in the Michalovce District and 1,038 in the Svidnik/Stropkov Districts) and 460 taken from children (231 in the Michalovce District and 229 in the Svidnik/Stropkov Districts),

Table 8.2. PCB and Some Organochlorine Pesticides in 2,047 Human Blood Serum Samples from Adults in Eastern Slovakia.

Compound	Content, ng/g, Lipid Weight Basis						
	Median	Mean	Minimum	10-Percentile	90-Percentile	Maximum	Percentage Detected
PCB-28	-	-	<1.7	-	-	565.0	38.7
PCB-52	-	-	<1.6	-	-	625.0	8.9
PCB-101	-	-	<1.3	-	-	253.0	15.6
PCB-105	-	-	<1.4	-	-	447	45.8
PCB-114	-	-	<1.3	-	-	152.0	27.8
PCB-118	31.4	62.6	<4.3	10.1	120.0	3,539.0	94.1
PCB-123[+149]	-	-	<1.2	-	-	75.9	9.8
PCB-138[+163]	205.0	366.0	8.5	92.4	691.0	14,050.0	100.0
PCB-153	335.0	585.0	38.8	156.0	1,112.0	25,089.0	100.0
PCB-156[+171]	38.5	69.3	<6.8	17.5	124.0	4,066.0	99.9
PCB-157	-	-	<1.3	-	-	385.0	44.8
PCB-167	9.9	18.0	<2.1	2.8	35.3	709.0	81.2
PCB-170	125.0	241.0	16.3	55.3	428.0	27,481.0	100.0
PCB-180	312.0	575.0	42.4	140.0	1,043.0	44,673.0	100.0
PCB-189	-	-	<1.7	-	-	1,497.0	55.6
PCBs (15 cong.)	1,087.0	1,972.0	149	504.0	3,452.0	101,413.0	
HCB	663.0	921.0	21.7	127.0	1,929.0	17,928.0	100.0
α-HCH	-	-	<0.8	-	-	17.5	6.7
β-HCH	46.2	57.3	<2.8	15.4	111.0	782.0	97.0
γ-HCH	-	-	<0.6	-	-	269.0	22.6
pp′-DDE	1,770.0	2,448.0	54.0	558.0	5,016.0	22,382.0	100.0
pp′-DDT	48.9	75.6	<3.6	20.0	157.0	940.0	99.5

Table 8.3. PCB and Some Organochlorine Pesticides in 434 Human Blood Serum Samples from Children (Age 8–9) in Eastern Slovakia.

Compound	Content, ng/g, Lipid Weight Basis						
	Median	Mean	Minimum	10-Percentile	90-Percentile	Maximum	Percentage Detected
PCB-28	-	-	<3.2	-	-	103.7	7.4
PCB-52	-	-	<2.9	-	-	153.4	3.5
PCB-101	-	-	<3.3	-	-	55.0	11.8
PCB-105	-	-	<1.6	-	-	19.3	7.4
PCB-114	-	-	<1.6	-	-	7.2	2.3
PCB-118	-	-	<4.8	-	-	118.0	49.8
PCB-123^{+149}	-	-	<3.1	-	-	13.2	3.7
PCB-138^{+163}	68.0	107.2	<9.9	24.4	223.6	1,011.2	98.8
PCB-153	109.4	172.2	<25	38.1	387.2	1,757.1	99.8
PCB-156^{+171}	-	-	<3.4	-	-	239.2	60.1
PCB-157	-	-	<1.5	-	-	20.3	6.2
PCB-167	-	-	<3.0	-	-	76.3	23.3
PCB-170	-	-	<4.1	-	-	1,056.0	91.9
PCB-180	88.5	150.5	<8.6	25.7	346.8	2,285.6	99.1
PCB-189	-	-	<1.6	-	-	38.9	9.4
PCBs (15 cong.)	377.0	568.0	66.6	150.0	1,208.0	6,495.0	
HCB	89.9	121.2	<12.8	44.9	219.3	2,279.0	99.3
α-HCH	-	-	<1.6	-	-	-	0.0
β-HCH	-	-	<5.3	-	-	121.0	49.3
γ-HCH	-	-	<1.6	-	-	49.0	4.4
pp′-DDE	473.0	677.2	31.0	157.7	1,378.0	11,733.8	100.0
pp′-DDT	-	-	<5.7	-	-	2,970.8	60.1

PCBs and selected organochlorine pesticides were determined in 2,047 adults (836 men and 1,211 women) and 434 children (221 boys and 213 girls), respectively.

Out of 328 blood samples taken from adults (143 in the Michalovce District and 185 in the Svidnik/Stropkov Districts), 2,378 substituted PCDD/Fs and non-*ortho* PCBs were determined in 320 (199 males and 121 females).

PCB congeners and organochlorine pesticides were quantified in more than two-thirds of the samples analyzed. Their median, mean, minimum, 10-percentile, 90-percentile, and maximum levels are reported in table 8.2 (adults) and table 8.3 (children). The arithmetic means and medians for any individual PCB congener were calculated with half-LOD values and LOD ones, respectively. The sum of all the 15 individual PCB congeners analyzed, including the half-LODs, represents "PCBs (15 cong)."

The most abundant PCB congeners quantified in all the samples from adults were PCB-153, 138^{+163}, 180, and 170. Out of the sum of all the 15 congeners analyzed, PCB-153 and the sum of PCB-153, 138^{+163} and 180 represented, on average, 30.6 percent ($s_x = 2.3\%$) and 77.5 percent ($s_x = 3.0\%$), respectively. The mean mutual ratio of PCB-153:138^{+163}:180

was 39:24:37. Similar numbers were found in the children's samples. The mean DDE:DDT ratio in the adults was 39.9 (min–max: 2.6–360). However, lower analyte levels in children's blood and, in many cases, insufficient sample volume caused that less abundant congeners could not be quantified (see table 8.3). Because of high correlation between PCB-153 levels and the sum of PCB congeners in this study (r = 0.94 and 0.99 in the case of adults and children respectively), one can work with the PCB-153 values as this congener was quantified in all the children's samples but one. It is interesting that mean PCB levels were statistically significantly higher in adult male blood lipids (2,383 vs. 1,688 ppb; p < 0.0001; Mann-Whitney test) while HCB levels were higher in adult female lipids (688 vs. 1,088 ppb; p < 0.0001). On the contrary, 8–9 year-old boys had higher mean HCB levels than girls (133 vs. 109 ppb; p = 0.004). There was no significant difference in mean p,p′-DDE content in adult male and female blood lipids (2,525 vs. 2,395 ppb).

The median, mean, minimum, 10-percentile, 90-percentile, and maximum levels of PCDDs, PCDFs, and non-*ortho* PCBs expressed as summed WHO-TEQs are given in table 8.4. If some congener was present at a concentration lower

Table 8.4. PCDDs, PCDFs, and Non-Ortho PCBs (77, 81, 126, and 169) in 320 Human Blood Serum Samples from Adults in Eastern Slovakia.

Compound	*Content, pg WHO TEQ/g, Lipid Weight Basis*					
	Median	*Mean*	*Minimum*	*10-Percentile*	*90-Percentile*	*Maximum*
PCDDs	3.1	3.8	0.9	1.5	7.1	15.8
PCDFs	9.3	11.2	1.4	5.3	17.7	88.6
Non-ortho PCBs	8.9	15.1	0.3	3.6	29.4	256.5
PCDDs + PCDFs	12.8	15.0	3.3	7.3	22.6	90.2
PCDDs + PCDFs + non-ortho PCBs	21.9	30.1	4.9	12.0	50.8	298.9

than its limit of detection half of the LOD was used for TEQ calculation. Whereas a mean ratio of TEQ_{PCDDs} and TEQ_{PCDFs} in many other countries is greater than 1, this ratio was 0.41 (min–max: 0.02–3.5) in the Slovak human general population studied; that is, the contribution of TEQ_{PCDFs} to the total TEQ was on average about 2.5 times higher than that of TEQ_{PCDDs}. This could be a result of higher PCB levels found in the Slovak population and, consequently, higher PCDF levels as PCDFs present in PCB formulations. On the other hand, lower mean TEQ_{PCDDs} can be caused by the reduced use of chlorinated phenols and other PCDD precursors in Slovakia in the past. There was no statistically significant difference between the levels of PCDDs, PCDFs, and coplanar PCBs expressed as TEQs in men and women.

As can be seen in figures 8.15 and 8.16, the levels of the POPs increased with the age of subjects, which proves that metabolic POPs degradation is lower than body intake.

The proximity of variables (TEQ_{PCDDs}, TEQ_{PCDFs}, TEQ_{cPCBs}, PCBs as a sum of 15 congeners, HCB and p,p′-DDE) is shown in figure 8.17, representing a principal component analysis biplot constructed using the statistical program SPSS 7.5 for Windows; simply said, the smaller angles are between vectors, the stronger connection between the variables and vice versa. Thus there is a good correlation between DDE and HCB human blood lipid levels as well as among PCB, coplanar PCB, and PCDF ones, unlike PCDDs, implying dissimilarity in the origin of the human exposure.

CONCLUSIONS

Twenty-five years of the manufacture of PCBs in eastern Slovakia have undoubtedly resulted in the increased environmental contamination of the surrounding area. It seems that the majority of contaminating PCBs have escaped from the Chemko factory through its effluent canal. This supposition

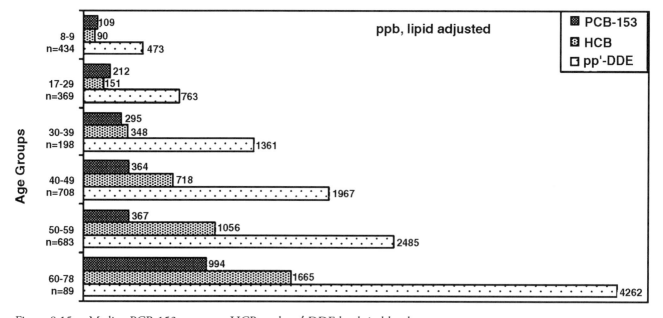

Figure 8.15.   Median PCB-153 congener, HCB, and p,p′-DDE levels in blood serum versus age groups.

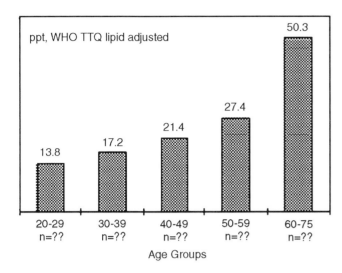

Figure 8.16.   Median WHO TEQ calculated from PCDDs, PCDFs, and non-ortho PCBs in blood serum versus age groups.

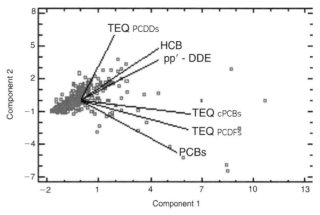

Figure 8.17.   A PCA biplot for PCDD, PCDF, cPCB, HCB, and DDE levels found in 320 adults living in eastern Slovakia.

is supported by extremely high PCB levels found in water and, especially, in sediment from the canal, as well as high levels in a river and lake that the canal empties into. The contamination of the watercourses has manifested itself in several hundredfold higher PCB levels in fish caught in those waters in comparison with control ones.

Substantially increased soil contamination was observed in the vicinity of sites where PCBs have been landfilled, stored, or used (as in heat exchange fluids in facilities preparing asphalted gravel). Increased PCB levels in air, water, and soil in the vicinity of the former manufacture site resulted in the higher exposure of wildlife to PCBs.

Environmental contamination with PCBs has caused increased PCB content in some foods of animal origin, in particular where animals were kept on small family farms and fed with feedstuffs grown in the polluted area.

Blood serum lipids taken from people consuming fish from the polluted area (Laborec River and Zemplinska Sirava Lake) contained the highest PCB levels (11.2 ppm). PCB content in blood lipids from workers working in the 1960s and 1970s at former PCB production was somewhat lower (8.6 ppm). Average PCB concentration in the human population of the polluted area (Michalovce District) was 4.2 ppm and 1.2 ppm in a control area (Stropkov District).

As in a previous human exposure study in the same areas in 1997–98 with five times fewer subjects, medium or mean PCB levels in female serum specimens from the PCBRISK project were 30 percent lower than those in male ones (mean 1,688 vs. 2,383 ppb; $p < 0.0001$); on the contrary, the HCB levels were almost 40 percent lower in male serum (mean 688 vs. 1,088 ppb; $p < 0.0001$). However, male children had higher HCB than female ones (mean 133 vs. 109 ppb;

$p = 0.004$). There was no statistically significant difference between the TEQ levels of PCDDs, PCDFs, and coplanar PCBs as well as DDE and DDT levels in men and women. Because of high correlation between PCB-153 levels and the sum of PCB congeners in the PCBRISK project ($r = 0.94$ and 0.99 in the case of adults and children, respectively), one can work with the PCB-153 values, since this most abundant PCB congener is quantifiable in a small volume of serum, as is a frequent case in children.

If the median PCB-153 congener, HCB and p,p′-DDE blood lipid levels found in the PCBRISK project are compared with the median levels of those pollutants found in blood lipids taken from 420 people living in the same area (Michalovce and Stropkov Districts) in 1997–98, one can observe a decline of 36 percent for PCB-153, 29 percent for p,p′-DDE, and even 50 percent for HCB over 3–4 years.

The higher contribution of TEQ PCDFs to the total TEQ compared to other countries, can be caused by higher PCB levels (PCDFs are present in PCB products) found in the Slovak population. On the other hand, lower mean TEQ PCDDs can be caused by the reduced use of chlorinated phenols and other PCDD precursors in Slovakia in the past.

REFERENCES

Chovancová, J., A. Kočan, and J. Petrík. 2004. "Concentrations of Dioxins and Dioxin-Like PCBs in Sediments and Foodstuffs from Eastern Slovakia." In *Proceedings of the 12th International Symposium: Advances and Applications of Chromatography in Industry,* June 29–July 1, 2004. Slovak University of Technology, Bratislava.

Chovancová, J., A. Kočan, S. Jursa, J. Petrík, and B. Drobná. 2003. "Dioxins and Dioxin-Like PCBs in Food Samples (Slovakia)." In *Book of Abstracts of the 1st International Symposium on Recent Advances on Food Analysis, 5–7 November 2003, Prague,*

Czech Republic, edited by the Institute of Chemical Technology, 178. Prague.

Drobná, B., J. Chovancová, S. Jursa, and J. Petrík. 2003. "PCBs in Food Samples of Animal Origin from Domestic Farms, Markets and Shops (Slovakia)." In *Book of Abstracts of the 1st International Symposium on Recent Advances on Food Analysis, 5–7 November 2003, Prague, Czech Republic,* edited by the Institute of Chemical Technology, 163. Prague.

Kočan, A., J. Petrík, B. Drobná, and J. Chovancová. 1994a. "Levels of PCBs and Some Organochlorine Pesticides in the Human Population of Selected Areas of the Slovak Republic." Pt. 1, "Blood." *Chemosphere* 29:2315–25.

———. 1994b. "Levels of PCBs and Some Organochlorine Pesticides in the Human Population from Selected Areas of the Slovak Republic." Pt. 2, "Adipose Tissue." *Organohalogen Compounds* 21:147–51.

Kočan, A., B. Drobná, J. Petrík, J. Chovancová, D. G. Patterson Jr., and L. L. Needham. 1995. "The Levels of PCBs and Selected Organochlorine Pesticides in Humans from Selected Areas of the Slovak Republic." Pt. 3, "Milk." *Organohalogen Compounds* 26:186–92.

Kočan, A., D. G. Patterson Jr., J. Petrík, W. E. Turner, J. Chovancová, and B. Drobná. 1996. "PCDD, PCDF and Coplanar PCB Levels in Blood from the Human Population of the Slovak Republic." *Organohalogen Compounds* 30:137–40.

Kočan, A., J. Petrík, H. Uhrinova, B. Drobná, and J. Chovancová. 1998. "The Occurrence of Semivolatile Persistent Organic Pollutants in Ambient Air in Selected Areas of Slovakia." *Toxicol. Environ. Chem.* 68:481–93.

Kočan, A., J. Petrík, S. Jursa, J. Chovancová, and B. Drobná. 2001. "Environmental Contamination with Polychlorinated Biphenyls in the Area of Their Manufacture in Slovakia." *Chemosphere* 43:595–600.

Kočan, A., B. Drobná, J. Petrík, S. Jursa, J. Chovancová, K. Čonka, B. Balla, E. Sovčíková, T. Trnovec. 2004. "Human Exposure to PCBs and Some Other Persistent Organochlorines in Eastern Slovakia as a Consequence of Former PCB Production." *Organohal. Compounds* 66:3539–46.

Pavuk, M., J. R. Cerhan, C. F. Lynch, A. Schecter, J. Petrík, J. Chovancová, and A. Kočan. 2003. "Environmental Exposure to PCBs and Cancer Incidence in Eastern Slovakia." *Chemosphere* 54:1509–20.

Petrík, J., J. Chovancová, A. Kočan, and I. Holoubek. 1991. "Project TOCOEN: The Fate of Selected Organic Pollutants in the Environment." Pt. 8, "PCBs in Human Adipose Tissues from Different Regions of Slovakia." *Toxicol. Environ. Chem.* 34:13–18.

Petrík, J., B. Drobná, A. Kočan, J. Chovancová, and M. Pavuk. 2001a. "Polychlorinated Biphenyls in Human Milk from Slovak Mothers." *Fresenius Environ. Bull.* 10:342–48.

Petrík, J., A. Kočan, S. Jursa, B. Drobná, J. Chovancová, and M. Pavuk. 2001b. "Polychlorinated Biphenyls in Sediments in Eastern Slovakia." *Fresenius Environ. Bull.* 10:375–80.

Petrík, J., J. Chovancová, and A. Kočan. 2003. "PCBs and Dioxins in Sediments and Surface Waters of the Michalovce Region." *Slovak Geol. Mag.* 9:173–76.

9

PCB Exposure, Neurobehavioral Performance, and Hearing Impairment in Children

Eva Sovčíková, *Slovak Medical University*

Tomáš Trnovec, *Slovak Medical University*

Ladislava Wsólová, *Slovak Medical University*

Soňja Wimmerová, *Slovak Medical University*

Beata Drobná, *Slovak Medical University*

Milan Husťák, *Air Force Military Hospital, Kosice*

Ján Petrík, *Slovak Medical University*

Anton Kočan, *Slovak Medical University*

Pavel Langer, *Slovak Academy of Sciences*

In many studies it has been shown that polychlorinated biphenyls (PCBs) can alter a number of developmental physiological processes in which the thyroid plays an essential role. Many investigators have reported associations of PCBs with neurodevelopmental delays. The aim of this study was to identify associations between exposure to PCBs and health outcomes as performance in neurobehavioral tests and thyroid hormones. Selected confounder factors, such as heavy metals and the health and social background of development in children, were also taken into account. Mothers of all the tested children had been living permanently in the area for at least five years before their children were born. Michalovce is in eastern Slovakia, where 435 children were examined. PCBs (ng/g serum lipids), TSH, T3 and T4, Pb, Hg, and Mn were determined in blood serum tests. Further tests assessed the thyroid volume, basic anthropometric measures, and neurobehavioral performance, which consisted of examination of (1) sensomotor functions and attention (Simple Reaction Time test, or SRT; the Vienna Discrimination Test, or VDT; tapping, with preferred and nonpreferred hand, and a memory test); and (2) complex mental processes (Benton Recognition test; digit span, forward and backward; digit symbols, or DS; cube-hand coordination; and the Raven nonverbal intelligence test). The "health and social background questionnaire" and the "scale of child's behavior at home" form were completed by the teacher and parents, respectively. The

study was cross-sectional, and associations between parameters could be deduced. Statistical analysis was performed using SPSS statistical software (p < 0.005). In the SRT test, there was a significant association between PCB serum concentration and gender and birth weight. Children in the fourth quartile of PCB serum concentration performed worse in the test, and children with lower birth weight (<3500 g) and girls had significantly longer reaction times. In the VDT (total number of hits), the PCB serum concentration and age of children were in association (significantly higher number of hits were attained by older children). In the Exacting Modification test (number of correct hits), a significant association was shown in children with higher PCBs levels (>600 ng/g serum lipids). Additionally, these children were from mothers with elementary school education. Girls performed a higher number of correct hits than boys. In the Benton Recognition test, the worst performance was associated with the higher level of serum concentration of FT_4 and higher PCB serum concentrations. In the scales of behavior test, at home and at school, as general indicators of behavior, there was a very important manifestation of hyperactive behavior of the children. At home, behavior was associated with both PCB serum concentration and blood Pb level in a positive way, and in the school, younger boys produced more hyperactive behavior, and PCB level in serum was positively associated with hyperactive behavior. In agreement with the findings of other authors, we found significant associations between PCB level and other predictors. PCBs may have effected the changes in neurobehavioral performances of this population of children environmentally exposed to PCBs in eastern Slovakia.

Polychlorinated biphenyls are lipophilic substances that easily penetrate biological membranes, including materno-fetal and blood-brain barriers, and appear in breast milk, exposing children prenatally and during early childhood. Based on reviews of both animal testing and human epidemiology studies, it is now generally accepted that PCBs are developmental neurotoxicants (Seegal and Schantz 1994; Tilson, Jacobson, and Rogan 1990).

The study of neuropsychological functions in children exposed to polychlorinated biphenyls was a subject of several excellent reviews (Hanneman et al. 1996; Jacobson and Jacobson 1997; Tilson and Kodavanti 1997; Schantz, Widholm, and Rice 2003). Most of the studies reviewed reported negative associations between prenatal PCB exposure and measures of cognitive functioning in infancy or childhood.

The Slovak study reviewed here (Sovčíková et al. 2004), when compared to the other studies, has the following characteristics: cross-sectional character of the study, direct PCB congener specific exposure assessment by serum analysis, wide range of environmental exposure, high PCB exposure level in comparison to other cohorts reported in literature, age of children relatively high (eight to nine years), dental and audiological investigations besides neurobehavioral examination, assessment of thyroid volume and hormones, and

assessment of potential confounders by means of a questionnaire or by additional analyses in children or local adult population (neurotoxic metals, chlorinated insecticides, and so on). The exposure data (means and medians) (Kočan et al. 2004) extracted from PCB serum concentration (sum of PCBs ng/g serum lipids) of children included in the cohort are shown in table 9.1. It can be seen from figure 9.1 that distribution of the PCB serum concentrations is skewed toward higher values. Of particular importance is the fact that in the cohort of 435 children, 309 were breast-fed and 26 were formula-fed.

In connection with the relatively high age of children of the Slovak cohort, it was evident from several relevant studies (Jacobson, Jacobson, and Humphrey 1990; Schantz 1996; Winneke 1995; Rogan and Gladen 1993; Vreugdenhil et al. 2002) that the age of children would be a very important factor for neurobehavioral outcomes after PCB exposure. In the youngest cohorts reported, mainly alterations of lower psychomotor skills were observed (Rogan and Gladen 1993; Chen and Hsu 1994); in older individuals, attention, vigilance, and memory functions were found to be compromised (Winneke et al. 1998; Vreugdenhil et al. 2004); and in 11-year-old children, prenatal exposure to PCBs was associated with lower full-scale and verbal IQ scores (Jacobson and Jacobson 1996). However, the latter paper was subjected to criticism (Schantz 1996). For the Slovak cohort of 8- to 9-year-old children it was important that neurotoxic effects of prenatal PCB and dioxin exposure might persist into school age, resulting in subtle cognitive and motor developmental delays. At the same time, more optimal intellectual stimula-

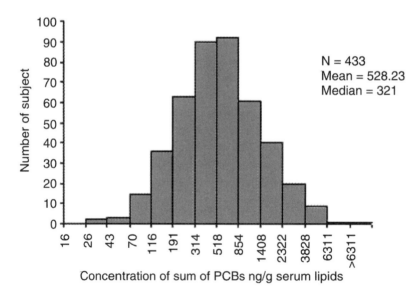

Figure 9.1.   Frequency diagram of the serum concentration of the sum of PCBs without LOD (ng/g lipids) in eight- to nine-year-old children recruited for the study.

Table 9.1. Means and Medians of Serum PCB Concentrations (ng/g Serum Lipids) of Children (Age 8–9).

Quartile	N	Lower Range	Upper Range	Mean	SD	Median	25 Percentile	75 Percentile
First	108	17.65	174.63	109.37	39.88	112.07	79.4	139.06
Second	109	175.44	320.93	244.28	42.5	238.36	208.22	285.28
Third	108	321.24	616.47	438.76	85.07	416.66	364.22	492.18
Fourth	108	618.54	6476.36	1323.2	900.78	1011.08	776.88	1550.1
Whole group	433	-	-	528.23	654.4	321.00	175.23	616.99

Note: Children were grouped into percentiles according to their PCB serum concentrations.

tion provided by a more advantageous parental and home environment could counteract the effects of prenatal exposure to PCBs and dioxins in the areas of cognitive skills and motor abilities (Vreugdenhil et al. 2002; Walkowiak et al. 2001).

In agreement with previous data (Chen and Hsu 1994; Winneke et al. 1998) the results of this Slovak study have shown a negative association between PCB serum concentrations and performance in psychomotor tests (SRT, VDT). With regard to experience from other studies (Vreugdenhil et al. 2004; Grandjean et al. 2001), an important role of age specific confounders in the cohort of children ages eight to nine was anticipated, and indeed, multifactor analysis of variance confirmed this assumption. It was also found that educational level of the mother and age of the child appeared as important counteracting confounders.

It is not clear in connection with thyroid hormones homeostasis if PCB-related subclinical alterations are responsible for later neurobehavioral deficit or delay (Winneke, Walkowiak, and Lilienthal 2002). It is remarkable in this respect that in the Benton Recognition test, a significant negative association between performance in the test and two independent variables, PCB serum concentrations and, surprisingly, free serum thyroxin (FT_4) concentration, was found. As far as is known, no data on association of thyroid hormones level and neurobehavioral performance in PCB exposed children have been published to date.

With regard to the importance of the age-specific effects of PCBs and related confounding factors in children, an evaluation of general behavior at home and at school was included in the examination schedule. Multifactor analysis of variance has shown an association between hyperactivity of children and PCB serum concentrations. The scaling of behavior at home was also associated with another predictor, the blood lead level. This may be related to the fact that the home environment is an important factor in children's lead exposure (Ursinyova and Hladikova 2000). Behavior at school was associated with hyperactivity only in young boys. Also, it can be seen that in mental tasks compared to the psychomotor tests, the outcome was more related to the intel-

ligence of the mother (outcomes in Raven test for adults) and the age of the child.

The next highly rated confounder was the particular school attended by the child, as the children were recruited from nine area schools. When the school was included into predictors for the multifactor analysis of variance, the previously found associations between performance in the SRT test and the VDT, and PCB serum concentrations remained significant. Results of two further tests, the tapping (left hand) and digit span tests, appeared to be significantly associated with PCB serum concentration (in contrast to lacking association before including school as a variable). The outcome of the Benton test was significantly associated with PCB serum concentration and FT_4 serum level before adjustment to school. After adjustment to school, T_3 appeared as another significant predictor. The performance in higher mental tasks (Raven test, symbols test, cubes) and scale of behavior at school were never significantly associated with PCB serum concentration. After adjustment for school, a significant association with education of mother, outcome of intelligence test in mother, and age of the child appeared.

Brain functions and hearing functions are often given under the same common denominator when discussing the mechanisms of their deficits as a sequence of an exposure to chlorinated organic compounds during pregnancy and early childhood. Goldey and colleagues (Goldey et al. 1995) were the first to publish that developmental exposure to PCBs (Aroclor 1254), in addition to reducing circulating thyroid hormone concentrations, causes hearing deficits in rats. Aroclor 1254 caused permanent auditory deficits (20–30 dB threshold shift) at the lowest frequency tested (1 kHz) in both the 4 and 8 mg/kg groups, whereas auditory thresholds were not significantly affected at higher frequencies (4, 16, 32, or 40 kHz). Brainstem auditory evoked responses in rats exposed to the same doses revealed dose-related decreased peak amplitudes at the lower frequencies (at 1 and 4 kHz but not at 16 or 32 kHz). This suggested a cochlear and/or auditory nerve site of Aroclor 1254 action (Herr, Goldey, and Crofton 1996). The hearing loss in animals exposed to Aroclor 1254 was later confirmed in several studies (Goldey and Crofton 1998; Crofton and Rice 1999; Crofton et al. 2000a; Herr, Goldey, and Crofton 1996). Importantly, the hearing deficit at 1 kHz in PCB-exposed animals was significantly attenuated by T4 replacement therapy (Goldey and Crofton 1998). Later studies have shown that the critical period for the ototoxicity of developmental A1254 exposure is within the first few postnatal weeks in the rat. This effect is consistent with the greater degree of postnatal hypothyroxinemia resulting from the greater magnitude of exposure that occurs postnatally via lactation (Crofton et al. 2000b).

In another study consistent with previous findings (auditory thresholds for 1-kHz tones were elevated by approximately 25 dB in the A1254-exposed animals, but thresholds for all higher frequencies were not different compared to controls), surface preparations of the organ of Corti revealed a mild to moderate loss of outer hair cells in the upper-middle and apical turns. Inner hair cells were not affected. These data clearly link the loss of low-frequency hearing caused by exposure during development to A1254 to a loss of outer hair cells in the organ of Corti. These data provide the first evidence of a structural deficit in the nervous system of adult animals exposed to PCBs during development (Crofton et al. 2000a).

Published data on the association between PCB exposure and hearing were based mainly on animal observations. Two papers were related to humans only. The first one reported PCB-associated increased thresholds at two out of eight frequencies on audiometry, but only on the left side, and no deficits on evoked potentials or contrast sensitivity in seven-year-old children prenatally exposed to seafood neurotoxicants (Grandjean et al. 2001). The other paper was focused on hearing impairments in boys of fish-eating mothers, but no individual PCB exposure data were available (Rylander and Hagmar 2000).

Before commenting on the results evaluating otoacoustic emissions in subjects exposed to PCBs, a few introductory remarks on this novel technique discovered by Dr. David Kemp (Kemp 1978) are appropriate. Studies have shown that when the sounds of different frequencies vibrate different sections of the organ of Corti, the higher frequencies vibrate the basal end and the lower frequencies vibrate the apex. When the organ of Corti vibrates, the outer hair cells create an electro-physical discharge (otoacoustic emission or OAE) that tunes and amplifies the stimulation for the benefit of the inner hair cells. When stimulated, the inner hair cells initiate the neural impulse that is carried up the auditory or eighth cranial nerve to the brain stem and on to the cortex of the brain where the signal is sent by way of association fibers to other parts of the brain for recognition, interpretation and understanding. OAE testing has developed into a significant addition in evaluating the auditory pathway. It

can be used for diagnostic and/or screening purposes (Hall 2000).

A study using this method has shown that the amplitudes of the distortion product otoacoustic emissions (DPOAE) (Lasky et al. 2002) were reduced and DPOAE thresholds elevated in adult rats exposed early in development to Aroclor 1254. Deficits were most pronounced at the lowest frequencies tested. These results, consistent with previous studies, implicate the cochlea and complement histological findings of Aroclor 1254–induced outer hair cell loss (Crofton et al. 2000a). In one-year-old rats born to females dosed with Aroclor 1254 (6 mg/kg) 28 days prior to mating and continuing until postnatal day 16, analysis of the DPOAE data revealed that exposure to PCBs resulted in consistent reductions in the signal-to-noise ratio between 2.0–6.0 kHz, and analysis of the ABR data revealed a significant reduction in peak II amplitude.

Important implications for humans were drawn (Lasky et al. 2002): Assuming that the critical period for PCB-induced auditory insult corresponds to the period of functional maturation of the cochlea, the human would be most vulnerable to PCB ototoxicity during the last half of pregnancy, when the only route of PCB exposure is via placental transfer. Based on these findings, further epidemiological studies of auditory function in children exposed to PCBs prenatally are warranted. Recording otoacoustic emissions could contribute significantly to that research.

The gap in knowledge on PCB ototoxicity in humans, especially in children, was at least partly filled by results of the 5th Framework EU PCBRISK project (Sovčíková et al. 2004; Trnovec et al. 2004). The same group of children that was tested for neurobehavioral performance was examined audiologically and for dental defects. The results (Sovčíková et al. 2004) of the examination of children by pure tone audiometry are presented in table 9.2. A significant association of the increases of the hearing thresholds at the low frequencies with quartiles of PCB serum concentrations can be seen. This finding is in agreement with many previous reports on increased hearing thresholds in rats exposed to PCB (Lasky et al. 2002). The sound pressure level (dB) of the transient evoked otoacoustic emissions was plotted against the medians of the sum of PCB serum concentrations. It can be seen that the decrease of the intensity of the otoacoustic emissions is associated with an increase of the PCB serum concentration. Because data on neurobehavioral performance and hearing were available for each child, it was possible to correlate them. From neurobehavioral parameters, the simple reaction time to visual stimulus was chosen and examined relative to hearing thresholds at the eight various tone frequencies tested. It was found (Trnovec et al. 2004) that for all frequencies the reaction time was positively associated

Table 9.2. Hearing Thresholds Determined by Pure Tone Audiometry in 433 Children (Age 8–9).

Quartile	kHz							
	0.25*	0.5*	1*	2	3	4	6	8
Mean	11.96	10.12	9.58	9.28	9.18	10.49	15.07	14.37
SD	3.54	3.34	2.39	3.01	2.97	3.39	6.11	6.53
Mean	12.12	10.17	9.95	9.83	10.07	10.9	14.58	14.15
SD	3.84	3.59	3.07	3.56	3.53	4.39	5.6	6.19
Mean	12.52	10.93	10.44	9.67	9.72	11.07	14.88	14.93
SD	4.55	4.47	4.47	3.64	3.94	4.52	5.38	6.56
Mean	13.26	11.18	10.81	10.19	10.39	11.62	14.7	14.49
SD	3.9	3.58	3.28	3.29	4.48	4.23	5.5	5.83

Note: The children were grouped according to the quartiles of their PCB serum concentrations. Significance at $p < 0.05$ in Kruskal-Wallis test is marked by an asterisk.

with the hearing threshold and that for five of the eight frequencies this association was significant.

The intensity of the sound pressure level of the transient evoked otoacoustic emissions was correlated with the battery of neurobehavioral tests targeted at complex mental processes: the Benton Recognition test, digit span (forward, backward), digit symbols, cube-hand coordination, SRT, the Raven test, and the scale of behavior (at home and at school). It was found in the Slovak cohort of children that the sound pressure level of transient, evoked otoacoustic emissions was significantly associated with results of all applied neurobehavioral tests with the exception of the Raven intelligence test. Thus it was shown for the first time that a decreased neurobehavioral performance may be associated with hearing impairment in children environmentally exposed to polychlorinated biphenyls.

ACKNOWLEDGMENTS

This work was conducted within the project Evaluating Human Health Risk from Low-Dose and Long-Term PCB Exposure (PCBRISK) supported by the European Union under the contract No. QLK-CT-2000-00488. However, the authors are solely responsible for this work, which does not represent the opinion of the EU, and the EU is not responsible for any use that might be made of data appearing therein.

REFERENCES

Chen, Y. J., and C. C. Hsu. 1994. "Effects of Prenatal Exposure to PCBs on the Neurological Function of Children: A Neuropsychological and Neurophysiological Study." *Develop. Med. Chil. Neurol.* 36:312–20.

Crofton, K. M., and D. C. Rice. 1999. "Low-Frequency Hearing Loss Following Perinatal Exposure to 3,3′,4,4′, 5-Pentachlorobiphenyl (PCB 126) in Rats." *Neurotoxicol. Teratol.* 21: 299–301.

Crofton, K. M., D. Ding, R. Padich, M. Taylor, and D. Henderson. 2000a. "Hearing Loss Following Exposure during Development to Polychlorinated Biphenyls: A Cochlear Site of Action." *Hear Res.* 144:196–204.

Crofton, K. M., P. R. Kodavanti, E. C. Derr-Yellin, A. C. Casey, and L. S. Kehn. 2000b. "PCBs, Thyroid Hormones, and Ototoxicity in Rats: Cross-Fostering Experiments Demonstrate the Impact of Postnatal Lactation Exposure." *Toxicol. Sci.* 57: 131–40.

Goldey, E. S., and K. M. Crofton. 1998. "Thyroxine Replacement Attenuates Hypothyroxinemia, Hearing Loss, and Motor Deficits Following Developmental Exposure to Aroclor 1254 in Rats." *Toxicol. Sci.* 45:94–105.

Goldey, E. S., L. S. Kehn, C. Lau, G. L. Rehnberg, and K. M. Crofton. 1995. "Developmental Exposure to Polychlorinated Biphenyls (Aroclor 1254) Reduces Circulating Thyroid Hormone Concentrations and Causes Hearing Deficits in Rats." *Toxicol. Appl. Pharmacol.* 135:77–88.

Grandjean, P., P. Weihe, V. W. Burse, L. L. Needham, E. Storr-Hansen, B. Heinzow, F. Debes, K. Murata, H. Simonsen, P. Ellefsen, E. Budtz-Jorgensen, N. Keiding, and R. White. 2001. "Neurobehavioral Deficits Associated with PCB in 7-Year-Old Children Prenatally Exposed to Seafood Neurotoxicants." *Neurotoxicol. Teratol.* 23:305–17.

Hall, J. W. 2000. *Handbook of Otoacoustic Emissions.* San Diego: Singular Thomson Learning.

Hanneman, W. H., M. E. Legare, E. Tiffany-Castiglioni, and S. H. Safe. 1996. "The Need for Cellular Biochemical, and Mechanistic Studies." *Neurotoxicol. Teratol.* 18:247–50; discussion, 271–76.

Herr, D. W., E. S. Goldey, and K. M. Crofton. 1996. "Developmental Exposure to Aroclor 1254 Produces Low-Frequency

Alterations in Adult Rat Brainstem Auditory Evoked Responses." *Fundam. Appl. Toxicol.* 33:120–28.

Jacobson, J. L., and S. W. Jacobson. 1996. "Intellectual Impairment in Children Exposed to Polychlorinated Biphenyls in Utero." *N. Engl. J. Med.* 335:783–89.

———. 1997. "Evidence for PCBs as Neurodevelopmental Toxicants in Humans." *Neurotoxicology* 18:415–24.

Jacobson, J. L., S. W. Jacobson, and H. E. Humphrey. 1990. "Effects of in Utero Exposure to Polychlorinated Biphenyls and Related Contaminants on Cognitive Functioning in Young Children." *J. Pediat.* 116:38–45.

Kemp, D. T. 1978. "Stimulated Acoustic Emissions from Within the Human Auditory System." *J. Acoust. Soc. Am.* 64: 1386–91.

Kočan, A., B. Drobná, J. Petrík, S. Jursa, J. Chovancová, K. Čonka, B. Balla, E. Sovčíková, T. Trnovec. 2004. "Human Exposure to PCBs and Some Other Persistent Organochlorines in Eastern Slovakia as a Consequence of Former PCB Production." *Organohal. Compounds* 66:3531–34.

Lasky, R. E., J. J. Widholm, K. M. Crofton, and S. L. Schantz. 2002. "Perinatal Exposure to Aroclor 1254 Impairs Distortion Product Otoacoustic Emissions (DPOAEs) in Rats." *Toxicol. Sci.* 68:458–64.

Rogan, W. J., and B. C. Gladen. 1993. "Breast-Feeding and Cognitive Development." *Early Human Develop.* 31:181–93.

Rylander, L., and L. Hagmar. 2000. "Medical and Psychometric Examinations of Conscripts Born to Mothers with a High Intake of Fish Contaminated with Persistent Organochlorines." *Scand. J. Work. Environ. Health* 26:207–12.

Schantz, S. L. 1996. "Developmental Neurotoxicity of PCBs in Humans: What Do We Know and Where Do We Go from Here?" *Neurotoxicol. Teratol.* 18:217–27.

Schantz, S. L., J. J. Widholm, and D. C. Rice. 2003. "Effects of PCB Exposure on Neuropsychological Function in Children." *Environ. Health Perspect.* 111:357–576.

Seegal, R. F., and S. L. Schantz. 1994. "Neurochemical and Behavioral Sequelae of Exposure to Dioxins and PCBs." In *Dioxins and Health,* edited by A. Schecter, 409–47. New York: Plenum Press.

Sovčíková, E., T. Trnovec, M. Husťák, J. Petrík, A. Kočan, B. Drobná, S. Wimmerová, and L. Wsólová. 2004. "Neurobehavioral Observation and Hearing Impairment in Children at School Age in Eastern Slovakia." *Organohal. Compounds* 66:3535–38.

Tilson, H. A., J. L. Jacobson, and W. J. Rogan. 1990. "Polychlorinated Biphenyls and the Developing Nervous System: Cross-Species Comparisons." *Neurotoxicol. Teratol.* 12:239–48.

Tilson, H. A., and P. R. Kodavanti. 1997. "Neurochemical Effects of Polychlorinated Biphenyls: An Overview and Identification of Research Needs." *Neurotoxicology* 18:727–43.

Trnovec, T., V. Bencko, P. Langer, M. van den Berg, A. Kočan, A. Bergman, and M. Husťák. 2004. "Study Design, Objectives, Hypotheses, Main Findings, Health Consequences for the Population Exposed, Rationale of Future Research." *Organohal. Compounds* 66:3523–30.

Ursinyova, M., and V. Hladikova. 2000. "Lead in the Environment of Central Europe." In *Trace Elements—Their Distribution and Effects in the Environment,* edited by B. Markert and K. Friese, 109–34. Oxford: Elsevier Science.

Vreugdenhil, H. J., C. I. Lanting, P. G. Mulder, E. R. Boersma, and N. Weisglas-Kuperus. 2002. "Effects of Prenatal PCB and Dioxin Background Exposure on Cognitive and Motor Abilities in Dutch Children at School Age." *J. Pediat.* 140: 48–56.

Vreugdenhil, H. J., P. G. Mulder, H. H. Emmen, and N. Weisglas-Kuperus. 2004. "Effects of Perinatal Exposure to PCBs on Neuropsychological Functions in the Rotterdam Cohort at 9 Years of Age." *Neuropsychology* 18:185–93.

Walkowiak, J., J. A. Wiener, A. Fastabend, B. Heinzow, U. Kramer, E. Schmidt, H.-J. Steingruber, S. Wundrum, and G. Winneke. 2001. "Environmental Exposure to Polychlorinated Biphenyls and Quality of the Home Environment: Effects on Psychodevelopment in Early Childhood." *Lancet* 358: 1602–7.

Winneke, G. 1995. "Endpoints of Developmental Neurotoxicity in Environmentally Exposed Children." *Toxicol. Letter* 77: 127–36.

Winneke, G., J. Walkowiak, and H. Lilienthal. 2002. "PCB-Induced Neurodevelopmental Toxicity in Human Infants and Its Potential Mediation by Endocrine Dysfunction." *Toxicology* 27:161–65.

Winneke, G., A. Bucholski, B. Heinzow, U. Kramer, E. Schmidt, J. Walkowiak, and J. A. Wiener. 1998. "Developmental Neurotoxicity of Polychlorinated Biphenyls (PCBS): Cognitive and Psychomotor Functions in 7-Month-Old Children." *Toxicol. Letter* 102–3:423–28.

10

Effects of PCBs on Tooth Enamel Development

Janja Jan, *University of Ljubljana*

Eva Sovčíková, *Slovak Medical University*

Anton Kočan, *Slovak Medical University*

Ladislava Wsólová, *Slovak Medical University*

Tomáš Trnovec, *Slovak Medical University*

The potential impact of polychlorinated biphenyls (PCBs) on humans has been a concern for decades due to their widespread distribution, persistence, bioaccumulation, and toxicity in animal studies. Developing tooth enamel is sensitive to a wide range of local and systemic disturbances. Because of the absolute metabolic stability of its structure, changes in enamel during its development are permanent in nature. PCBs have been shown to disturb tooth development in experimental animals, but only limited amounts of data exist on their adverse effects in humans. In this study we have aimed to evaluate the effects of long-term exposure to PCBs on developmental enamel defects in children in eastern Slovakia. We examined 432 children aged eight to nine years. Developmental defects of enamel were assessed using the Developmental Defects of Enamel Index on buccal surfaces of permanent teeth. The data set from the PCBRISK project has provided information on various confounding factors and modifiers. Analyses of blood samples for PCBs were made by high-resolution gas chromatography using electron capture detection. The proportion of teeth affected with demarcated opacities and/or hypoplasia in the children in higher exposed groups was significantly higher (Kruskal-Wallis χ^2 = 9.985; p = 0.007). The extent of the enamel defects was also greater in the higher exposed children (Kruskal-Wallis χ^2 = 10.714; p = 0.005). This study has demonstrated a dose-response relationship between PCB exposure and developmental dental defects in children. Further evaluation of the mechanism of this toxicity is needed.

INTRODUCTION

Polychlorinated biphenyls are highly persistent and ubiquitous organochlorine environmental pollutants. Being lipophilic, they increase in concentration up the food chain and bioconcentrate in animal and human tissues. Humans are exposed to PCBs mainly via diet, but PCBs are also transferred to the fetus and infants transplacentally and lactationally (Ahlborg et al. 1992). PCB-induced developmental defects that may be irreversible are raising concern due to high environmental exposure and possible greater sensitivity of infants (Brouwer, Ahlborg, and Berg 1995). Evidence is building that human tooth development may be a sensitive endpoint of PCB toxicity.

ENAMEL DEVELOPMENT

Tooth development, from the initiation of morphogenesis to the mineralization of tooth-specific matrices, is governed by a chain of epithelial-mesenchymal tissue interactions, similar to other developing organs (Jernvall and Thesleff 2000). In humans, the timing of tooth development lasts from the second month of gestation until adolescence. Once formed, the dental hard tissues are not remodeled, and, therefore, any major disturbances in the function of odontoblasts (dentin-forming cells) and/or ameloblasts (enamel-forming cells) are permanent. Given the slow rate of development and eruption of permanent teeth, the defects cannot be detected until six to seven years following the exposure.

Causes of developmental defects of enamel may be environmental or genetic. There are prenatal, neonatal, and postnatal environmental influences on enamel development and these include trauma, infection, hypoxia, and toxins (Brook, Fearne, and Smith 1997). However, most of the lesions are idiopathic.

The clinical features of developmental defects of enamel vary considerably. They range from a small circumscribed area of one surface of a tooth to widespread defects affecting all surfaces throughout the full thickness of the enamel. Similarly, the defects may be localized to one or two teeth or they may be generalized, involving many teeth or even the whole dentition. The clinical appearance may be of an opacity or of hypoplasia. In opacities there may be full thickness of enamel but the appearance and texture of the surface are altered due to deficient mineralization. In hypoplasia, pitting, grooving or an area of absent enamel represents deficiencies in the amount of enamel formed.

However, rather than a particular appearance being related to a specific cause, the nature and extent of the enamel defects are influenced by factors such as the severity and duration of the insult, the development stage, and the host's repose to the injury. Insults occurring during the earliest stages of enamel development, that is, matrix formation, will result in enamel hypoplasia. In contrast, insults occurring during the calcification and maturation stages of enamel development may lead to deficiency of mineralization and usually manifest as enamel opacities.

The clinical concern with mild developmental defects of enamel is its effect upon aesthetics. In severe cases the hypomineralized enamel surface may break down and the dentine is then exposed, with hypersensitivity, caries, and sometimes even loss of teeth as the result (Ellwood and O'Mullane 1994; Leppäniemi, Lukinmaa, and Alaluusua 2001).

HUMAN AND EXPERIMENTAL STUDIES ON THE ADVERSE EFFECTS OF PCBS ON ENAMEL DEVELOPMENT

Developmental dental defects were first described in two episodes in Asia of epidemic PCB poisoning, Yusho and YuCheng. Excess of ectodermal defects and developmental delay (Ahlborg et al. 1992), including a variety of dental changes, such as mottled, chipped, and carious deciduous teeth, perinatally erupted teeth, distortion of tooth roots, retarded eruption, and lack of permanent teeth, have been reported (Fukuyama, Anan, Akamine, and Aono 1979; Rogan et al. 1988; Lan et al. 1989). These children born to mothers who had consumed the contaminated oil would have had transplacental exposure and possibly exposure through breast milk (Rogan et al. 1988). Occupational exposure of the mother to PCBs also has been reported to result in mottled,

chipped, and carious primary teeth in the child exposed via placenta (Hara 1985). When following up the cohort of Yusho children, developmental dental defects were also observed in permanent teeth (Wang et al. 2003). Nevertheless, co-contamination with polychlorinated dibenzo-furans (PCDFs) was largely responsible for the overall toxicity (Safe 1994; Masuda 1996). PCBs encountered by the general population are not nearly so contaminated.

A Finnish study confirmed that developing human teeth are vulnerable to these compounds. They found that developmental dental defects were correlated with the total exposure to polychlorinated aromatics via mother´s milk (Alaluusua et al. 1996a; Hölttä et al. 2001). The correlation was strong with exposure to prevailing levels of polychlorinated dibenzo-p-dioxins (PCDDs) and PCDFs but weak with exposure to PCBs alone (Alaluusua et al. 1999). No association was found between the prevailing levels of PCBs, PCDDs, and PCDFs in mother's milk and the occurrence of perinatal tooth eruption (Alaluusua et al. 2002). The prevailing levels of those pollutants were apparently below the threshold to be a cause. When compared to YuCheng milk samples, the levels were 10–20 times lower for PCBs and 100–1,000 times lower for PCDFs.

It was suggested that enamel defects may be the best available biological indicator of an early dioxin exposure in the child population (Alaluusua et al. 1999).

Support for this observation is provided by experimental studies that have shown that the specific cells forming the teeth are sensitive to those compounds which lead to permanent changes in dental hard tissues. Rhesus macaques exposed to PCDD and dioxin-like compounds, including PCBs, via food showed squamous metaplasia of ameloblasts and the development of jaw cysts (McNulty 1985), and the teeth appeared to have a yellowish-brown tinge (Yoshihara et al. 1979). Studies on continuously growing rat incisors reported the selective toxic effects on ameloblasts in adult rats treated with the PCB mixture KC-400 (Hashiguchi et al. 1985) and impaired dentin and enamel formation after exposure to PCDD (Alaluusua et al. 1993; Kiukkonen et al. 2002; Simanainen et al. 2003). Similar defects have been found in rat pups after their breast feeding dam was exposed to PCDD (Lukinmaa et al. 2001).

As shown in different experimental conditions, the sensitivity of rat and mouse teeth to PCDD extends from early morphogenetic stages to the completion of tooth development, with the consequences ranging from the blockage of tooth development (Partanen et al. 1998; Kattainen et al. 2001; Lukinmaa et al. 2001; Miettinen et al. 2002) to defective dentinogenesis and amelogenesis (Alaluusua et al. 1993; Lukinmaa et al. 2001; Gao et al. 2004).

Developmental enamel defects were also seen after direct long-term exposure of otherwise healthy children to PCBs in a contaminated region (Jan and Vrbič 2000). This environmental epidemiological study provided an opportunity to increase our knowledge on the risk of PCBs alone in humans, as only negligible amounts of toxic PCDFs were present there. Exposure to dioxin-like and nonplanar PCBs was assessed. Enamel defects occurred at an estimated daily intake of TEQ only four times higher than the recommended World Health Organization tolerable daily intake of 10 pg TEQ/kg body weight/day (Masuda 1996) and thus proved to be a sensitive marker of PCB toxicity in man.

The exact pathogenetic mechanism explaining how PCBs cause enamel defects is currently not clearly understood. The alteration of retinoid metabolism, inhibition of thyroid hormone activity, and reduced blood flow have been suggested as mechanisms leading to dental defects (Hashiguchi et al. 1985). Changes in enzyme levels, hormones, growth factors, and their receptors are the principal known biochemical consequences of exposure to dioxin-like coplanar PCBs (Birnbaum 1994) that are thought to be mediated by the aryl hydrocarbon receptor (AhR) (Safe 1994). The role of AhR in dioxin-induced tooth defects has been shown (Sahlberg et al. 2002; Gao et al. 2004). The epidermal growth factor receptor (EGFR) seems to be involved, because the dioxin-induced failure of enamel matrix deposition and mineralization of dentin observed in cultured molars of mouse embryos were not seen in molars of mice lacking EGFR (Partanen et al. 1998). Dioxin also interferes with tooth development by enhancing and accelerating apoptosis in the dental epithelium (Partanen et al. 2004). However, nonplanar PCBs are likely to act through different mechanisms of action (Giesy and Kannan 1998). There are insufficient data on their role in dental development disturbance.

Recent studies have revealed especially high PCB contamination in eastern Slovakia, in the Michalovce region, where PCBs from a chemical plant manufacturing Delors contaminated the surrounding district (Kočan et al. 2001). The total serum PCB levels in samples from the general population there exceeded the background levels by several times. PCB levels in breast milk samples were the highest in Slovakia (Petrík et al. 2001).

The goal of this study was to investigate the effect of long-term PCB exposure on the development of the dental enamel of children in eastern Slovakia. Exposure was measured at the individual level, with a close control on other possible confounding factors.

MATERIALS AND METHODS

The description of the child population examined and study methodology are presented elsewhere (Jan et al. 2004; Trnovec et al. 2004). In short, 432 children aged eight to nine years were examined in September 2002 by two calibrated dentists. Developmental defects of enamel were assessed using the Developmental Defects of Enamel (DDE) Index (FDI 1992) on buccal surfaces of permanent teeth. Three main types of developmental defects of enamel were recorded: demarcated opacities, diffuse opacities, and hypoplasia. The extent of the defects was recorded in thirds of the surface area. Questionnaires that were completed by the parents and other data from the PCBRISK project data set provided information on various confounding factors and modifiers (e.g. place of residence during tooth development, parity, duration of breast feeding, eating fish from rivers and/or lakes, children's diseases, medications, fluoride, and metal exposure). Analyses of blood samples for organochlorines were made by high-resolution gas chromatography using electron capture detection.

Children were categorized into three groups according to their serum total PCB concentration (group 1 < 200; group 2 > 200 < 600; group 3 > 600 ng PCBs/g serum lipids).

All the data were analyzed using the SPSS 9.0 statistical software package.

RESULTS

Of the 432 children examined, 57.5 percent had at least one tooth affected with an enamel defect. 65.0 percent examined permanent teeth were affected with enamel defects, 49.4 percent with demarcated opacites and/or hypoplasia, and 33.4 percent with diffuse opacities.

The proportion of teeth with different kinds and extensions of enamel defects correlated with serum PCB concentration (table 10.1).

The proportion of teeth affected with any developmental enamel defect was significantly higher in more highly exposed children (Kruskal-Wallis $\chi^2 = 7.237$; p = 0.027), according to their serum total PCB concentration categorized in the

Table 10.1. Spearman's Rank Correlation Coefficient between Serum PCB Concentration and the Proportion of Teeth with Enamel Defects.

	Any Defect	Demarcated Opacity and/or Hypoplasia	Extension 2[a] or More
Correlation coefficient	0.134	0.134	0.148
Sig. (2-tailed)	0.007	0.007	0.002
N	401	401	433

[a]Affected teeth had defects greater in size than one-third the buccal area.

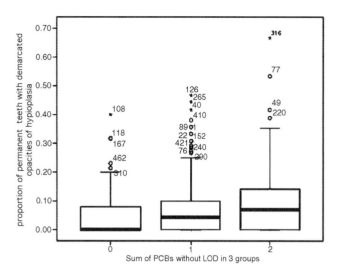

Figure 10.1.   Distribution of enamel defects among differently exposed children.

DISCUSSION

This study demonstrated a dose-response relationship between long-term PCB exposure and developmental enamel defects in children.

The difference in the prevalence of defects among differently exposed children was mostly due to demarcated opacities and hypoplasia and thus confirmed the results of our previous study (Jan and Vrbič 2000). The observation is in accordance with studies on PCB-treated rats and nonhuman primates (Hashiguchi et al. 1985; McNulty 1985), where selective toxic effects on ameloblasts and cells of stratum intermedium in the secretory and maturation stage of enamel development were reported.

The proportion of teeth with diffuse opacities that are likely to be of fluoride etiology (Clarkson and O'Mullane 1989) was relatively low and could not have masked demarcated lesions during clinical examination.

Children with enamel defects had both early- and late-forming teeth affected, suggesting that these defects resulted from a systemic factor acting over a long period of time. Since children were lifelong exposed to PCBs through the food chain, the dietary intake of PCBs had a larger effect on their total body burden than the exposure in utero or via milk (Patandin et al. 1999). This is in accordance with our findings that the first permanent molars did not prove to be more sensitive to polyhalogenated aromatic hydrocarbons, as was suggested (Brook et al. 1997) from the results of the Finnish studies on PCDD (Alaluusua et al. 1996a, 1996b; Hölttä et al. 2001). In those studies, permanent first molars were chosen as target teeth so as to indicate lactational exposure to prevailing levels of PCDD/Fs and PCBs. In our study no associations between the duration of breast feeding and enamel defects were found. A possible explanation is that in early life, accumulated PCB levels did not reach the critical levels necessary to cause distinctive adverse effects on

three concentration groups. Furthermore, the extent of the defects was also greater (Kruskal-Wallis $\chi^2 = 10.714$; p = 0.005). When excluding diffuse opacities from the final statistical analysis, there was a significantly higher proportion of teeth affected with demarcated opacities and/or hypoplasia in the children in higher exposed groups (Kruskal-Wallis $c^2 = 9.985$; p = 0.007) (fig. 10.1).

In order to evaluate the relative importance of the various possible etiological reasons for enamel defects, we have utilized univariate analysis of variance (table 10.2). The proportion of permanent teeth with enamel defects was statistically significantly affected by serum PCB concentration (as categorized in three concentration groups), hereditary illnesses, and interaction between three parameters (serum PCB concentration, hereditary illnesses, and frequency of eating fish from rivers and/or lakes).

Table 10.2.   Univariate Analysis of Variance, with the Proportion of Permanent Teeth with Any Defect as Dependent Variable.

Source	Type III Sum of Squares	df	Mean Square	F	p
Corrected model	1.246*	11	0.113	5.229	0.000
Intercept	1.652	1	1.652	76.246	0.000
Serum PCB concentration	0.256	2	0.143	6.583	0.002
Hereditary illness	0.556	1	0.556	25.657	0.000
Interaction	0.482	8	0.060	2.781	0.005
Error	8.125	375	0.022	-	-
Total	16.609	387	-	-	-
Corrected total	9.372	386			

*R-square = 0.133 (adjusted R-square = 0.108).

enamel development. Also, the duration of breast feeding was much shorter than in the dioxin studies. These findings are consistent with results from PCB-exposed children in Slovenia (Jan and Vrbič 2000). Unfortunately, dioxin-like PCB compounds have not been measured at the individual level, so direct comparisons with previous studies are difficult to make.

Analysis also assessed many of the covariates suspected to affect estimates of the PCB–enamel defects relation, including those from the complex data set from the PCBRISK project (Trnovec et al. 2004), but found little evidence of confounding or effect modification. Only hereditary illnesses and frequency of eating fish from rivers and/or lakes proved to be statistically significant (table 10.2).

CONCLUSIONS

In conclusion, our results demonstrated a dose-response relationship between long-term PCB exposure and developmental enamel defects in children in eastern Slovakia, indicating that enamel development is one of the sensitive endpoints of PCB toxicity. Given the widespread dissemination of PCBs in the environment and the food chain, the possibility that it causes developmental defects is a public health concern.

There is a need for studies in exposed animal models on the role of individual PCB congeners and their mixtures on enamel development impairment. It is essential to be aware of potential pathways other than AhR-mediated toxicity.

ACKNOWLEDGMENTS

This study was supported by a European Union project (PCBRISK) (No. QLK4-CT-2000-00488) and by the Slovenian Ministry of Science, Education and Sport (No. J3-8713-0381-99).

REFERENCES

Ahlborg, U. G., A. Brouwer, M. A. Fingerhut, J. L. Jacobson, S. W. Jacobson, S. W. Kennedy, A. F. Kettrup, J. H. Koeman, H. Poiger, C. Rappe, S. H. Safe, R. F. Seegal, J. Tuomisto, and M. van den Berg. 1992. "Impact of Polychlorinated Dibenzo-*p*-Dioxins, Dibenzofurans, and Biphenyls on Human and Environmental Health, with Special Emphasis on Application of the Toxic Equivalency Factor Concept." *Eur. J. Pharmacol.* 228:179–99.

Alaluusua, S., P. L. Lukinmaa, R. Pohjanvirta, M. Unkila, and J. Tuomisto. 1993. "Exposure to 2,3,7,8-Tetrachlorodibenzo-para-Dioxin Leads to Defective Dentin Formation and Pulpal Perforation in Rat Incisor Tooth." *Toxicology* 81:1–13.

Alaluusua, S., P. L. Lukinmaa, M. Koskimies, S. Pirinen, P. Hölttä, M. Kallio, T. Holttinen, and L. Salmenperä. 1996a. "Developmental Dental Defects Associated with Long Breast Feeding." *Eur. J. Oral. Sci.* 104:493–97.

Alaluusua, S., P. L. Lukinmaa, T. Vartiainen, M. Partanen, J. Torppa, and J. Tuomisto. 1996b. "Polychlorinated Dibenzo-*p*-Dioxins and Dibenzofurans via Mother's Milk May Cause Developmental Defects in the Child's Teeth." *Environ. Toxicol. Pharmacol.* 1:193–97.

Alaluusua, S., P. L. Lukinmaa, J. Torppa, J. Tuomisto, and T. Vartiainen. 1999. "Developing Teeth as Biomarker of Dioxin Exposure." *Lancet* 353:206.

Alaluusua, S., H. Kiviranta, A. Leppäniemi, P. Hölttä, P. L. Lukinmaa, L. Lope, A. L. Järvenpää, M. Renlund, J. Toppari, H. Virtanen, M. Kaleva, and T. Vartiainen. 2002. "Natal and Neonatal Teeth in Relation to Environmental Toxicants." *Pediatr. Res.* 52:652–55.

Birnbaum, L. S. 1994. "Endocrine Effects of Prenatal Exposure to PCBs, Dioxins, and Other Xenobiotics: Implications for Policy and Future Research." *Environ. Health Perspect.* 102:676–79.

Brook, A. H., J. M. Fearne, J. M. Smith. 1997. "Environmental Causes of Enamel Defects." In *Dental Enamel,* edited by D. Chadwick and G. Cardew, 212–20. Chichester: Wiley.

Brouwer, A., U. G. Ahlborg, and M. Berg. 1995. "Functional Aspects of Developmental Toxicity of Polyhalogenated Aromatic Hydrocarbons in Experimental Animals and Human Infants." *Eur. J. Pharmacol. Environ. Toxicol. Pharmacol.* Section 293:1–40.

Clarkson, J., and D. O'Mullane. 1989. "A Modified DDE Index for Use in Epidemiological Studies of Enamel Defects." *J. Dent. Res.* 68:445–50.

Ellwood, R. P., and D. M. O'Mullane. 1994. "Association Between Dental Enamel Opacities and Dental Caries in a North Wales Population." *Caries Res.* 28:383–87.

FDI Commission on Oral Health, Research and Epidemiology. 1992. "A Review of the Developmental Defects of Enamel Index (DDE Index)." *Int. Dent. J.* 42:411–26.

Fukuyama, H., Y. Anan, A. Akamine, and M. Aono. 1979. "Alteration in Stomatological Findings of Patients with Yusho (PCB Poisoning) in the General Examination." In Japanese. *Fukuoka Igaku Zasshi* 70:187–98.

Gao, Y., C. Sahlberg, A. Kiukkonen, S. Alaluusua, R. Pohjanvirta, J. Tuomisto, and P. L. Lukinmaa. 2004. "Lactational Exposure of Han/Wistar Rats to 2,3,7,8-Tetrachlorodibenzo-*p*-Dioxin Interferes with Enamel Maturation and Retards Dentin Mineralization." *J. Dent. Res.* 83:139–44.

Giesy, J. P., and K. Kannan. 1998. "Dioxin-like and Non-Dioxin-like Toxic Effects of Polychlorinated Biphenyls (PCBs): Implications for Risk Assessment." *Crit. Rev. Toxicol.* 28:511–69.

Hara, I. 1985. "Health Status and PCBs in Blood of Workers Exposed to PCBs and of Their Children." *Environ. Health Perspect.* 59:85–90.

Hashiguchi, I., A. Akamine, Y. Hara, K. Maeda, H. Anan, T. Abe, M. Aono, and H. Fukuyama. 1985. "Effects on the Hard

Tissue of Teeth in PCB Poisoned Rats." In Japanese. *Fukuoka Igaku Zasshi* 76:221–28.

Hölttä, P., H. Kiviranta, A. Leppäniemi, T. Vartiainen, P. L. Lukinmaa, and S. Alaluusua. 2001. "Developmental Dental Defects in Children Who Reside by a River Polluted by Dioxins and Furans." *Arch. Environ. Health* 56:522–28.

Jan, J., and V. Vrbič. 2000. "Polychlorinated Biphenyls Cause Developmental Enamel Defects in Children." *Caries Res.* 34:469–73.

Jan, J., E. Sovčíková, I. Kovrizhnykh, A. Kočan, S. Wimmerová, and T. Trnovec. 2004. "Developmental Dental Defects in Children Exposed to PCBs." *Organohal. Compounds* 66: 3544–47.

Jernvall, J., and I. Thesleff. 2000. "Reiterative Signaling and Patterning during Mammalian Tooth Morphogenesis." *Mech. Dev.* 92:19–29.

Kattainen, H., J. Tuukkanen, U. Simanainen, J. T. Tuomisto, O. Kovero, P. L. Lukinmaa, S. Alaluusua, J. Tuomisto, and M. Viluksela. 2001. "In Utero/lactational 2,3,7,8-Tetrachlorodibenzo-*p*-Dioxin Exposure Impairs Molar Tooth Development in Rats." *Toxicol. Appl. Pharmacol.* 174:216–24.

Kiukkonen, A., M. Viluksela, C. Sahlberg, S. Alaluusua, J. T. Tuomisto, J. Tuomisto, and P. L. Lukinmaa. 2002. "Response of the Incisor Tooth to 2,3,7,8-Tetrachlorodibenzo-*p*-Dioxin in a Dioxin-Resistant and a Dioxin-Sensitive Rat Strain." *Toxicol. Sci.* 69:482–89.

Kočan, A., J. Petrík, S. Jursa, J. Chovancová, and B. Drobná. 2001. "Environmental Contamination with Polychlorinated Biphenyls in the Area of Their Manufacture in Slovakia." *Chemosphere* 43:595–600.

Lan, S. J., Y. Y. Yen, Y. C. Ko, and E. R. Chen. 1989. "Growth and Development of Permanent Teeth Germ of Transplacental Yu-Cheng Babies in Taiwan." *Bull. Environ. Contam. Toxicol.* 42:931–34.

Leppäniemi, A., P. L. Lukinmaa, and S. Alaluusua. 2001. "Non-fluoride Hypomineralizations in the Permanent First Molars and Their Impact on the Treatment Need." *Caries Res.* 35: 36–40.

Lukinmaa, P. L., C. Sahlberg, A. Leppäniemi, A. M. Partanen, O. Kovero, R. Pohjanvirta, J. Tuomisto, and S. Alaluusua. 2001. "Arrest of Rat Molar Tooth Development by Lactational Exposure to 2,3,7,8-Tetrachlorodibenzo-*p*-Dioxin." *Toxicol. Appl. Pharmacol.* 173:38–47.

Masuda, Y. 1996. "Approach to Risk Assessment of Chlorinated Dioxins from Yusho PCB Poisoning." *Chemosphere* 32: 583–94.

McNulty, W. P. 1985. "Toxicity and Fetotoxicity of TCDD, TCDF and PCB Isomers in Rhesus Macaques (*Macaca mulatta*)." *Environ. Health Perspect.* 60:77–88.

Miettinen, H. M., S. Alaluusua, J. Tuomisto, and M. Viluksela. 2002. "Effect of in Utero and Lactational 2,3,7,8-Tetrachlorodibenzo-*p*-Dioxin Exposure on Rat Molar Development: The Role of Exposure Time." *Toxicol. Appl. Pharmacol.* 184:57–66.

Partanen, A. M., S. Alaluusua, P. J. Miettinen, I. Thesleff, J. Tuom-

isto, R. Pohjanvirta, and P. L. Lukinmaa. 1998. "Epidermal Growth Factor Receptor as a Mediator of Developmental Toxicity of Dioxin in Mouse Embryonic Teeth." *Lab. Invest.* 78:1473–81.

Partanen, A. M., A. Kiukkonen, C. Sahlberg, S. Alaluusua, I. Thesleff, R. Pohjanvirta, and P. L. Lukinmaa. 2004. "Developmental Toxicity of Dioxin to Mouse Embryonic Teeth in Vitro: Arrest of Tooth Morphogenesis Involves Stimulation of Apoptotic Program in the Dental Epithelium." *Toxicol. Appl. Pharmacol.* 194:24–33.

Patandin, S., P. C. Dagnelie, P. G. Mulder, E. Op de Coul, J. E. van der Veen, N. Weisglas-Kuperus, and P. J. Sauer. 1999. "Dietary Exposure to Polychlorinated Biphenyls and Dioxins from Infancy until Adulthood: A Comparison between Breast-Feeding, Toddler, and Long-Term Exposure." *Environ. Health Perspect.* 107:45–51.

Petrík, J., B. Drobná, A. Kočan, J. Chovancová, and M. Pavuk. 2001. "Polychlorinated Biphenyls in Human Milk from Slovak Mothers." *Freestones Environ. Bull.* 10:342–48.

Rogan, W. J., B. C. Gladen, K. L. Hung, S. L. Koong, L. Y. Shih, J. S. Taylor, Y. C. Wu, D. Yang, N. B. Ragan, and C. C. Hsu. 1988. "Congenital Poisoning by Polychlorinated Biphenyls and Their Contaminants in Taiwan." *Science* 241:334–36.

Safe, S. H. 1994. "Polychlorinated Biphenyls (PCBs): Environmental Impact, Biochemical and Toxic Responses, and Implications for Risk Assessment." *Crit. Rev. Toxicol.* 24:87–149.

Sahlberg, C., R. Pohjanvirta, Y. Gao, S. Alaluusua, J. Tuomisto, and P. L. Lukinmaa. 2002. "Expression of the Mediators of Dioxin Toxicity, Aryl Hydrocarbon Receptor (AHR) and the AHR Nuclear Translocator (ARNT) Is Developmentally Regulated in Mouse Teeth." *Int. J. Dev. Biol.* 46:295–300.

Simanainen, U., J. T. Tuomisto, J. Tuomisto, and M. Viluksela. 2003. "Dose-Response Analysis of Short-Term Effects of 2,3,7,8-Tetrachlorodibenzo-*p*-Dioxin in Three Differentially Susceptible Rat Lines." *Toxicol. Appl. Pharmacol.* 187:128–36.

Trnovec, T., V. Bencko, P. Langer, M. van den Berg, A. Kočan, Å. Bergman, and M. Husťák. 2004. "Study Design, Objectives, Hypotheses, Main Findings, Health Consequences for the Population Exposed, Rationale of Future Research." *Organohal. Compounds* 66:3523–30.

Wang, S. L., T. T. Chen, J. F. Hsu, C. C. Hsu, L. W. Chang, J. J. Ryan, Y. L. Guo, and G. H. Lambert. 2003. "Neonatal and Childhood Teeth in Relation to Perinatal Exposure to Polychlorinated Biphenyls and Dibenzofurans: Observations of the Yucheng Children in Taiwan." *Environ. Res.* 93:131–37.

Yoshihara, S., N. Ozawa, H. Yoshimura, Y. Masuda, T. Yamaryo, H. Kuroki, K. Murai, K. Akagi, M. Yamanaka, T. Omae, M. Okumura, M. Fujita, T. Yamamoto, A. Ohnishi, H. Iwashita, T. Kohno, Y. Ohnishi, T. Ishibashi, M. Kikuchi, H. Fukuyama, Y. Anan, A. Akamine, and M. Aono. 1979. "Preliminary Studies on the Experimental PCB Poisoning in Rhesus Monkeys." In Japanese. *Fukuoka Igaku Zasshi* 70: 135–71.

11

Environmental and Human Contamination with Persistent Organochlorine Pollutants and Development of Diabetes Mellitus

Žofia Rádiková, *Slovak Academy of Sciences*
Anton Kočan, *Slovak Medical University*
Miloslava Hučková, *Slovak Academy of Sciences*
Pavel Langer, *Slovak Academy of Sciences*
Tomáš Trnovec, *Slovak Medical University*
Elena Šebӧková, *Slovak Academy of Sciences*
Iwar Klimeš, *Slovak Academy of Sciences*

Some studies suggest that exposure to persistent organochlorine pollutants (POPs) may affect glucose metabolism. The aim of our study was to search for further interrelations between long-term organochlorine pollution and disturbances in glucose homeostasis in large cohorts of populations from three districts of eastern Slovakia. A total of 2,050 adults, 835 males (44.9 ± 0.4 years, BMI = 27.4 ± 0.1 kg/m^2), and 1,215 females (44.7 ± 0.3 years, BMI = 26.6 ± 0.1 kg/m^2) from three districts of eastern Slovakia variably polluted with POPs were examined. Fasting blood samples were obtained from all 2,050 participating subjects. A standard (0, 60, 120 minutes) oral glucose tolerance test (75g glucose) was performed in 1,222 willing subjects without previous evidence of diabetes or other dysglycemia. Concentrations of glucose, insulin, and organochlorines (α, β, γ–HCH, HCB, DDE, DDT, and 15 PCB-congeners) were determined.

Variables determining the glucose homeostasis status, such as fasting and 120 min glucose and insulin concentrations correlated with the levels of HCB, βHCH, DDE, and DDT in males and in females (p = 0.02–0.000, r = 0.11–0.28). In females, predictive parameters for fasting plasma glucose are BMI, levels of HCB, βHCH, DDT, and age. In males, the predictive parameters are BMI, βHCH, and age. The predictive parameters for fasting plasma insulin, besides BMI, are DDT and DDE for females and males, respectively.

A relation between environmental pollution and disorders of glucose metabolism has been found in a population chronically exposed to POPs. The interactive effects of many compounds on human health cannot be excluded. Therefore the dissociation of the impact of individual pollutants might be difficult.

In this work we revealed evidence for a causal relationship between disturbances of glucose metabolism and environmental pollution with POPs. For definitive statements, further analysis is necessary. This chapter reviews the current status of knowledge regarding organochlorines and diabetes.

Persistent organic pollutants are man-made chemical products ubiquitously present as complex mixtures in the environment (Jensen 1987). Organochlorine compounds include polychlorinated biphenyls (PCBs), organochlorine insecticides (such as dichloro-diphenyl-trichloroethane, or DDT, and its metabolite dichloro-diphenyl-dichloroethylene, or DDE), pesticides (i.e., hexachlorocyclohexanes, or HCHs), fungicides (i.e., hexachlorobenzene, or HCB), dioxins (i.e., 2,3,7,8-tetrachlorodibenzodioxin, or TCDD), furans, and so on (Jensen 1987). Human exposure occurs transdermally as well as via inhalation and ingestion. POPs accumulate in the human body, predominantly in adipose tissues (Kutz, Wood, and Bottimore 1991), cross the placenta and are transferred into human breast milk fat (Ando, Saito, and Wakisaka 1985; Kočan et al. 1994). Parturition and lactation are the main routes of elimination and the main sources of contamination for fetus and newborns (Korrick and Altshul 1998). Human populations have been exposed to the various pollutants accidentally (O'Keefe et al. 1985; Pesatori et al. 1998), occupationally (Wolff 1985; Calvert et al. 1999;

Langer et al. 2002), and environmentally (Kočan et al. 1994; Glynn et al. 2003). Blood and tissue levels of organochlorines following accidental and occupational exposure are significantly higher than in the latter category (Wolff 1985; Jensen 1987). Environmental exposure occurs predominantly by consumption of contaminated food.

Organochlorine compounds include many endocrine disruptors. These are exogenous substances interfering with production, secretion, transport, metabolism, binding, action, and elimination of natural hormones (Kavlock et al. 1996). Thus organochlorines have several adverse effects on human health (Safe 1994; Hansen 1998), including their oncogenic (Fingerhut et al. 1991; Kogevinas et al. 1995; Warner et al. 2002), immunotoxic (Lu and Wu 1985; Nordstrom et al. 2000; Langer et al. 2002), neurotoxic (Tilson and Kodavanti 1998; Boersma and Lanting 2000; van Wendel de Joode et al. 2001), hepatotoxic (Safe 1993), teratogenic, and embryotoxic (Miller 2004) effects, negative effects on gonads (Giwercman et al. 1993; Swan, Elkin, and Fenster 2000; Den Hond et al. 2002) and the thyroid gland (Hagmar et al. 2001; Hansen 1998; Langer et al. 2003).

Various studies indicate that organochlorines may also affect glucose metabolism (Morgan, Lin, and Saikaly 1980; Henriksen et al. 1997; Bertazzi et al. 1998; Pesatori et al. 1998; Calvert et al. 1999; Michálek, Akhtar, and Kiel 1999; Steenland et al. 1999; Cranmer et al. 2000; Longnecker and Daniels 2001; Longnecker et al. 2001). Longnecker and Daniels (2001) reviewed epidemiological data on associations of environmental contaminants to diabetes and found, however, the data limited, inconsistent, or inconclusive. Our team has been also involved in a series of field surveys aimed at investigations of environmental pollution with PCBs in specific areas of the Slovak Republic (Kočan et al. 1994, 2001; Chovancová, Kočan, and Jursa 2005; Jursa et al. 2006; Hovander et al. 2006; Langer et al. 1998, 2002, 2003, 2007, in press) and their various health consequences. This chapter, which includes several pieces of our data, is focused on the analysis of the relationship between environmental contamination with organochlorines and development of diabetes. A synoptic overview of the main findings is shown in table 11.1.

An accident in a trichlorophenol plant near Seveso, Italy, in 1976 caused an exposure of a large population to TCDD (Bertazzi et al. 1998). Besides acute toxic effects of TCDD, including chloracne, nausea, headache, and eye irritation, long-term toxic effects were studied (Bertazzi et al. 1998; Pesatori et al. 1998). The classification of the exposed and unexposed population was based only on monitoring of soil levels, place of residence, and measurements of a limited number of human blood samples (Bertazzi et al. 1998). Increased mortality from cardiovascular diseases, cancer, and diabetes was reported. An impact of psychosocial stressor (accident experience), the lack of exact data on diabetes (diagnoses were obtained from medical records and from death records), and a small number of subjects in the most contaminated area were the limitations, which make the results of this study inconclusive.

Several other studies were conducted to examine long-term health effects of occupational exposure to chemicals and materials contaminated with TCDD (Sweeney et al. 1997–98; Steenland et al. 1999; Calvert et al. 1999). A slight but significant increase in the risk of diabetes (OR = 1.12, $p < 0.003$) and fasting serum glucose ($p < 0.001$) was observed with increasing serum concentrations of TCDD in a study on 281 plant workers and 260 unexposed controls (Sweeney et al. 1997–98). However, typical risk factors for diabetes (obesity, age, familiar background) seemed to be more dominant than TCDD in diabetes development. Increased serum glucose levels in workers with the highest half-life extrapolated lipid adjusted serum TCDD concentrations were observed, suggesting an effect of TCDD exposure on increased risk of diabetes mellitus in all workers (OR = 1.49, 95% CI 0.77–2.91). No dose-response relationship was observed between TCDD levels and incidence of diabetes (Calvert et al. 1999). Diagnosis of diabetes was based on fasting serum glucose levels only, or on self-reported history of diabetes diagnosed by a physician without confirmation by medical records. No other disturbances in glucose homeostasis (i.e., impaired fasting glucose, impaired glucose tolerance) were considered. In another study, no excess mortality from diabetes was found in a cohort of 5,132 male workers from U.S. plants that produced TCDD-contaminated products (Steenland et al. 1999). However, diagnosis of diabetes in this study was based only on death certificate data, which seem to be unsatisfactory for studying diabetes.

Data on increased morbidity from diabetes were found in about 900 veterans of Operation Ranch Hand, responsible for spraying the herbicidal mixture Agent Orange contaminated with TCDD, when compared with about 1,200 unexposed control individuals (Henriksen et al. 1997; Michálek, Akhtar, and Kiel 1999). Increased risk of diabetes (relative risk = 1.5, 95% CL = 1.2–2.0), decreased time-to-disease onset, and increased severity of diabetes were found in these veterans (Henriksen et al. 1997). In nondiabetic Ranch Hand veterans, increased risk of abnormally high insulin levels (relative risk = 3.4, 95% CL = 1.9–6.1) with dioxin exposure was observed.

In a study by Cranmer et al. (2000) 69 subjects living close to a plant formerly producing pesticides in Jacksonville, Arkansas, were examined. All subjects had normal fasting glucose and normal glucose tolerance; among them 7 sub-

Table 11.1. Overview of Studies Investigating the Relation between Environmental Pollution with Organochlorines and Diabetes Mellitus.

Study Design	Number of Exposed Cases	Pollutant	Parameter	Association	First Author and Year
Mortality/ follow-up	2,620	DDT	DM from death records or questionnaire	0 0	Morgan et al. 1980
Cross-sectional	281	TCDD	Fasting glucose	0	Sweeney et al. 1997–98
Mortality study	6,748[a]	TCDD	DM from death records	0	Pesatori et al. 1998
Mortality study	5,132	TCDD	DM from death certificate	Æ	Steenland et al. 1999
Cross-sectional	281	TCDD	Fasting glucose or self-reported DM	+/−	Calvert et al. 1999
Follow-up	871	TCDD	DM from medical records, fasting glucose, and insulin	0	Michálek et al. 1999
Cross-sectional	69	TCDD	Insulin in non-DM subjects	0	Cranmer et al. 2000
Cross-sectional	2,245	PCBs	DM from project records	+/−	Longnecker et al. 2001
Cross-sectional	205	PCBs, DDE, HCB, b-HCH	DM from questionnaire	0	Glynn et al. 2003

[a] Number of subjects with high/substantial exposure.

Note: DM = diabetes mellitus. Association between pollution and diabetes: Æ = no association, + = positive association, +/− = positive association uncertain. TCDD concentrations measured only in 289 cases.

jects with the highest (top 10%) TCDD concentrations demonstrated hyperinsulinemia during an oral glucose tolerance test (Cranmer et al. 2000). Hyperinsulinemia in the presence of normal glucose levels strongly suggests insulin resistance. The small number of the study participants limits the conclusions of TCDD-mediated insulin resistance.

Higher blood levels of DDT and its metabolite DDE were found in subjects with diabetes compared to those without diabetes in a study of a group of pesticide users (Morgan et al. 1980). Not only was there a causal relationship between the toxic effects of organochlorines and diabetes, but slower elimination of organochlorines in diabetes could be hypothesized. Glynn et al. (2003) studied associations between lifestyle/medical factors and serum concentrations of 7 PCB congeners, DDE, HCB, βHCH, oxychlordane, and trans-nonachlor in 205 Swedish women living in areas with high availability of organochlorine-contaminated fishes. They found association of organochlorines with age, BMI, weight change, diabetes mellitus, consumption of fish, and place of residence. Data on diabetes were obtained only from questionnaires; fasting glucose levels were not estimated. Interestingly, the authors suggest that the positive association between diabetes and HCB concentrations (p = 0.008) may be the result of the alteration of the toxicokinetics of various compounds by a disease (in this case diabetes). The paper highlights the assumption that some parameters studied (age, gender, BMI, etc.) may confound associations between organochlorine concentrations and a disease, especially if these factors are related to the organochlorine concentrations and are risk factors for several diseases (Glynn et al. 2003).

An association between serum PCB levels and diabetes was reported previously in a large birth cohort study, where 44 pregnant women with diabetes had 30% higher PCB levels (p < 0.0002) than 2201 pregnant women without diabetes (Longnecker et al. 2001). However, this association did not show an origin of diabetes in women with the highest PCB levels. Furthermore, possible influences of diabetes and impaired glucose homeostasis on toxicokinetics of organo-chlorines were discussed. The drawback of this study is the lack of exact measurements of glucose concentrations; data on diabetes were obtained just from project records.

Between 1959 and 1985, the chemical plant Chemko, in Strazske in the Michalovce District in eastern Slovakia, produced approximately 22,000 tons of PCBs. Open-air waste sites and liquid waste dumped into a nearby river and lakes resulted in massive environmental pollution of soil, water, food, fauna, and flora, which is still persisting (Kočan et al. 2001; Chovancová, Kočan, and Jursa 2005; Jursa et al. 2006). The Michalovce District was recognized as one of the world's most heavily polluted areas with PCBs (Trnovec et al. 2004). Also pollution by pesticides and fungicides (HCHs, HCB, DDT) in this region occurred due to their extensive agricultural use in the past. Increased prevalence of thyroid disorders with high titers of anti-thyreoperoxidase, anti-thyreoglobulins, and anti-TSH-receptor antibodies (Langer et al. 1998, 2003, in press) as well as significantly increased frequency of glutamic acid decarboxylase anti-bodies (anti-GAD) were found in employees of Chemko compared to controls from a background area (Langer et al. 1998, 2002). However, in these studies, organochlorine compounds were not estimated and direct toxic effects of PCBs on metabolism were only anticipated. The data obtained from the next survey in this area confirmed the association of increased thyroid volume and indicators of potential thyroid dysfunction with high PCB levels (Langer et al. 2003, in press).

Therefore, the most recent project PCBRISK was focused on evaluating human health risk from low-dose and long-term PCB exposure in 2,050 subjects with variable degrees of contamination with organochlorine pollutants. Variables determining the glucose homeostasis status (fasting and 120 min post-oral-glucose-load glucose and insulin concentrations) correlated with the levels of HCB, βHCH, DDE, and DDT ($p = 0.02$–0.000, $r = 0.11$–0.28) (Rádiková et al. 2004). Moreover, an increased frequency of diabetes and pre-diabetes (i.e., impaired fasting glucose, impaired glucose tolerance) was found in subjects with high levels of PCBs, HCB, βHCH, DDT and DDE (Klimeš et al. 2003). In contrast to other studies (i.e., Bertazzi et al. 1998; Steenland et al. 1999; Longnecker et al. 2001), in our survey, diagnoses of diabetes and other dysglycemias were based on direct sampling and measurement of fasting and oral glucose tolerance test-derived plasma glucose concentrations and classified according to the ADA criteria (ADA 2004).

There are several hypotheses for potential mechanisms by which organochlorines might cause diabetes. TCDD may reduce cellular glucose uptake, possibly due to alterations in the activity of glucose transporters (Enan et al. 1996). Higher dioxin levels are associated with hyperinsulinemia and insulin resistance (Cranmer et al. 2000). Another speculative mechanism is that dioxin affects the activity of pancreatic nitric oxide synthase, responsible for the regulation of insulin release from the pancreatic β-cells (Michálek, Akhtar, and Kiel 1999). Coplanar PCBs act through the Ah-receptor like dioxins, and therefore toxic effects of PCBs similar to dioxins regarding decreased insulin sensitivity cannot be excluded. Remillard and Bunce (2002) suggested that functions of Ah-receptor might antagonize the PPARg (peroxisome proliferator-activated receptor) functions and thus promote development of diabetes as PPARg regulates the expression of several enzymes involved in lipid and glucose metabolism (Arner 2003). A mechanism of action for PCBs independent of the binding of PCB congeners to the Ah receptor was suggested by Fischer, Zhou, and Wagner (1996) and Fischer, Wagner, and Madhukar (1999) in a study on rat insulinoma cells, where release of insulin was induced by PCBs. This mechanism could lead to hyperinsulinemia and to exhausting of insulin production in pancreatic β-cells and/or to compensatory desensitization of peripheral tissues to insulin and disturbances in glucose homeostasis. The insulin release is preceded by a rise in intracellular calcium (Fischer, Wagner, and Madhukar 1999) which may be mediated by ryanodine-sensitive calcium channels in pancreatic beta cells (Islam 2002). These calcium channel complexes are highly sensitive to perturbation by *ortho*-rich PCBs, which are not active at the AhR (Pessah and Wong 2001) and probably, along with other perturbations in calcium regulation (Tilson and Kodavanti 1998), disrupt endocrine as well as nervous system function (Imsilp 2004).

In all epidemiological studies investigating the relationships between organochlorines and development of a disease, these associations remain still an open matter of discussion. Exposure to pollutants (accidental) occurred many years before the epidemiological studies or occurred during a longer period of time (occupational, environmental) and the individual variability of other influencing factors is very high (Remillard and Bunce 2002). At present, no completely "pollutants-free" control population is available due to the ubiquitous presence of persistent organic pollutants and because of a background exposure of both studied subjects and controls. Furthermore, all subjects are exposed to a large number of other chemical substances (not only to those studied) and their interactive effects on human health cannot be excluded (Glynn et al. 2003). Also, a possibility that a disease itself may modify the toxicokinetics, distribution, metabolism, and/or elimination of organochlorines, has to be considered (Glynn et al. 2003). Finally, parameters such as gender, age, BMI, weight loss are known risk factors for diabetes but are also associated with organochlorine levels (Pelletier, Imbeault, and Tremblay 2003).

This review chapter showed, using data of several studies, associations between environmental organochlorine pollution and disturbances in glucose metabolism. Beside other proven risk factors for pre-diabetes and diabetes, environmental pollution may play an important role in disease development. However, stronger evidence for a causal relationship between organochlorines contamination and diabetes development remains necessary.

ACKNOWLEDGMENTS

This work was conducted within the project PCBRISK supported by the European Union under contract No. QLK-CT-2000-00488. However, the authors are solely responsible for this work, which does not represent the opinion of the EU, and the EU is not responsible for any use that might be made of data appearing herein.

REFERENCES

American Diabetes Association (ADA). 2004. "Diagnosis and Classification of Diabetes Mellitus." *Diabetes Care* 27:S5–S10.

Ando, M., H. Saito, and I. Wakisaka. 1985. "Transfer of Polychlorinated Biphenyls (PCBs) to Newborn Infants through the Placenta and Mothers' Milk." *Arch. Environ. Contam. Toxicol.* 14:51–57.

Arner, P. 2003. "The Adipocyte in Insulin Resistance: Key Molecules and the Impact of the Thiazolidinediones." *Trends Endocrinol. Metab.* 14:137–45.

Bertazzi, P. A., I. Bernucci, G. Brambilla, D. Consonni, and A. C. Pesatori. 1998. "The Seveso Studies in Early and Long-Term Effects of Dioxin Exposure: A Review." *Environ. Health Perspect.* 106 (Suppl. 2): 625–33.

Boersma, E. R., and C. I. Lanting. 2000. "Environmental Exposure to Polychlorinated Biphenyls (PCBs) and Dioxins. Consequences for Long-term Neurological and Cognitive Development of the Child Lactation." *Adv. Exp. Med. Biol.* 478:271–87.

Calvert, G. M., M. H. Sweeney, J. Deddens, and D. K. Wall. 1999. "Evaluation of Diabetes Mellitus, Serum Glucose, and Thyroid Function among United States Workers Exposed to 2,3,7,8-Tetrachlorodibenzo-*p*-Dioxin." *Occup. Environ. Med.* 56:270–76.

Chovancová J., A. Kočan, and S. Jursa. 2005. "PCDDs, PCDFs and Dioxin-like PCBs in Food of Animal Origin (Slovakia)." *Chemosphere* 61:1305–11.

Cranmer, M., S. Louie, R. H. Kennedy, P. A. Kern, and V. A. Fonseca. 2000. "Exposure to 2,3,7,8-Tetrachlorodibenzo-*p*-Dioxin (TCDD) Is Associated with Hyperinsulinemia and Insulin Resistance." *Toxicol. Sci.* 56:431–36.

Den Hond, E., H. A. Roels, K. Hoppenbrouwers, T. Nawrot, L. Thijs, C. Vandermeulen, G. Winneke, D. Vanderschueren, and J. A. Staessenet. 2002. "Sexual Maturation in Relation to Polychlorinated Aromatic Hydrocarbons: Sharpe and Skakkebaek's Hypothesis Revisited." *Environ. Health Perspect.* 110: 771–76.

Enan, E., B. Lasley, D. Stewart, J. Overstreet, and C. A. Vandevoort. 1996. "2,3,7,8-Tetrachlorodibenzo-*p*-Dioxin (TCDD) Modulates Function of Human Luteinizing Granulosa Cells via cAMP Signaling and Early Reduction of Glucose Transporting Activity." *Reprod. Toxicol.* 10:191–98.

Fingerhut, M. A., W. E. Halperin, D. A. Marlow, L. A. Piacitelli, P. A. Honchar, M. H. Sweeney, A. L. Greife, P. A. Dill, K. Steenland, and A. J. Suruda. 1991. "Cancer Mortality in Workers Exposed to 2,3,7,8-Tetrachlorodibenzo-*p*-Dioxin." *N. Eng. J. Med.* 324:212–18.

Fischer, L. J., M. A. Wagner, and B. V. Madhukar. 1999. "Potential Involvement of Calcium CaM Kinase II, and MAP Kinase in PCB-Stimulated Insulin Release from RINm5F Cells." *Toxicol. Appl. Pharmacol.* 159:194–203.

Fischer, L. J., H-R. Zhou, and M. A. Wagner. 1996. "Polychlorinated Biphenyls Release Insulin from RINm5F Cells." *Life Sci.* 59:2041–49.

Giwercman, A., E. Carlsen, N. Keiding, and N. E. Skakkebaek. 1993. "Evidence for Increasing Incidence of Abnormalities of the Human Testis: A Review." *Environ. Health Perspect.* 101 (Suppl. 2): 65–71.

Glynn, A. W., F. Granath, M. Aune, S. Atuma, P. O. Darnerud, R. Bjerselius, H. Vainio, and E. Weiderpass. 2003. "Organochlorines in Swedish Women: Determinants of Serum Concentrations." *Environ. Health Perspect.* 111:349–55.

Hagmar, L., L. Rylander, E. Dyremark, E. Klasson-Wehler, and E. M. Erfurth. 2001. "Plasma Concentrations of Persistent Organochlorines in Relation to Thyrotropin and Thyroid Hormone Levels in Women." *Int. Arch. Environ. Health* 74:184–88.

Hansen, L. G. 1998. "Stepping Backward to Improve Assessment of PCB Congener Toxicities." *Environ. Health Perspect.* 106 (Suppl. 1): 171–89.

Henriksen, G. L., N. S. Ketchum, J. E. Michálek, and J. A. Swaby. 1997. "Serum Dioxin and Diabetes in Veterans of Operation Ranch Hand." *Epidemiology* 8:252–58.

Hovander, L., L. Linderholm, M. Athanasiadou, I. Athanassiadis, A. Bignert, B. Fangstrom, J. Kočan, J. Petrík, T. Trnovec, and A. Bergman. 2006. "Levels of PCBs and Their Metabolites in the Serum of Residents of a Highly Contaminated Area in Eastern Slovakia." *Environ. Sci. Technol.* 40:3696–703.

Imsilp, K. 2004. "Disposition and Toxic Effects of Polychlorinated Biphenyls Following Different Routes of Exposure." Ph.D. diss., University of Illinois at Urbana-Champaign.

Islam, S. 2002. "The Ryanodine Receptor Calcium Channel of Beta-Cells: Molecular Regulation and Physiological Significance." *Diabetes* 51:1299–1309.

Jensen, A. A. 1987. "Polychlorinated Biphenyls (PCBs), Polychlorodibenzo-*p*-Dioxins (PCDDs) and Polychlorodibenzofurans (PCDFs) in Human Milk, Blood, and Adipose Tissue." *Sci. Total. Environ.* 64:259–93.

Jursa S., J. Chovancová, J. Petrík, and J. Lokša. 2006. "Dioxin-like and Non-Dioxin-Like PCBs in Human Serum of Slovak Population." *Chemosphere* 64: 686–91.

Kavlock, R. J., G. P. Daston, C. DeRosa, P. Fenner-Crisp, L. E. Gray, S. Kaattari, G. Lucier, M. Luster, M. J. Mac, C. Maczka, R. Miller, J. Moore, R. Rolland, G. Scott, D. M. Sheehan, T. Sinks, and H. A. Tilson. 1996. "Research Needs for the Risk Assessment of Health and Environmental Effects of Endocrine Disruptors: A Report of the U.S. EPA-Sponsored Workshop." *Environ. Health Perspect.* 104 (Suppl. 4): 715–40.

Klimeš, I., J. Koška, L. Kšinantová, K. Bučková, Z. Červenáková, R. Imrich, J. Petrík, S. Jursa, M. Tajtáková, P. Suchánek, M. Vigaš, P. Langer, A. Kočan, T. Trnovec, and E. Šeböková. 2003. "Increased Frequency of Glucose Intolerance in the Population of Specific Areas of Eastern Slovakia Chronically Exposed to Contamination with Polychlorinated Biphenyls (PCB)." Paper given at 63rd American Diabetes Association's Scientific Sessions, June 13–17, New Orleans. In *Diabetes* 52 (Suppl. 1): 955-P.

Kočan, A., J. Petrík, B. Drobná, and J. Chovancová. 1994. "Levels of PCBs and Some Organochlorine Pesticides in the Human Population of Selected Areas of the Slovak Republic." Pt. 1, "Blood." *Chemosphere* 29:2315–25.

Kočan, A., J. Petrík, S. Jursa, J. Chovancová, and B. Drobná. 2001. "Environmental Contamination with Polychlorinated Biphenyls in the Area of Their Former Manufacture in Slovakia." *Chemosphere* 43:595–600.

Kogevinas, M., T. Kauppinen, R. Winkelmann, H. Becher, P. A. Bertazzi, H. B. Bueno-de-Mesquita, D. Coggon, L. Green, E. Johnson, and M. Littorin. 1995. "Soft Tissue Sarcoma and Non-Hodgkin Lymphoma in Workers Exposed to Phenoxyherbicides, Chlorophenols and Dioxins: Two Nested Case-Control Studies." *Epidemiology* 6:396–402.

Korrick, S. A., and L. Altshul. 1998. "High Breast Milk Levels of Polychlorinated Biphenyls (PCBs) among Four Women Living Adjacent to a PCB-Contaminated Waste Site." *Environ. Health Perspect.* 106:513–18.

Kutz, F. W., P. H. Wood, D. P. Bottimore. 1991. "Organochlorine Pesticides and Polychlorinated Biphenyls in Human Adipose Tissue." *Rev. Environ. Contam. Toxicol.* 120:1–82.

Langer, P., M. Tajtáková, G. Fodor, A. Kočan, P. Bohov, J. Michálek, and A. Kreze. 1998. "Increased Thyroid Volume and Prevalence of Thyroid Disorders in an Area Heavily Polluted by Polychlorinated Biphenyls." *Eur. J. Endocrinol.* 139:402–9.

Langer, P., M. Tajtáková, H. J. Guretzki, A. Kočan, J. Petrík, J. Chovancová, B. Drobná, S. Jursa, M. Pavúk, T. Trnovec, E. Šeböková, and I. Klimeš. 2002. "High Prevalence of Anti-Glutamic Acid Decarboxylase (anti-GAD) in Employees at a Polychlorinated Biphenyl Production Factory." *Arch. Environ. Health* 57:412–15.

Langer, P., A. Kočan, M. Tajtáková, J. Petrík, J. Chovancová, B. Drobná, S. Jursa, M. Pavúk, J. Koška, T. Trnovec, E. Šeböková, and I. Klimeš. 2003. "Possible Effects of Polychlorinated Biphenyls and Organochlorinated Pesticides on the Thyroid after Long-Term Exposure to Heavy Environmental Pollution." *J. Occup. Environ. Med.* 45:526–32.

Langer, P., A. Kočan, M. Tajtáková, J. Petrík, J. Chovancová, B. Drobná, S. Jursa, Ž. Rádiková, J. Koška, L. Kšinantová, M. Hučková, R. Imrich, S. Wimmerová, D. Gašperíková, Y. Shishiba, T. Trnovec, E. Šeböková, and I. Klimeš. 2007. "Fish from Industrially Polluted Freshwater as the Main Source of Organochlorinated Pollutants and Increased Frequency of Thyroid Disorders and Dysglycemia." *Chemosphere* 67:S379–85.

Langer, P., M. Tajtáková, A. Kočan, J. Petrík, J. Koška, L. Kšinantová, Ž. Rádiková, J. Ukropec, R. Imrich, M. Hučková, J. Chovancová, B. Drobná, S. Jursa, M. Vlček, A. Bergman, M. Athanasiadou, L. Hovander, Y. Shishiba, T. Trnovec, E. Šeböková, and I. Klimeš. In press. "Thyroid Ultrasound Volume, Structure and Function After Long-term High Exposure of Large Population to Polychlorinated Biphenyls, Pesticides and Dioxin." *Chemosphere.*

Longnecker, M. P., and J. L. Daniels. 2001. "Environmental Contaminants as Etiologic Factors for Diabetes." *Environ. Health Perspect.* 109 (Suppl. 6): 871–76.

Longnecker, M. P., M. A. Klebanoff, J. W. Brock, and H. Zhou. 2001. "Polychlorinated Biphenyl Serum Levels in Pregnant Subjects with Diabetes." *Diabetes Care* 24:1099–1101.

Lu, Y. C., and Y. C. Wu. 1985. "Clinical Findings and Immunological Abnormalities in Yu-Cheng Patients." *Environ. Health Perspect.* 59:17–29.

Michálek, J. E., F. Z. Akhtar, and J. L. Kiel. 1999. "Serum Dioxin, Insulin, Fasting Glucose, and Sex Hormone-Binding Globulin in Veterans of Operation Ranch Hand." *J. Clin. Endocrinol. Metab.* 84:1540–43.

Miller, R. W. 2004. "How Environmental Hazards in Childhood Have Been Discovered: Carcinogens, Teratogens, Neurotoxicants, and Others." *Pediatrics* 113 (Suppl. 4): 945–51.

Morgan, D. P., L. I. Lin, and H. H. Saikaly. 1980. "Morbidity and Mortality in Workers Occupationally Exposed to Pesticides." *Arch. Environ. Contam. Toxicol.* 9:349–82.

Nordstrom, M., L. Hardell, G. Lindstrom, H. Wingfors, K. Hardell, and A. Linde. 2000. "Concentrations of Organochlorines Related to Titers to Epstein-Barr Virus Early Antigen IgG as Risk Factors for Hairy Cell Leukemia." *Environ. Health Perspect.* 108:441–45.

O'Keefe, P. W., J. B. Silkworth, J. F. Gierthy, R. M. Smith, A. P. DeCaprio, J. N. Turner, G. Eadon, D. R. Hilker, K. M. Aldous, L. S. Kaminsky, and D. N. Collins. 1985. "Chemical and Biological Investigations of a Transformer Accident at Binghamton, N.Y." *Environ. Health Perspect.* 60:201–9.

Pelletier, C., P. Imbeault, and A. Tremblay. 2003. "Energy Balance and Pollution by Organochlorines and Polychlorinated Biphenyls." *Obes. Rev.* 4:17–24.

Pesatori, A. C., C. Zocchetti, S. Guercilena, D. Consonni, D. Turrini, and P. A. Bertazzi. 1998. "Dioxin Exposure and Non-Malignant Health Effects: A Mortality Study." *Occup. Environ. Med.* 55:126–31.

Pessah, I. N., and P. W. Wong. 2001. "Etiology of PCB Neurotoxicity: From Molecules to Cellular Dysfunction." In *Recent Advances in the Environmental Toxicology and Health Effects of PCBs,* edited by L. W. Robertson and L. G. Hansen, 179–184. Lexington: University of Kentucky Press.

Rádiková, Ž., J. Koška, L. Kšinantová, R. Imrich, A. Kočan, J. Petrík, M. Hučková, J. Chovancová, B. Drobná, S. Jursa, S. Wimmerová, L. Wsolová, P. Langer, T. Trnovec, E. Šeböková, and I. Klimeš. 2004. "Increased Prevalence of Diabetes Mellitus and Other Dysglycemias in a Population Chronically Exposed to Polychlorinated Biphenyls and Other Persistent Organochlorine Pollutants." Abstract, PCB Workshop, June 13–15, Champaign, Ill.

Remillard, R. B., and N. J. Bunce. 2002. "Linking Dioxins to Diabetes: Epidemiology and Biologic Plausibility." *Environ. Health Perspect.* 110:853–58.

Safe, S. H. 1993. "Toxicology, Structure-Function Relationship, and Human and Environmental Health Impacts of Polychlorinated Biphenyls: Progress and Problems." *Environ. Health Perspect.* 100:259–68.

———. 1994. "Polychlorinated biphenyls (PCBs): Environmental Impact, Biochemical and Toxic Responses, and Implications for Risk Assessment." *Crit. Rev. Toxicol.* 24:87–149.

Steenland, K., L. Piacitelli, J. Deddens, M. Fingerhut, and L. I. Chang. 1999. "Cancer, Heart Disease, and Diabetes in Workers Exposed to 2,3,7,8-Tetrachlorodibenzo-*p*-Dioxin." *J. Natl. Cancer Inst.* 91:779–86.

Swan, S. H., E. P. Elkin, and L. Fenster. 2000. "The Question of Declining Sperm Density Revisited: An Analysis of 101 Studies Published 1934–1996." *Environ. Health Perspect.* 108:961–66.

Sweeney, M. H., G. M. Calvert, G. A. Egeland, M. A. Fingerhut, W. E. Halperin, and L. A. Piacitelli. 1997–98. "Review and Update of the Results of the NIOSH Medical Study of Workers Exposed to Chemicals Contaminated with 2,3,7,8-Tetrachlorodibenzodioxin." *Teratog. Carcinog. Mutagen.* 17:241–47.

Tilson, H. A., and P. R. Kodavanti. 1998. "The Neurotoxicity of Polychlorinated Biphenyls." *Neurotoxicology* 19:517–25.

Trnovec, T., A. Kočan, J. Petrík, J. Chovancová, B. Drobná, S. Jursa, and K. Čonka. 2004. "PCBs in East Slovakia and the Structure and Function of the PCBRISK Project." Abstract, PCB Workshop, June 13–15, Champaign, Ill.

van Wendel de Joode, B., C. Wesseling, H. Kromhout, P. Monge, M. Garcia, and D. Mergler. 2001. "Chronic Nervous-System Effects of Long-Term Occupational Exposure to DDT." *Lancet* 357:1014–16.

Warner, M., B. Eskenazi, P. Mocarelli, P. M. Gerthoux, S. Samuels, L. Needham, D. Patterson, and P. Brambilla. 2002. "Serum Dioxin Concentrations and Breast Cancer Risk in the Seveso Women's Health Study." *Environ. Health Perspect.* 110:625–28.

Wolff, M. S. 1985. "Occupational Exposure to Polychlorinated Biphenyls (PCBs)." *Environ. Health Perspect.* 60:133–38.

12

Long-Term Exposure to Polychlorinated Organic Environmental Pollutants and the Human Thyroid

Pavel Langer, *Slovak Academy of Sciences*

Maria Tajtáková, *Pavel Jozef Šafárik University*

Anton Kočan, *Slovak Medical University*

Ján Petrík, *Slovak Medical University*

Juraj Koška, *Slovak Academy of Sciences*

Lucia Kšinantová, *Slovak Academy of Sciences*

Žofia Rádiková, *Slovak Academy of Sciences*

Richard Imrich, *Slovak Academy of Sciences*

Miloslava Hučková, *Slovak Academy of Sciences*

Jana Chovancová, *Slovak Medical University*

Beata Drobná, *Slovak Medical University*

Stanislav Jursa, *Slovak Medical University*

Yoshimasa Shishiba, *Mishuku Hospital*

Tomáš Trnovec, *Slovak Medical University*

Elena Šeböková, *Slovak Academy of Sciences*

Iwar Klimeš, *Slovak Academy of Sciences*

In three districts of eastern Slovakia (a part of one being heavily polluted by PCB), 2,046 adults (834 males and 1,212 females, 459 <35 yrs and 1,590 >35 yrs) were examined. Thyroid volume (ThV) was measured by ultrasound, free thyroxine (FT4), total triiodothyronine (TT3), thyrotropin (TSH), and thyroperoxidase antibodies (anti-TPO) in serum were measured by sensitive electrochemiluminiscent immunoassay, urinary iodine in 996 subjects was estimated by Hitachi laboratories (Tokyo) using sensitive microcroplate method. Organochlorinated pollutants (PCB = polychlorinated biphenyls, HCB = hexachlorobenzene, DDE = dichloro-p-chlorophenyl-ethene) were measured by congener specific analysis.

After all data were sorted according to PCB level, considerable differences in organochlorine levels were found between medians of lower and upper quintile (e.g., 504 vs. 3,663 ng/g serum lipids for a sum of 18 PCB congeners, 784 vs. 3,222 ng/g for DDE and 320 vs. 959 ng/g for HCB).

Respective ThV (mean ± SD) was 8.76 ± 3.63 vs. 11.15 ± 5.16 mL (p < 0.001 by t-test), FT4 level was 16.04 ± 2.63 vs. 17.06 ± 3.15 pmol/L (p < 0.001), and TT3 level was 1.91 ± 3.32 vs. 2.01 ± 0.37 nmol/L (p < 0.001). In addition, respective frequency of high FT4 (e.g., >22.0 pmol/L) was 13/408 (3.1 %) vs. 23/406 (5.7 %) (p > 0.05 by chi-square), that of TT3 (e.g. >2.50 nmol/L) was 21/408 (5.1 %) vs. 45/406 (11.1 %) (p < 0.001), and that of positive anti-TPO (e.g. >37 IU/mL) was 83/408 (20.3 %) vs. 116/406 (28.6 %) (p < 0.001).

In this and our previous studies, thyroid ultrasound volumetry was used for the first time to assess possible effects of PCB on ThV. The mean level of urinary iodine was 137 μg/L, and the median was 116 μg/L, which, according to World Health Organization (WHO) criteria, is a sign of sufficient iodine intake. Thus any interfering effect of iodine deficiency on ThV can be excluded and increased ThV could be explained mainly by possible effect of organochlorines.

ThV in all subjects was significantly correlated with PCB level (r = 0.208; p <0.001 by Spearman rank correlation test). Nevertheless, the mechanism of ThV increase remains to be elucidated.

Moreover, PCB levels were significantly correlated (p < 0.001) with those of FT4 (r = 0.114) and TT3 (r = 0.101) and highly significant correlation between FT4 and TT3 was also found (r = 0.270; p < 0.0001). This could be explained by the impaired equilibrium between bound and free fractions of such hormones in blood due to well-known effects of PCB on their displacement from specific binding proteins in plasma. Finally, significant negative correlation was found between FT4 and TSH (r = 0.234; p < 0.001).

Although the above described interrelations were demonstrated by evaluating the whole cohort of examined population, some relatively small groups of subjects were found with most pronounced changes, which suggests possible role of hereditary background in the development of the observed impairments.

Among several persistent polychlorinated organic substances that are either used as pesticides, insecticides, defoliants, multipurpose industrial chemicals and plasticizers, and so on or originate as byproducts of large-scale chemical industrial production, perhaps the most widespread global pollutants are polychlorinated biphenyls (PCBs). Because of their striking structural similarity to thyroid hormones, attention has been repeatedly focused on their possible effects on thyroid function and on the metabolism of thyroid hormones.

Recently, several authors attempted to summarise numerous data on organochlorines and thyroid function obtained mainly in animal experiments as well as sporadic findings in human adult and adolescent populations (Brucker-Davis 1998; Brouwer et al. 1998; Kimbrough and Krouskas 2001; Karmaus 2001; Hagmar 2003; Khan and Hansen 2003).

Among the main repeatedly observed and thus definitely accepted experimental findings on short-term effects of PCBs on animal thyroid and thyroid hormone metabolism should be listed, first, repeatedly reported displacement of thyroxine (T4) from specific plasma transport proteins (transthyretin, thyroxine binding globulin and albumin; Chauhan, Kodavanti, and McKinney 2000). Second, enhanced biliary excretion of T4 due to the induction of UDP-glucuronosyltransferase which results in increased production of T4- and T3-glucuronide and their increased disposal via the intestinal route, although a certain fraction of such conjugates is subsequently subjected to hydrolysis by glucuronidase of intestinal bacteria and then to recirculation via enterohepatic cycle. And third, interactions of organochlorines with nuclear receptors of triiodothyronine (T3), which is the active intracellular form of thyroid hormone resulting mainly from intracellular deiodination of circulating prohormone (T4). Since such receptors, after being activated by ligands, act as specific transcription activators and thus as gene expression enhancers, organochlorines as possible ligands could interfere with DNA-related actions of thyroid hormone (Feng et al. 2000; Miyazaki et al. 2004; Yamada-Okabe et al. 2004, 2005). All such effects may be considered valid also for human beings.

In addition, there are a few suspected effects which are not yet confirmed and accepted. One of them is a possible mechanism of direct action of organochlorines on the thyroid tissue structure and thus on the inhibition of thyroid hormone biosynthesis *in situ* as based on the data obtained earlier by Capen and Martin (1989) by electron microscope studies of thyroid ultrastructure. Second, lipophilic organochlorines could accumulate in the thyroid cell membrane and thus increase its fluidity and facilitate the cross-talk between thyroid antigens and lymphocytes of the PCB-modulated immune system which could contribute to the pathogenesis of autoimmune thyroid disorders (Tan et al. 2004). Third, estrogen receptors have been repeatedly found in normal and disordered thyroid glands (Kawabata et al. 2003); furthermore, estrogen-promoted growth of human thyroid tumor cells by different molecular mechanisms has been reported (Manole et al. 2001), increased thyroid function in ovariectomized ewes exposed to phytoestrogens has been found (Madej et al. 2002), and estrogen-like activity of some PCB congeners, their metabolites, and DDE has also been observed (Hansen 1998; Shiraishi et al. 2003). Thus it appears that such mechanism could also contribute to the thyroid growth. This view is supported by repeated reports on moderately higher risk of thyroid cancer after the use of oral contraceptives (La Vecchia et al. 1999). Finally, observations on possible interference of organochlorines with hypothalamic-pituitary-thyroid-peripheral feedback loops should be mentioned, among them recent findings on the inhibitory effect of PCB on thyrotropin release from the pituitary after the administration of exogenous hypothalamic hormone TRH (thyrotropin releasing hormone) by Khan and Hansen (2003). Since there are still several doubts and contradictory data about these mechanisms, they should be considered plausible but not yet accepted.

Perhaps the first report on the effect of organochlorines on human thyroid was that by Bahn et al. (1980), who found 11 percent prevalence of hypothyroidism and elevated levels of autoantibodies against thyroid microsomes (which are now considered identical with those against thyroperoxidase) in workers from a PCB-producing factory. Later, Saracci et al. (1991) collected data on 13,482 workers exposed to chlorophenoxy herbicides or chlorophenols who were followed for 17 years and found that 4 of them died from thyroid cancer, significantly more than no case of thyroid cancer in 4,908 workers either with unknown exposure or probably exposed. Zober, Ott, and Messerer (1994) found a significantly higher percentage of thyroid disease (e.g., 7.0 % vs. 1.2 %, respectively) in 158 men exposed to TCDD (tetrachloro-dibenzo-dioxin) due to the explosion in a chemical factory in Germany 36 years ago as compared to 161 unexposed employees of the same factory. In a Spanish village located in the vicinity of an organochlorine factory, Grimalt et al. (1994) found significantly increased incidence of thyroid cancer (3 new cases among 5,003 inhabitants within 10 years). Data of Norway Cancer Registry (Frich, Akslen, and Glattre 1997) showed significantly elevated risk of thyroid cancer in 40,939 women married to fishery workers, among which 174 cases were found during 1,210,683 person-years. Since fish food from northern seas belongs to the most polluted by PCB, some interrelations perhaps cannot be excluded. In contrast, no attention has been paid to the thyroid in a number of other epidemiological studies of industry workers exposed to PCBs and TCDD as reviewed by Nicholson and Landrigan (1994) and Swanson, Ratcliffe, and Fischer (1995) or individual studies by Kogevinas et al. (1995) and Flesh-Janys et al. (1995). From this brief overview it is apparent that, until the mid-1990s, reports on the interrelations between organochlorines and human thyroid disease were mostly sporadic and inconsistent.

POSSIBLE EFFECTS OF PCBS ON HUMAN THYROID VOLUME AND STRUCTURE

Some authors reported increased frequency of goiter of undefined thyroid volume (ThV) as evaluated by semi-quantitative palpation either in subjects affected by the Yusho (Tsuji et al. 1997) or YuCheng accident (Guo et al. 1999). Recently, Karmaus (2001) reviewed 13 thus far published papers on PCBs and human thyroid and reported data from one of his previous papers showing increased ThV in adulthood in a number of individuals who had been exposed to organochlorines and heavy metals during adolescence.

As far as we are aware, in our studies the precise thyroid ultrasound volumetry has been used for the first time in the evaluation of possible effects of organochlorinated substances on the ThV. Ultrasound examinations were performed by an observer with two decades experience in the field (Mária Tajtáková); the intra-observer variation as estimated by three subsequent measurements of 50 ThV ranging between 3.0 and 20.5 ml (median 6.2 ml) was 3.9 ± 3.5 %. ThV was measured by real-time sonography using the ellipsoid method (Brunn et al. 1981) with the aid of Sonoline (Siemens) apparatus and 7.5 mHz linear transducer. Thyroid hypoechogenicity was estimated by ultrasound in all thyroids. This

phenomenon is based on the decreased density of ultrasound waves reflected by a decreased number of cell-colloid interfaces of thyroid tissue which was impaired by an autoimmune process to various extents. It was evaluated as positive or negative as based on the comparison of the density of echonormal thyroid image with that of normally low-echogenic surrounding muscles (fig. 12.1). Three dimensions of nodules were measured by ultrasound in anterior-posterior and lateral projection and were sorted according to the highest dimension.

In the pilot study (Langer et al. 1998) significantly increased ThV was found in 238 long-term employees of a former, but still contaminated, PCB-producing factory (median 17.3 ml, 90th percentile 22.8 ml) as compared to 486 referents (median 11.3 ml, 90th percentile 15.5 ml; p < 0.001) and also in 454 17-year-old adolescents living in the polluted area (median 9.0 ml, 90th percentile 11.0 ml) compared to 965 referents (median 7.7 ml, 90th percentile 9.8 ml; p < 0.001).

In that pilot study the blood levels of PCB were not measured. However, the next survey (Langer et al. 2003b) showed very high levels of total PCB (consisting of 15 main congeners) in a group of 101 employees of that chemical factory and subjects living in a polluted area (mean ± SE = 7,300 ± 871 ng/g serum lipids) compared to that found in 360 referents living in a much less polluted area (2,045 ± 147 ng/g). All thyroid volumes in the polluted area were significantly higher than those in referents (Spearman rank correlation coefficient r = 0.208; p < 0.001). In the polluted area, the thyroid volume in the upper quartile was significantly higher than that found in the upper quartile of referents (Mann-Whitney rank sum test; p < 0.016). The highest ThV (18.7 ± 2.32 ml; mean ± SE) was clustered among 23 subjects with a PCB level in the range of 10,000–58,667 ng/g lipids (median 14,932 ng/g), the frequency of ThV >20.0 ml in that group being 9/23 (39.1 %), while in the remaining 438 subjects with a PCB level of 290–9952 ng/g (median 1,629 ng/g) the mean ThV was 14.2 ± 0.29 ml (mean ± SE) and the frequency of ThV >20.0 ml was 42/438 (9.6 %; p > 0.001 as compared to the above group of 23 by Yates chi-square).

In the recent PCBRISK survey (Langer et al. 2007a, 2007b) more than 2,000 adults were examined in three districts of eastern Slovakia, a part of one of those districts being heavily polluted by PCBs for almost five decades; nevertheless, background exposure of neighboring districts cannot be neglected (Kočan et al. 1994, 2001). In all subjects, ThV was found significantly related to the level of total PCBs (r = 0.208; p < 0.001). In 407 subjects in the upper quintile of PCB levels (range of 2349–101413 ng/g, median = 3663 ng/g), the ThV (11.15 ± 5.16 ml, mean ± SD,

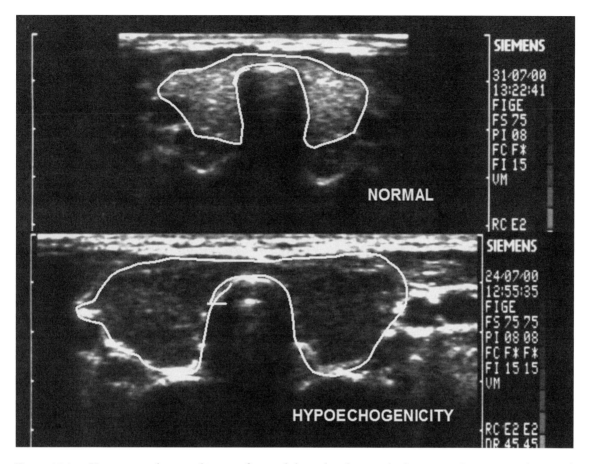

Figure 12.1.   Upper part: ultrasound image of normal thyroid with normal echogenicity. Lower part: ultrasound image of enlarged thyroid with severe homogenously diffuse hypoechogenicity (it should be noted that several degrees of hypoechogenicity may be distinguished such as homogenously diffuse, non homogenous, etc.).

median = 10.1 ml, 90 % = 17.9 ml) was significantly higher (p < 0.001 by t-test) than that in 1640 subjects in four lower quintiles of PCB levels (e.g., 148–2346 ng/g, median = 1640 ng/g) in which the mean ThV was 9.55 ± 4.53 ml (median = 8.5 ml, 90 % = 14.4 ml).

However, since aging appeared to be one of the major confounding factors positively associated with both ThV and PCB level, we recently compared the participation of both and showed a positive effect of PCB other than that of age (Langer et al. 2007b). Such interrelation among age, PCB, and ThV is also depicted in figure 12.2, which shows more remarkable and PCB level–dependent ThV increases in subjects aged 36–50 or >50 years compared to the group aged >35 years. However, such interrelation between ThV and PCB level is apparently nonlinear, since a majority of subjects with even very high PCB levels show ThV in the normal range. Several explanations of this phenomenon may be suggested. One of the most plausible appears to be hereditary background resulting in increased susceptibility of some individuals. In general, some role could be also played by the threshold level of PCB (Rádiková et al. in press) neces-

sary to trigger or facilitate the toxic action, which could be related to the actual and previous longtime blood level, time of exposure, total body burden, and so on, and which could be different for each subject. Although such a threshold for PCB blood level may show a wide range, may be individual, and may either exist or not, this problem deserves attention and should be further elucidated, since several studies reported relatively low organochlorine levels in serum of case patients which must not reach the critical threshold and, thus, may not provide conclusive data.

In the general population, a great majority of thyroid disorders are of autoimmune nature and are called AITD (autoimmune thyroid disease). This concerns both the hypothyroidism or autoimmune thyroiditis (resulting from the presence of cytotoxic anti-thyroperoxidase and anti-thyroglobulin autoantibodies) and the hyperthyroidism (resulting from autoantibodies stimulating the thyrotropin receptor). They have similar pathogenetic mechanisms. They can co-occur in families and the transition from one clinical picture to the other within the same individual over time was repeatedly observed. Since all types of such antibodies

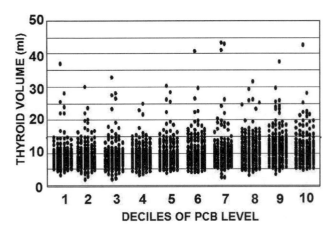

Figure 12.2.   Distribution of thyroid volume (ThV) in a total of 2,046 adults (1,212 females and 834 males) divided in three age groups (gray points = <35 years, black points = 36–50 years, black triangles = 51–75 years) as stratified in deciles of PCB level. The upper range limits of serum PCB level (ng/g serum lipid) in individual deciles are 504, 627, 767, 906, 1,087, 1,387, 1,711, 2,343, 3,664, and 101,413, respectively. At 90 percent of the 10th decile (i.e., 99th percentile) the value is 15,103 ng/g serum lipid. This figure shows that the increase of ThV in the young group (gray points) with increasing PCB level is almost negligible, while in the oldest group (black triangles) ThV increases as serum PCB increases. Thus the subjects within the same age range have significantly smaller thyroids at the lower PCB level than at a higher one.

could be present in blood of the same subject and the final status of thyroid function depends on the prevailing type (Kung and Jones 1998), it may be that organochlorines aggravate the development of such disorders in hereditary predisposed individuals.

From figure 12.2 it may be seen that the range of ThV distribution even in the first decile of PCB levels, for example, in a virtually normal and almost negligibly PCB-affected population is relatively large (cca between 4 and 20 ml), which is due to the enormous growth potential of the thyroid orchestrated by numerous growth factors regulated by TSH (Studer and Derwahl 1996; Roger et al. 1997). Thus even a normal thyroid in some individuals is about two to five times larger than that in others, which is one of the new aspects resulting from ultrasound volumetry as also supported by our previous studies (Tajtáková et al. 1998; Langer et al. 1999, 2003a). It is very likely that this is due to hereditary factors, which also could play a role in the acceleration of growth rate under exogenous toxic influences which simply results in the increase of range span of thyroid volumes.

However, it should be underlined that the most important exogenous factor influencing the thyroid volume is the intake of iodine. Within a recent European study (Delange

et al. 1997) it was found that Slovakia is one of a few countries with high urinary iodine levels and small thyroid volumes in schoolchildren. This results from 50 years of well-organized and well-monitored mandatory consumption of iodized salt and an embargo on the importation of any other salt. This was encouraged by the fact that Slovakia is a landlocked country located far from the seas, resulting in severe iodine deficiency until about the end of World War II. Such views also are supported by generally low ThV in this and our previous studies related to the ThV in general adolescent population in Slovakia (Langer et al. 1999, 2003b; Tajtáková et al. 1988). Urinary iodine also has been measured in about 1,000 subjects participating in the PCBRISK study (thanks to a generous support of ICCIDD Committee of Japan and Hitachi Laboratories in Tokyo), which showed a median value of 137 μg/L, corresponding to optimal iodine intake according to the recommendations of the WHO (WHO/UNICEF/ICCIDD 1993). Our opinion is that such long-term sufficient iodine intake by the whole population eliminates the interfering effects of different iodine intake on ThV and thus offers the unique opportunity to study the antithyroid effects different from iodine deficiency, such as environmental and genetic ones, which could not be valid for some countries with regional or even familial differences in iodine intake (e.g., if one family prefers iodized salt and the other does not). This is also true for some countries with very high iodine intake, mostly due to high consumption of seafood (e.g., Japan, Hudson Bay, Faroe Islands). Thus the differences in ThV in Slovakia described above do not appear to be caused by any regional or individual differences in iodine intake but to the participation of exogenous toxic substances in regular thyroid growth resulting from the process of aging (Langer et al. 2007b) as also shown in figure 12.2.

In addition to increased ThV in groups with high PCB levels, there was increased frequency of ultrasound hypoechogenicity (HYE) and nodules (NOD) as defined above (see also fig. 12.3), such difference being much more pronounced in males than in females. Thus, among 461 subjects examined in 1998 (Langer et al. 2003b), no difference in HYE was found in females (e.g., 83/179 = 46.4 % in background area vs. 18/42 = 42.8 % in polluted area), while in males the respective data were 21/181 (= 11.6 %) vs. 26/59 (= 44.1 % – p < 0.001 by Yates chi-square). Similarly, no difference in females was found in the frequency of solitary NOD with the diameter >10 mm (10.6 % vs. 11.9 %), while that in males (2.7 % vs. 10.1 %) was again highly significant (p < 0.001).

Such findings were further supported by the data obtained in 2001 within the PCBRISK survey in which increased frequency of HYE has been also found in females. Thus in

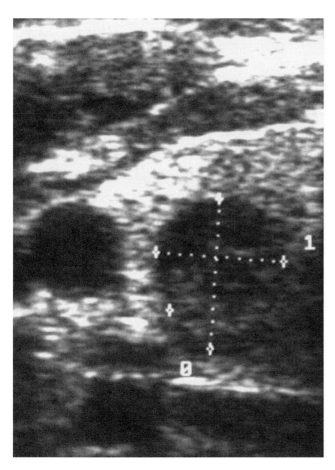

Figure 12.3.   Example of ultrasound image of hypoechogenic large thyroid nodule.

frequencies of the above indices; when separated by serum PCB group, both genders differed in the same direction. In addition, table 12.1 shows that the levels of DDE and HCB were also considerably higher in the subjects with a high PCB level, possibly due to negligence in unlimited use of pesticides, which was apparently much higher in the agricultural plus PCB-polluted area than in the mountainous area of background pollution. Although participation of those pesticides in the toxic effects cannot be excluded, from the present data the particular role of individual substances in individual adverse effects cannot be definitively quantified. Although various statistical methods have been used to achieve this aim, a considerable black box remains between the serum level and specific endpoints. Such increased frequency of NOD in groups of subjects with increased PCB levels supports the views on impaired thyroid growth mechanisms.

POSSIBLE IMMUNO-MODULATORY EFFECTS OF PCBS AS EVALUATED BY THE FREQUENCY OF AUTOANTIBODIES

Immunotoxic and immunomodulatory effects of various organochlorines are well known and repeatedly reported. Recently, Jung et al. (1998) observed impaired leucocyte function in workers formerly exposed to dioxins, Dewailly et al. (2000) found increased frequency of otitis media in Inuit infants prenatally exposed to organochlorines (DDE, HCB and PCB), and similarly, Heilmann, Grandjean, and Weihe (2003) found decreased antidiphtheria vaccine response in children from Faroe Islands related to the PCB level in maternal milk. Twenty years after a dioxin accident in Seveso, Italy, Baccarelli et al. (2002) found decreased IgG levels in serum of the most highly exposed population, and Bernhoft et al. (2000) reported similar findings of IgG decrease in the plasma of polar bears in the Norwegian Arctic.

As mentioned above in the previous section, in 1980 Bahn et al. reported a high frequency of antimicrosomal antibodies in workers of a PCB-producing factory. Since such antibodies are now considered identical with anti-thyroperoxidase antibodies (TPOab), this work actually stimulated us to search for those. Thus, within the first survey in 1994 (Langer et al. 1998) we found increased frequency of anti-thyroperoxidase antibodies in 190 examined females from a former PCB-producing factory compared to 482 referent females from the background polluted area, for example, 54/190 (28.4 %) versus 99/482 (20.5 % – p < 0.05 by chi-square). Similar respective differences of the frequency of anti-thyroglobulin antibodies were found in a group of females aged 31–60 years (36/169 = 21.3 % vs. 50/342 = 14.6 % – p < 0.05) as well as in the frequency of TSH-receptor antibodies in 238

females the frequency of HYE in the first versus fifth quintile of PCB levels (for such levels, see above) was 46/308 (14.9 %) versus 47/184 (25.5 % – p < 0.01 by chi-square), while that in males was 1/101 (1.0 %) versus 23/223 (10.3 % – p < 0.01 by chi-square).

The frequency of NOD >10 mm was also related to PCB level. Thus when comparing that in the first to third quintile (range of PCB levels 148 to 1,329 ng/g serum lipids, median = 762 ng/g, 90 % = 1198 ng/g) with fourth plus fifth quintile (range 1,330 to 101,413 ng/g, median 2,308 ng/g, 90% = 6,300) in females, the frequency was 27/796 (3.4 %) versus 29/416 (6.9 %, p < 0.01 by chi-square) and in males 4/433 (0.9 %) vs. 10/401 (2.5 % – p ≤ 0.05).

Pooled data obtained within 1998 and 2001 (PCBRISK) surveys from the previous four paragraphs are summarized in table 12.1, indicating clear differences between exposures and between genders. In the general population, the higher frequency of NOD, HYE, and positive TPOab in females is common and well known; it appears that the males from the polluted region were "feminized," at least in regard to the

Table 12.1. Summary of Data on Thyroid Nodules, Hypoechogenicity, and Anti-TPO in Individuals with High vs. Low PCB, DDE, and HCB Exposures.

	PCB < 3000 ng/g		PCB > 3000 ng/g	
	Females	Males	Females	Males
Number of subjects	1,253	848	178	221
Median PCB (ng/g)	883	1,129	4,481	5,087
Median DDE (ng/g)	1,685	1,691	3,880	3,339
Median HCB (ng/g)	784	523	1,556	862
Nodules	81 (6.4 %)	15 (1.8 %)	16 (9.0 %)	11 (5.0 %)[b]
Hypoechogenicity	267 (21.3 %)	68 (8.0 %)	61 (34.2 %)[a]	39 (17.6 %)[b]
Anti-TPO	363 (29.0 %)	112 (13.2 %)	51 (28.6 %)	47 (21.2 %)[b]

[a]Significant difference from low PCB at $p < 0.05$.

[b]Signficant difference at $p < 0.01$.

age- and sex-matched subjects from respective areas (10.5 % vs. 2.5 % – $p < 0.001$). In the employees of the chemical factory we also found increased frequency of anti-glutamic acid decarboxylase antibodies considered a marker of diabetes mellitus type 1 (Langer et al. 2002), but this was not convincingly confirmed in the PCBRISK survey.

In the 1998 survey no difference was found between the frequency of positive TPOab in females (e.g., 37/179 = 20.7 % in background area vs. 7/42 = 16.6 % in polluted area), but in males the respective difference (e.g., 4/181 = 2.2 % vs. 12/59 = 20.3 %) was highly significant ($p < 0.001$). Within the PCBRISK survey in 2001, significant increase in the frequency of positive TPOab was found between the first and fifth quintile of PCB levels (for such levels, see above) in both females (e.g., 73/308 = 23.7 % vs. 67/184 = 36.4 % – $p < 0.01$) and, especially, in males (e.g., 10/101 = 9.9 % vs. 49/223 = 21.9 % – $p < 0.01$).

In conclusion, increased frequency of positive anti-TPO as related to high PCBs levels was found repeatedly in relatively large numbers of the population. Since the increase of anti-TPO level is a sign of the same autoimmune process as ultrasound hypoechogenicity, we further searched for a coincidence of those two signs. It should be explained that hypoechogenicity was evaluated during the field examination, while anti-TPO was estimated several months later. The frequency of such coincidence in females was 35/308 (11.3 %) in the first quintile of PCB levels, while that in the fifth quintile was 27/184 (14.9 %, or not significant), but the respective frequency in males was significant, 0/101 (0.0 %) versus 3.6% ($p < 0.01$), which supports the interrelations between PCB levels and frequency of autoimmunity signs.

INTERRELATIONS BETWEEN THYROID HORMONE AND PCB LEVELS

The thyroid produces about 80–90 percent of whole-body thyroxine (T4) and 10–20 percent of triiodothryonine (T3). Both are circulating in blood, bound with various affinities and binding capacities, to three carrier proteins (transthyretin, thyroxine binding globulin, or TBG, and albumin). T4 is a circulating prohormone, about 75 percent of which is bound to TBG in humans, and its half-life in plasma is about five days, while T3 is the genuine active hormone which finally binds to specific nuclear receptors in target cells and thus acts as a transcription enhancer at the DNA level. Whereas the thyroid is the sole producer of T4, only about 20 percent of circulating T3 is produced by thyroid, but about 80 percent of T3 is produced by intracellular deiodination of T4 in the liver and kidneys. The turnover of T3 is much faster than that of T4, the half-life of T3 being only about 1.8 days.

Peripheral tissues are fed by T3 from two sources, one being the circulating prohormone T4, which after entering the cells in free form is deiodinated to T3 by specific enzymes. This fraction of intracellular T3 is being called T3(T4). The second source of intracellular T3 is the circulating T3 originating either from the thyroid or from peripheral deiodination of T4 in other organs and then released to blood. This one is called T3(T3). Considering this carousel, peripheral organs could be divided into predominantly T3-producing (liver and kidneys) and predominantly T3-consuming (e.g., brain, skeletal muscles, etc.).

Since T4 in plasma is strongly bound to carrier proteins, only about 0.03 percent of the total amount (TT4) appears in free (unbound) form (FT4). There exists a dynamic equilibrium between TT4 and FT4, individual T4 molecules oscillating between bound and free state within milliseconds which depends on local temperature, pH, concentration of various ions, proteins, free organic substances, and so on. From this it follows that the level of FT4, for instance, will be considerably different in the warm liver and in the cold skin. In our opinion, organochlorines are influencing particularly this process.

A negative feedback loop exists between the level of circulating thyroid hormone, the decisive regulatory step being the rate of T4 to T3 conversion by deiodinase in pituitary cells. The data reported so far on the effects of organochlorines on serum levels of thyroid hormones in human subjects are scanty and contradictory. Thus Koopman-Esseboom et al. (1994) found a decrease of TT4 and an increase of TSH level in two-week-old infants of high-dioxin-exposed mothers compared to those of low-dioxin-exposed mothers, while such difference could no longer be found at the age of three months. Pluim, Koppe, and Olie (1993) found an increase of TT4 and TSH in neonates of mothers with high levels of dioxin (TCDD) in milk, while at the age of 11 weeks only increased TSH persisted. In middle-aged and elderly Swedish fishermen Rylander et al. (2006) found a positive association between DDE and TSH.

In adult Great Lakes fish consumers, Persky et al. (2001) and Turyk et al. (2006) found a negative correlation of T4 and inconsistent interrelations of TSH with PCB. Calvert et al. (1999) reported increased FT4 index in 278 workers employed in the manufacture of 2,4,5,-trichlorophenol contaminated with TCDD. Osius et al. (1999) found increased 95 percent value of TSH and FT3 in 57 second-grade schoolchildren living 20 km from the waste incinerator as compared to 583 referent children, while no difference in FT4 level was observed. No differences in TT4, TT3, and TSH were found in German workers with low (486 ppt) and high (526 ppt) dioxin exposure to dioxin (Triebig et al. 1998). Nagayama et al. (2001) did not find any changes of serum levels of TT4, TT3, and TSH in 16 Yusho survivors with TEQ concentrations of PCDDs, PCDFs (polychlorinated dibenzo-furans), and PCBs in serum about seven times higher than in the normal Japanese population. Sala et al. (2001) reported a negative correlation between hexachlorobenzene and TT4 levels in serum, while no interrelation was found with the TSH level. Hagmar et al. (2001a) described a negative correlation between plasma level of PCB 153 and TT3 in 182 fishermen's wives (which actually means a decrease of TT3 level), while no similar interrelations were found in males, nor were there any interrelations between PCB and TSH, TT4, FT4, and FT3 in males or females (Hagmar

et al. 2001b). The same author recently reviewed 13 studies (among them 6 in neonates and infants) that showed contradictory data on the increase, decrease, or no change of FT4, TT3, and TSH levels. Negative association of TT4 with PCB, DDE, and HCB was also found in Mohawk youth (Schell et al. 2004), but no such association was observed in New York anglers (Bloom et al. 2003). Rylander et al. (2006) did not find any association between DDE and TT4, while Meeker, Altshul, and Hauser (2007) reported positive correlation of DDE with FT4 and TT3. In pregnant women Takser et al. (2005) found negative correlation of PCB with TT3, but no correlation with FT4.

From such sporadic data obtained in different countries and under different circumstances it seems not so easy to find some common denominator in the search for the interrelations between PCBs and thyroid hormone metabolism. However, it should be noted that the subjects examined by different authors had different (and in some cases relatively low) levels of PCBs and other organochlorines, which could be considered as one of the main causes of discrepant results. Actually, beginning in the early 1950s the estimation of total T4 was used almost for three decades as the predominant laboratory tool for diagnosis of thyroid diseases, since it has been considered a sole circulating thyroid hormone, while T3 was considered only its strange modification. However, numerous papers showed that about one-third of patients with hyperthyroidism actually have a normal T4 level, while vice versa, about one-third of patients with a normal T4 level have hyperthyroidism. Finally, it was definitely shown that the level of total T4 more closely depends on the level of the main binding protein (e.g., TBG) than on the thyroid function, and thus total T4 estimation was definitely abandoned from the diagnostic armament for thyroid disease. From this it follows that it apparently is not of any special significance either for the evaluation of PCB effect on thyroid function. This may be supported by our finding of no difference in TT4 level between the polluted (N = 238) and background (N = 498) area (e.g., 116.1 ± 31.2 versus 112.2 ± 37.0 nmol/L), Nevertheless, some present studies on the effects of organochlorines use thyroxine as the sole marker of thyroid function (Bloom et al. 2003).

Circulating T3 was measured in blood later, in the early seventies, and subsequently its true role has been elucidated. Nevertheless, since about 20 years ago the estimation of TSH by sensitive IRMA method based on the use of two monoclonal antibodies has been considered the first step for reliable laboratory diagnosis of thyroid disease. However, it should be underlined that the interrelation between TSH level and thyroid function is discontinuous. Thus in hyperthyroidism the TSH level is paradoxically suppressed (below about 0.3 mU/L in overt clinical cases) by extremely high levels of thyroid hormones released from the thyroid due to the effect of TSH receptor stimulating antibodies. In contrast, in hypothyroidism characterised by a decreased production of thyroid hormones due to the damage of thyroid tissue by autoimmune process, TSH level is increasing step by step. Since this is a long-term process taking several decades and progressing by various rates in various individuals, TSH level is increasing slowly. It is generally accepted the TSH level >4.0 mU/L shows the lower limit for subclinical hypothyroidism, which, according to prevailing opinions, does not require exogenous thyroid hormone. Nevertheless, this point is being frequently discussed. It was observed that a certain number of subjects with "high normal" TSH levels (e.g., between 3.0 and 4.0 mU/L) will develop hypothyroidism within 10 years (Vanderpump et al. 1995). Nevertheless, it should be underlined that in short term experiments in experimental animals decreased responses of the TSH level to almost immediate decrease of T4 due to the effect of (sometimes high doses of) organochlorines were found (e.g., Hood and Klaasen 2000; Khan and Hansen 2003), but the attempts to apply such interrelations to human beings without any caution seems to be oversimplified.

Finally, subjects with a TSH level between about 0.3 and 4.0 mU/L are considered euthyroid, and thus the average TSH level, particularly in large groups of subjects, will oscillate between about 1.5 and 2.5 mU/L, which was also the case in our PCB surveys. Thus in our first survey (1994) no difference was found in the mean level of TSH between the polluted (N = 238) and background (N = 460) area (e.g., 1.56 ± 0.86 mU/l vs. 1.51 ± 0.84 mU/l). Within the PCBRISK study in 1,873 subjects with TSH levels in the range of 0.5–5.0 mU/L, the average value was 2.20 ± 1.00. Similarly, in the U.S. survey NHANES III (Hollowell et al. 2002) the average level in 16,533 Americans was 1.47 ± 0.02 mU/L (SE), about 90 percent of values being between 0.5 and 2.5 mU/L, while 3.2 percent were <0.4 mU/L and 4.7 percent >4.5 mU/L. For the purpose of evaluating the effects of PCBs on the thyroid status in a population, there is no special reason to search for differences in the average TSH level; a more appropriate approach would be to evaluate differences in the frequency of decreased and increased levels to see how many people actually suffer from thyroid disorders.

Within the PCBRISK survey the frequency of increased TSH levels that are considered as either a sign of subclinical or even clinically overt hypothyroidism (e.g. >4.0 mU/L) did not show any considerable difference between the fifth and first quintile either for combined males and females (e.g., 45/409 = 11.0% vs. 55/410 = 13.4% – n.s.), or for females only (e.g., 31/185 = 16.7% vs. 48/308 = 15.6% –

n.s) and for males only (e.g., 14/224 = 6.2% vs. 6/101 = 5.9% – n.s.). In contrast, the frequency of decreased TSH levels that are considered a sign of subclinical or clinically overt hyperthyroidism (e.g., <0.5 mU/L) was higher in the fifth quintile than in the first quintile in females (e.g., 9/185 = 4.9% vs. 3/308 = 1.0%, p < 0.01) and also in males (11/224 = 4.9% vs. 1/101 = 1.0%, p ≤0.05).

As shown in figure 12.5, within the PCBRISK survey increased frequency of high FT4 levels (e.g., >22.0 pmol/L) was found in the fifth quintile of PCB levels (e.g., 62/409 = 15.2%) as compared to the first quintile (e.g., 22/410 = 5.4%, p < 0.001). The levels of total T3 were positively correlated with those of FT4 (fig. 12.5) and a similar difference appeared between the frequency of high TT3 levels (e.g., >2.2 nmol/L) in the fifth quintile (e.g., 45/409 = 11.0%) and the first quintile (e.g., 20/410 = 4.9%, p < 0.001). These findings are similar to those reported by Calvert et al. (1999), who found a mean FT4 index of 27.8 ± 0.27 pmol/L in 278 workers employed for >15 years in the manufacture of 2,4,5,-trichlorophenol contaminated with TCDD. This was significantly higher than that found in 257 age-, sex-, and race-adjusted controls (for example, 26.8 ± 0.28 pmol/L).

Our findings on increased FT4 as related to increased PCB levels (fig. 12.4) support the hypothesis that the persistent increase of PCB for decades perhaps resulted in persistent impairment of equilibrium between bound and free T4 which further resulted in increased flow of free T4 to the cells of target tissues and increased rate of intracellular T4 to T3 conversion. From this perhaps further followed increased production of T3 which could be supported by the correlation between FT4 and TT3 (fig. 12.5). Although this process is apparently somewhat different from immunomodulatory

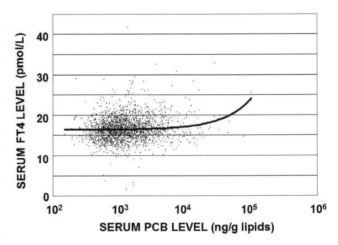

Figure 12.4.   Interrelation between serum PCB level and serum free T4 level in 586 males age >40 years (Pearson's correlation coefficient r = 0.079, p < 0.05).

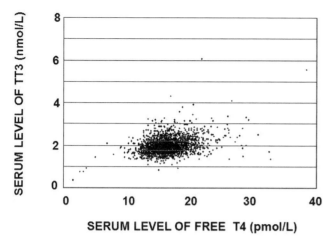

Figure 12.5.   Correlation between free T4 and total T3 levels (Spearman rank correlation coefficient r = 0.270, p < 0.0001).

actions of PCBs, they both may be considered as fractions of multiple toxic activities of organochlorines.

In conclusion, some fundamental interrelations between high serum PCB levels and markers of impaired thyroid growth and function were observed in a large population exposed to heavy environmental pollution for nearly a half century as compared to the population with background, although not quite negligible, exposure from neighboring districts.

ACKNOWLEDGMENTS

This study was supported by the 5th Framework Program on Quality of Life and Management of Living Resources as part of the EU's PCBRISK project (QLK4-2000-00488). The authors wish to express their appreciation also to the ICCIDD Committee of Japan and Hitachi Laboratories (Tokyo); to Alica Mitková for her excellent technical assistance; to 28 district practitioners from Michalovce, Svidník, and Stropkov Districts; and to the Public Health authorities of these districts for their invaluable help in the planning, arrangement, and carrying out of the repeated field survey. This work was also partly supported by the grant from VEGA No. 2-309923.

REFERENCES

Baccarelli, A., P. Mocarelli, D. G. Patterson, M. Bonzini, A. C. Pesatori, N. Caporaso, and M. T. Landi. 2002. "Immunologic Effects of Dioxin: New Results from Seveso and Comparison with Other Studies." *Environ. Health Perspect.* 110: 1169–73.

Bahn, A. K., J. L. Mills, P. J. Snyder, P. H. Gann, L. Houten, O. Bialik, L. Hollman, and R. D. Utiger. 1980. "Hypothy-roidism in Workers Exposed to Polybrominated Biphenyls." *New Eng. J. Med.* 302:31–33.

Ballatori, N., J. L. Boyer, and J. C. Rockett. 2003. "Exploiting Genome Data to Understand the Function, Regulation, and Evolutionary Origins of Toxicologically Relevant Genes." *Environ. Health Perspect.* 111:871–75.

Bernhoft, A., J. U. Skaare, O. Woog, A. E. Derocher, H. J. S. Larsen. 2000. "Possible Immunotoxic Effects of Organochlorines in Polar Bears (*Ursus maritimus*) at Svalbard." *J. Toxicol. Environ. Health* 59:561–74.

Bloom, M. S., J. M. Weiner, J. E. Vena, and G. P. Bechler. 2003. "Exploring Associations between Serum Levels of Select Organochlorines and Thyroxine in a Sample of New York State Sportsmen: The New York State Angler Cohort Study." *Environ. Res.* 93:52–66.

Brouwer, A., D. C. Morse, M. C. Lans, A. G. Schuur, A. J. Murk, E. Klasson-Wehler, A. Bergman, and T. J. Visser. 1998. "Interactions of Persistent Environmental Organohalogens with the Thyroid Hormone System: Mechanisms and Possible Consequences for Animal and Human Health." *Toxicol. Ind. Health* 14:59–84.

Brucker-Davis, F. 1998. "Effects of Environmental Synthetic Chemicals on Thyroid Function." *Thyroid* 8:827–56.

Brunn, J., U. Block, G. Ruf, I. Bos, W. P. Kunye, and P. C. Scriba. 1981. "Volumetrie der Schilddrüsenlappen Mittels Real-Time Sonographie." *Dtsche. Med. Wschr.* 106:1138–40.

Calvert, G. M., M. H. Sweeney, J. Deddens, and D. K. Wall. 1999. "Evaluation of Diabetes Mellitus, Serum Glucose, and Thyroid Function among United States Workers Exposed to 2,3,7,8,-Tetrachlorodibenzo-p-Dioxin." *Occup. Environ. Med.* 56:270–76.

Capen, C. C., and S. L. Martin. 1989. "The Effects of Xenobiotics on the Structure and Function of Thyroid Follicular and C-Cells." *Toxicol. Pathol.* 17:266–93.

Chauhan, K. R., P. R. S. Kodavanti, and J. D. McKinney. 2000. "Assessing the Role of Ortho-Substitution on Polychlorinated Biphenyl Binding to Transthyretin, a Thyroxine Transport Protein." *Toxicol. Appl. Pharmacol.* 162:10–21.

Delange, F., G. Benker, Ph. Caron, O. Eber, W. Ott, F. Peter, J. Podoba, M. Simescu, Z. Szybinski, F. Vertongen, P. Vitti, W. Wiersinga, and V. Zamrazil. 1997. "Thyroid Volume and Urinary Iodine in European Schoolchildren Standardization of Values for Assessment of Iodine Deficiency." *Eur. J. Endocrinol.* 136:180–87.

Dewailly, E., P. Ayotte, S. Bruneau, S. Gingras, M. Belles-Isles, R. Roy. 2000. "Susceptibility to Infections and Immune Status in Inuit Infants Exposed to Organochlorines." *Environ. Health Perspect.* 108:205–11.

Feng, X., Y. Jiang, P. Maltzer, and M. P. Yen. 2000. "Thyroid Hormone Regulation of Hepatic Genes in Vivo Detected by Complementary DNA Microarray." *Mol. Endocrinol.* 14: 947–55.

Flesh-Janys, D., J. Berger, P. Gurn, A. Manz, S. Nagel, H. Waltsgott, and J. H. Dwyer. 1995. "Exposure to Polychlorinated Dioxins and Furans (PCDD/F) and Mortality in a Cohort of

Workers from Herbicide-Producing Plant in Hamburg, Federal Republic of Germany." *Am. J. Epidemiol.* 142:356–65.

Frich, L., L. A. Akslen, and E. Glattre. 1997. "Increased Risk of Thyroid Cancer among Norwegian Women Married to Fishery Workers—A Retrospective Cohort Study." *Brit. J. Cancer* 76:385–89.

Grimalt, J. O., J. Sunyer, V. Moreno, O. C. Amaral, M. Sala, A. Rosell, J. M. Anto, and J. Albaiges. 1994. "Risk Excess of Soft-Tissue Sarcoma and Thyroid Cancer in a Community Exposed to Airborne Organochlorinated Compound Mixtures with a High Hexachlorobenzene Content." *Int. J. Cancer* 56:200–203.

Guo, Y. L., M-L. Yu, L-Y. Lin, C. C. Hsu, and W. J. Rogan. 1999. "Chloracne, Goiter, Arthritis and Anemia after Polychlorinated Biphenyl Poisoning: 14-Year Follow-up of the Taiwan Yucheng Cohort." *Environ. Health Perspect.* 107:715–19.

Hagmar, L. 2003. "Polychlorinated Biophenyls and Thyroid Status in Humans." *Thyroid* 13:1021–28.

Hagmar, L., L. Rylander, E. Dyremark, E. Klaasson-Wehler, A. Bergman, and E. M. Erfurth. 2001a. "Plasma Concentrations of Persistent Organochlorines in Relation to Thyrotropin and Thyroid Hormone Levels in Women." *Int. Arch. Occup. Environ. Health* 74:184–88.

Hagmar, L., J. Björk, A. Sjödin, A. Bergman, and E. M. Erfurth. 2001b. "Plasma Levels of Persistent Organohalogens and Hormone Levels in Adult Male Humans." *Arch. Environ. Health* 56:138–43.

Hansen, L. G. 1998. "Stepping Backward to Improve Assessment of PCB Congener Toxicities." *Environ. Health Perspect.* 106 (Suppl. 1): 171–79.

Heilmann, C., P. Grandjean, P. Weihe. 2003. "Decreased Childhood Vaccine Response in Children Exposed to PCBs from Maternal Seafood Diet." *Organohalogen Compounds* 60–65: 1–4.

Hollowell, J. G., N. W. Staehling, W. D. Flanders, W. H. Hannon, E. W. Gunter, C. A. Spencer, and L. E. Braverman. 2002. "Serum TSH, T(4), and Thyroid Antibodies in the United States Population (1988 to 1994): National Health and Nutrition Examination Survey (NHANES III)." *J. Clin. Endocrinol. Metab.* 87:489–99.

Hood, A., and C. D. Klaasen. 2000. "Differential Effects of Microsomal Enzyme Inducers on in Vitro Thyroxine (T4) and Triiodothyronine (T3) Glucuronidation." *Toxicol. Sci.* 55:78–84.

Hovander, L., L. Linderholm, M. Athanasiadou, L. Athanassiadis, A. Bignert, B. Fangström, A. Kočan, J. Petrík, T. Trnovec, and Å. Bergman. 2006. "Levels of PCB and Their Metabolites in the Serum of Residents of a Highly Contaminated Area in Eastern Slovakia." *Environ. Sci. Technol.* 40:3696–3703.

Jung, D., P. A. Berg, L. Edler, W. Ehrenthal, D. Fenner, D. Flesh-Janys, C. Huber, R. Klein, C. Koitka, G. Lucier, A. Manz, A. Muttray, L. Needham, O. Päpke, M. Pietsch, C. Portier, D. Patterson, W. Prellwitz, D. M. Rose, A. Thews, and J. Konietzko. 1998. "Immunologic Findings in Workers Formerly Exposed to 2,3,7,8,-Tetrachlorodibenzo-p-dioxin and Its Congeners." *Environ. Health Perspect.* 106:689–94.

Jursa, S., J. Chovancová, J. Petrík, and J. Lokša. 2006. "Dioxin-Like and Non-Dioxin-Like PCBs in Human Serum of Slovak Population." *Chemosphere* 64:686–91.

Karmaus, W. 2001. "Of Jugglers, Mechanics, Communities, and the Thyroid Gland: How Do We Achieve Good Quality Data to Improve Public Health?" *Environ. Health Perspect.* 109 (Suppl. 6): 863–69.

Kawabata, W., T. Suzuki, T. Moriya, K. Fujimori, H. Naganuma, S. Inoue, Y. Kinouchi, K. Kameyama, H. Takami, T. Shimosegawa, and H. Sasano. 2003. "Estrogen Receptors (Alpha and Beta) and 17beta-Hydroxysteroid Dehydrogenase Type 1 and 2 in Thyroid Disorders: Possible in Situ Estrogen Synthesis and Actions." *Mod. Pathol.* 16:437–44.

Khan, M. A., and L. G. Hansen. 2003. "Ortho-Substituted Polychlorinated Biphenyl (PCB) Congeners (95 or 101) Decrease Pituitary Response to Thyrotropin Releasing Hormone." *Toxicol. Lett.* 144:173–82.

Kimbrough, R. D., and C. A. Krouskas. 2001. "Polychlorinated Biphenyls, Dibenzo-p-Dioxins, and Dibenzofurans and Birth Weight and Immune and Thyroid Function in Children." *Regul. Toxicol. Pharmacol.* 34:42–52.

Kočan, A., J. Petrík, B. Drobná, and J. Chovancoá. 1994. "Levels of PCBs and Some Organochlorine Pesticides in the Human Population of Selected Areas of the Slovak Republic." Pt. 1, "Blood." *Chemosphere* 29:2315–25.

Kočan, A., J. Petrík, S. Jursa, J. Chovancová, and B. Drobná. 2001. "Environmental Contamination with Polychlorinated Biphenyls in the Area of Their Former Manufacture in Slovakia." *Chemosphere* 43:595–600.

Kogevinas, M., T. Kauppinen, R. Winkelmann, H. Becher, P. A. Bertazzi, H. B. Bueno-de-Mesquita, D. Coggon, L. Green, E. Johnson, and M. Litton. 1995. "Soft Tissue Sarcoma and Non-Hodgkin's Lymphoma in Workers Exposed to Phenoxy Herbicides, Chlorophenols, and Dioxins: Two Nested Case-Control Studies." *Epidemiology* 6:386–402.

Koopman-Esseboom, C., D. C. Morse, N. Weisglas-Kuperus, I. J. Lutkechipholt, C. G. van der Pauw, L. G. M. Tuinstra, A. Brouwer, and P. J. J. Sauer. 1994. "Effects of Dioxins and Polychlorinated Biphenyls on Thyroid Hormone Status of Pregnant Women and Their Infants." *Pediatric Research* 36: 468–76.

Kung, A. W., and B. M. Jones. 1998. "A Change from Stimulatory to Blocking Antibody Activity in Graves' Disease During Pregnancy." *J. Clin. Endocrinol. Metab.* 83:514–18.

Langer, P., M. Tajtáková, G. Fodor, A. Kočan, P. Bohov, J. Michálek, and A. Kreze. 1998. "Increased Thyroid Volume and Prevalence of Thyroid Disorders in an Area Heavily Polluted by Polychlorinated Biphenyls." *Eur. J. Endocrinol.* 139:402–9.

Langer, P., M. Tajtáková, P. Bohov, and I. Klimeš. 1999. "Possible Role of Genetic Factors in Thyroid Growth Rate and in the Assessment of Upper Limit of Normal Thyroid Volume in Iodine-Replete Adolescents." *Thyroid* 9:557–62.

Langer, P., M. Tajtáková, J. Koška, P. Bohov, E. Šeböková, and I. Klimeš. 2003a. "Multimodal Distribution versus Logarithmic Transformation of Thyroid Volumes in Adolescents:

Detection of Subgroup with Subclinical Thyroid Disorders and Its Impact on the Assessment of the Upper Limit of Normal Thyroid Volumes." *Endocrine Journal* 50:117–25.

Langer, P., A. Kočan, M. Tajtáková, J. Petrík, J. Chovancová, B. Drobná, S. Jursa, M. Pavúk, J. Koška, T. Trnovec, E. Šeböková, and I. Klimeš. 2003b. "Possible Effects of Polychlorinated Piphenyls and Organochlorinated Pesticides on the Thyroid after Long-Term Exposure to Heavy Environmental Pollution." *J. Occup. Environ. Med.* 45:526–32.

Langer, P., A. Kočan, M. Tajtáková, J. Petrík, J. Chovancová, B. Drobná, S. Jursa, Ž. Rádiková, J. Koška, L. Kšinantová, M. Hučková, R. Imrich, S. Wimmerová, D. Gašperíková, Y. Shishiba, T. Trnovec, E. Šeböková, and I. Klimeš. 2007a. "Fish from Industrially Polluted Freshwater as the Main Source of Organochlorinated Pollutants and Increased Frequency of Thyroid Disorders and Dysglycemia." *Chemosphere* 67:S379–85.

Langer, P., M. Tajtáková, A. Kočan, J. Petrík, J. Koška, L. Kšinantová, Ž. Rádiková, J. Ukropec, R. Imrich, M. Hučková, J. Chovancová, B. Drobná, S. Jursa, M. Vlček, A. Bergman, M. Athanasiadou, L. Hovander, Y. Shishiba, T. Trnovec, E. Šeböková, and I. Klimeš. 2007b. "Thyroid Ultrasound Volume, Structure and Function after Long-term High Exposure of Large Population to Polychlorinated Biphenyls, Pesticides and Dioxin." *Chemosphere* 69:118–27 doi.10.1016/j.chemosphere.2007.04.039.

La Vecchia, C., E. Ron, S. Franceschi, L. Dal Maso, S. D. Mark, L. Chatenoud, C. Braga, S. Preston-Martin, A. McTiernan, L. Kolonel, K. Mabuchi, F. Jin, G. Wingren, M. R. Galanti, A. Hallquist, E. Lund, F. Levi, D. Linos, and E. Negri. 1999. "A Pooled Analysis of Case-Control Studies on Thyroid Cancer." Part 3, "Oral Contraceptives, Menopausal Replacement Therapy and Other Female Hormones." *Cancer Causes Control* 10:157–66.

Ma, R. and D. A. Sasoon. 2005. "PCBs Exert an Estrogenic Effect through Repression of the *Wnt7a* Signaling Pathway in the Female Reproductive Tract." *Environ. Health Perspect.* 114:898–904.

Madej, D., E. Persson, T. Lundh, and Y. Ridderstrale. 2002. "Thyroid Gland Function in Ovariectomized Ewes Exposed to Phytoestrogens." *J. Chromatogr. B.* 777:281–87.

Manole, D., B. Schildknecht, B. Gosnell, E. Adams, and M. Derwahl. 2001. "Estrogen Promotes Growth of Human Thyroid Tumor Cells by Different Molecular Mechanisms." *J. Clin. Endocrinol. Metab.* 86:1072–77.

Meeker, J. D., L. Altshul, and R. Hauser. 2007. "Serum PCBs, *p,p'*-DDE and HCB Predict Thyroid Hormone Levels in Men." *Environ. Res.* 104:296–304.

Miyazaki, W., T. Iwasaki, A. Takeshita, Y. Kuroda, and N. Koibuchi. 2004. "Polychlorinated Biophenyls Suppress Thyroid Hormone Receptor-Mediated Transcription through a Novel Mechanism." *J. Biol. Chem.* 279:18195–18202.

Nagayama, J., H. Tsuji, T. Iida, H. Hirakawa, T. Matsueda, and M. Ohki. 2001. "Effects of Contamination Level of Dioxins and Related Chemicals on Thyroid Hormone and Immune Response System in Patients with 'Yusho.'" *Chemosphere* 43:1005–10.

Nicholson, W. J., and P. J. Landrigan. 1994. "Human Health Effects of Polychlorinated Biphenyls." In *Dioxines and Health,* edited by A. Schecter, 487–523. New York: Plenum Press.

Osius, N., W. Karmaus, H. Kruse, and J. Witten. 1999. "Exposure to Polychlorinated Biphenyls and Levels of Thyroid Hormones in Children." *Environ. Health Perspect.* 107:843–49.

Persky, V., M. Turyk, A. Anderson, L. P. Hanrahan, C. Falk, D. N. Steenport, J. R. Chatterton, S. Freels, and the Great Lakes Consortium. 2001. "The Effects of PCB Exposure and Fish Consumption on Endogenous Hormones." *Environ. Health Perspect.* 109:1275–83.

Pluim, H. J., J. G. Koppe, and K. Olie. 1993. "Effects of Dioxins and Furans on Thyroid Hormone Regulation in the Human Newborn." *Chemosphere* 27:391–94.

Rádiková, Ž, M. Tajtáková, A. Kočan, T. Trnovec, E. Šeböková, I. Klimeš, and P. Langer. In press. "Possible Effects of Environmental Nitrates and Toxic Organochlorines on Human Thyroid in Highly Polluted Areas of Slovakia." *Thyroid.*

Roger, P. P., D. Christophe, J. E. Dumont, and I. Pirson. 1997. "The Dog Thyroid Primary Culture System: A Model of the Regulation of Function, Growth and Differentiation Expression by cAMP and other Well-Defined Signalling Cascades." Review. *Eur. J. Endocrinol.* 137:579–98.

Rylander, L., E. Wallin, B. A. G. Jönssson, M. Stridberg, E. M. Erfurth, and L. Hagmar. 2006. "Associations between CB-153 and *p,p'*-DDE and Hormone Levels in Serum in Middle-aged and Elderly Men." *Chemosphere* 65:275–81.

Sala, M., J. Sunyer, C. Herrero, J. To-Figueras, and J. Grimalt. 2001. "Association between Serum Concentrations of Hexachlorobenzene and Polychlorobiphenyls with Thyroid Hormone and Liver Enzyme in a Sample of the General Population." *Occup. Environ. Med.* 58:172–77.

Saracci, R., M. Kogevinas, P-A. Bertazzi, B. H. B. Mesquita, D. Coggon, L. M. Green, T. Kauppinen, K. A. L'Abbée, M. Littorin, and E. Lynge. 1991. "Cancer Mortality in Workers Exposed to Cholorphenoxy Herbicides and Chlorophenols." *Lancet* 338:1027–31.

Schell, L. M., M. Gallo, A. P. DeCaprio, L. Hubicki, M. Denham, and L. Ravenscroft. 2004. "Thyroid Function in Relation to Burden of PCB, *p,p'*-DDE, HCB, Mirex and Lead Among Akwesasne Mohawk Youth: A Preliminary Study." *Environ. Toxicol. Pharmacol.* 18:91–99.

Shiraishi, F., T. Okumura, M. Nomachi, S. Serizawa, J. Nishikawa, J. S. Edmonds, H. Shiraishi, and M. Morita. 2003. "Estrogenic and Thyroid Hormone Activity of a Series of Hydroxy-Polychlorinated Biphenyls." *Chemosphere* 52:33–42.

Studer, H., and M. Derwahl. 1996. "Mechanisms of Nonneoplastic Endocrine Hyperplasia—A Changing Concept: A Review Focused on the Thyroid Gland." *Endocrine Rev.* 16:411–26.

Swanson, G. M., H. E. Ratcliffe, and L. J. Fischer. 1995. "Human Exposure to Polychlorinated Biphenyls(PCBs): A Critical Assessment of the Evidence for Adverse Health Effects." *Regul. Toxicol. Pharmacol.* 21:136–50.

Tajtáková, M., D. Hančinová, P. Langer, J. Tajták, E. Malinovský, and J. Varga. 1988. "Thyroid Volume in East Slovakian Adolescents Determined by Ultrasound 40 Years after the Introduction of Iodized Salt." *Klin Wochenschr* 66:749–51.

Tajtáková, M., P. Langer, V. Gonsorčíková, P. Bohov, and D. Hančinová. 1998. "Recognizing a Subgroup with Rapidly Growing Thyroids among Adolescents under Iodine Replete Conditions: Seven Years Follow-up." *Eur. J. Endocrinol.* 138: 674–80.

Takser, L., D. Mergler, M. Baldwin, S. Grosbols de, A. Smarglassi, and J. Lafond. 2005. "Thyroid Hormone in Pregnancy in Relation to Environmental Exposure to Organochlorine Compounds and Mercury." *Environ. Health Perspect.* 113: 1039–45.

Tan, Y., Ch-H. Chen, D. Lawrence, and D. O. Carpenter. 2004. "Ortho-Substituted PCBs Kill Cells by Altering Membrane Structure." *Toxicol. Sci.* 80:54–59.

Triebig, G., E. Werle, O. Päpke, G. Heim, C. Broding, and H. Ludwig. 1998. "Effects of Dioxins and Furans on Liver Enzymes, Lipid Parameters, and Thyroid Hormones in Former Thermal Metal Recycling Workers." *Environ. Health Perspect.* 106 (Suppl. 2): 313–16.

Tsuji, H., K. Sato, J. Shimono, K. Azuma, M. Hashiguchi, and M. Fujishima. 1997. "Thyroid Function in Patients with Yusho: 28 Year Follow-up Study." *Fukuoka Yagukaku Zashi* 88:231–35.

Turyk, M. E., H. A. Anderson, S. Freels, R. Chatterton Jr., L. L. Needham, D. G. Patterson, D. N. Steenport, L. Knobeloch, P. Imm, V. W. Persky, and the Great Lakes Consortium. 2006. "Associations of Organochlorines with Androgenous Hormones in Male Great Lakes Fish Consumers and Nonconsumers." *Environ. Res.* 102:299–307.

Vanderpump, M. P. J., W. M. G. Tunbridge, D. M. French, D. Appleton, D. Bates, F. Clark, J. Grimley Evans, D. M. Hasan, H. Rodgers, F. Tinbridge, and E. T. Young. 1995. "The Incidence of Thyroid Disorders in the Community: A Twenty-Year Follow-up of the Whickham Survey." *Clin. Endocrinol.* 43:655–89.

WHO/UNICEF/ICCIDD. 1993. "Indicators for Assessing Iodine Deficiency Disorders and Their Control Programmes." Report of a joint WHO/UNICEF/ICCIDD consultation. Review version. September.

Yamada-Okabe, T., T. Aono, H. Sakai, Y. Kashima, H. Yamada. 2004. "2,3,7,8-Tetrachlorodibenzo-*p*-Dioxin Augments the Modulation of Gene Expression Mediated by the Thyroid Hormone Receptor." *Toxicol. Appl. Pharmacol.* 194:201–10.

Yamada-Okabe, T., H. Sakai, Y. Kashima, and H. Yamada-Okabe. 2005. "Modulation at a Cellular Level of the Thyroid Hormone Receptor-mediated Gene Expression by 1,2,5,6,9,10-Hexabromocyclododecane (HBCD), 4,4′-Diiodobiphenyl (DIB), and Nitrofen (NIP)." *Toxicol. Lett.* 155:127–33.

Zober, A., M. G. Ott, and P. Messerer. 1994. "Morbidity Follow Up Study of BASF Employees Exposed to 2,3,6,8-Tetrachlorodiabenzo-p-Dioxin (TCDD) after a 1953 Chemical Reactor Incident." *Occup. Environ. Med.* 51:479–86.

13

Estrogenic, Antiestrogenic, and Dioxin-Like Activities in Human Male Serum Samples from Eastern Slovakia

Miroslav Machala, *Veterinary Research Institute, Brno, Czech Republic*

Martina Plisková, *Veterinary Research Institute, Brno, Czech Republic*

Jan Vondrácek, *CAS, Brno, Czech Republic*

Rocio Fernandez Canton, *University of Utrecht*

Jiri Neca, *Veterinary Research Institute, Brno, Czech Republic*

Anton Kočan, *Slovak Medical University*

Ján Petrík, *Slovak Medical University*

Tomáš Trnovec, *Slovak Medical University*

Thomas Sanderson, *University of Utrecht*

Martin van den Berg, *University of Utrecht*

ABBREVIATIONS

AhR	Aryl hydrocarbon receptor
CYP1A1	Cytochrome P4501A1
CYP1B1	Cytochrome P4501B1
E2	17beta-estradiol
EEQ	17beta-estradiol equivalents
ER	Human estrogen receptor
PCA	Principal component analysis
PCBs	Polychlorinated biphenyls
PCDD/Fs	Polychlorinated dibenzo-*p*-dioxins and dibenzo-furans
POPs	Persistent organic pollutants
TCDD	2,3,7,8-tetrachlorodibenzo-*p*-dioxin
TEQs	2,3,7,8-tetrachlorodibenzo-*p*-dioxin (TCDD) equivalents

The aim of this chapter is to review the current understanding of (anti)estrogenic effects of polychlorinated biphenyls (PCBs) and to describe modulations of estrogenic activity by PCBs present in human male blood. In this study, a decreased total estrogen receptor (ER)-mediated activity (the ER-CALUX assay) and increased aryl hydrocarbon receptor (AhR)-mediated or dioxin-like activity (the DR-CALUX assay) were induced by high levels of PCBs present in male serum samples. The concentration of 17b-estradiol (E2) was lower in serum samples of subjects highly exposed to PCBs. Principal component analysis revealed a significant negative correlation between expression of CYP1A1 and CYP1B1 mRNAs in lymphocytes, as well as dioxin-like activity, and estrogenic activity. A small part of the sulfuric acid/silica–fractionated samples, containing only persistent organic pollutants, elicited either weak estrogenic or anti-estrogenic potencies. The increased frequency of anti-estrogenic samples was associated with a higher sum of PCBs and the presence of dioxin-like compounds. While the lower-chlorinated PCBs were estrogenic, the prevalent higher-chlorinated PCBs 138, 153, 170, 180, 187, 194, 199, and 203, as well as major PCB metabolites, behaved as anti-estrogens in the ER-CALUX assay. Our data suggested that the effect of prevalent higher-chlorinated non-dioxin-like PCBs is responsible for the weak anti-estrogenicity of serum samples. However, the reduction of E2 levels appeared to be a more important effect of high PCB exposure. Suppression of E2 concentrations by PCB-induced metabolism or modulation of estrogen synthesis might play a more significant role in endocrine disruption than the direct effects of PCBs on ER-mediated activity.

INTRODUCTION

Induction of aryl hydrocarbon receptor (AhR)-mediated cellular events and modulation of estrogen signaling are recognized as two important modes of action of xenobiotics that may lead to both developmental and reproductive toxicity or to carcinogenesis (van den Berg et al. 1998; Kavlock et al. 1996; Ankley et al. 1998). A number of polychlorinated biphenyls have been reported to affect activation of AhR, ERs, or both. However, the effects of "real-life" mixtures of PCBs at levels found, for example, in human blood, are only rarely studied. The purpose of this work is to give a brief overview of current methodology available for such studies and to present an example of data obtained when investigating effects of serum samples of human males highly contaminated with persistent organic pollutants (POPs), especially with PCBs. Such studies might provide a valuable insight into effects of complex mixtures of pollutants on estrogenic activity in human blood.

In general, activation of the AhR and induction of AhR-dependent gene expression correspond well with the dioxin-like toxicity of xenobiotics, and the methodology of determination of AhR-mediated activity is well established (Safe 1994; Behnisch, Hosoe, and Sakai 2001). A majority of *in vitro* bioassays for this type of toxic activity use rodent or human hepatoma cell lines transfected with a reporter gene under control of activated AhR (Zacharewski, Berhane, and Gillesby 1995; Garrison et al. 1996; Murk et al. 1996; Anderson et al. 1999; Seidel et al. 2000) or a detection of AhR-dependent cytochrome P4501A (CYP1A) expression/ activity in the same cellular model as an endpoint (Gale et al. 2000; Brack et al. 2000; Nilsen, Berg, and Goksoyr 1998). In the past, a number of *in vitro* and *in vivo* assays for estrogenicity/anti-estrogenicity have also been successfully introduced, which use varying experimental design or endpoints. Unlike the AhR activity bioassays, they may sometimes give contradictory results for individual compounds, making their interpretation challenging. (Anti)estrogenicity assays usually detect the following processes associated with modulation of estrogen signaling by xenobiotics:

1. *In vitro* cell models for the determination of ER activation include breast cancer epithelial T47D or MCF-7 cell line, transfected with estrogen-response elements (ERE)-regulated reporter constructs, such as pERE-tk-CAT (Nesaretnam and Darbre 1997; Bonenfeld-Jorgensen et al. 2001) or pERE-tata-Luc gene (Jobling et al. 1995; Legler et al. 1999). Alternatively, a chimeric receptor-based gene expression is used as the endpoint (Zacharewski 1997). The yeast-based estrogenic assays use *Saccharomyces cerevisiae* strains transformed with the human ER cDNA and an ERE-regulated *LacZ* reporter gene that encodes for the b-galactosidase enzyme (ref. in Zacharewski 1997).

2. Alternative methods use reporter genes controlled by promoters derived from endogenous ER-inducible genes, such as luciferase gene driven by ERE in front of the vitellogenin-tyrosine kinase as a promoter (Pons et al. 1990; Moore et al. 1997) and 17b-estradiol (E2)-responsive DNA constructs derived from the *creatine kinase B, cathepsin D* (Ramamoorthy et al. 1999) or *pS2* gene promoters (Daly and Darbre 1990; Gillesby and Zacharewski 1998). Endogenous gene products known to be produced in response to E2 treatment, such as vitellogenin and pS2, can also be determined as suitable endpoints (Ankley et al. 1998).

3. Estrogen-dependent growth responses can be detected using either the E-SCREEN cell proliferation assay with termination on day six in the late exponential phase (Soto et al. 1997; Gutendorf and Westendorf 2001) or the postconfluent cell proliferation and the focus MCF-7 assays (Gierthy, Arcaro, and Floyd 1997; Arcaro et al. 1999).

4. In immature mouse or rat uterotropic bioassays, the wet uterine weight or uterine epithelial mitogenic and secretory activity are determined after three-day exposure to a tested compound (Kanno et al. 2001; Markey et al. 2001; Yamasaki et al. 2002). Alternatively, acute 24- or 48-hour effects of less persistent xenobiotics such as lower-chlorinated PCBs can be detected in the female rat integrated endocrine disruption assay introduced by Hansen and coworkers (Jansen et al. 1993; Li and Hansen 1996; Rose et al. 2002).

5. A number of assays have been reported that detect additional possible mechanisms of estrogenic/anti-estrogenic effects of xenobiotics, such as modulation of steroidogenesis (Letcher et al. 1999) or decreased level of ER in a target tissue (ref. in Safe and Wormke 2003).

Polychlorinated biphenyls have been reported to elicit both dioxin-like and estrogenic or anti-estrogenic effects in various *in vitro* and *in vivo* models (van den Berg et al. 1998; Hansen 1998; Cooke, Sato, and Buchanan 2001). In general, the biological activities of individual PCB congeners depend on planarity of a molecule (Safe 1994), but also on its molecular weight and biotransformation rate (Rose et al. 2002). Coplanar non-ortho-substituted PCBs activate AhR and the AhR-dependent signaling pathways in a way similar to 2,3,7,8-tetrachlorodibenzo-*p*-dioxin (TCDD) (Safe 1994), and a majority of their adverse effects is thought to be mediated through activation of AhR. TCDD and other AhR agonists, including coplanar PCBs, have been frequently reported to elicit anti-estrogenic activity (Safe and Wormke

2003; Buchanan et al. 2000; Buchanan et al. 2002). Several modes of anti-estrogenic action of AhR agonists have been suggested, including direct repression of E2-dependent gene expression by interactions of activated AhR with specific promoter regions of E2 responsive genes (inhibitory dioxin response elements, iDRE) (ref. Safe and Wormke 2003), inhibition of E2-induced cell cycle proteins and uterine epithelial mitogenesis (Wang, Smith, and Safe 1998; Buchanan et al. 2002) and effects of PCBs on E2 metabolism (Pang et al. 1999; van Duursen et al. 2003).

Distinct AhR-independent neurotoxic, (anti)estrogenic, tumor-promoting and other adverse effects have been reported after exposure to noncoplanar ortho-substituted PCBs (Hansen 1998; Robertson and Hansen 2001; Machala et al. 2003). Nevertheless, the exact mechanisms of estrogenic or anti-estrogenic activities of noncoplanar PCBs remain to be established. The reported results are often contradictory, derived from data obtained in different *in vitro* or *in vivo* models (Hansen 1998; Brouwer et al. 1999).

One of the major goals of research on environmental endocrine disrupters, such as PCBs and other POPs, is to answer the question, whether these compounds have effects on chronically exposed human populations (Daston, Cook, and Kavlock 2003). Determination of *in vitro* estrogenic/anti-estrogenic activities of extracts of human male blood samples collected from a PCB-contaminated area is one possible approach for assessment of impact of PCBs and/or other POPs on estrogen-dependent signaling. Human male blood contains significant levels of both nonpersistent endogenous estrogens and POPs with potentially estrogenic or anti-estrogenic effects, including PCBs and their metabolites, DDT residues and polychlorinated dibenzo-*p*-dioxins and dibenzofurans (PCDD/Fs). In this study, the estrogen receptor (ER)-mediated activities of 150 human male samples were investigated by the ER-CALUX assay using the T47D.Luc breast cancer cell line stably transfected with pERE-tata-Luc detecting the direct activation of ER. The samples were collected from eastern Slovakia, where high levels of PCB contamination have been reported previously (Kočan et al. 2001), and from the background region in the same geographic area. The aim of this study was to determine effects of chronic PCB exposure on anti/estrogenic and dioxin-like activities exerted by extracts of human male sera, and to compare concentrations of major POPs and levels of E2 in serum with *in vitro* bioassay data.

MATERIALS AND METHODS

Blood Sampling, Extraction, and Cleanup

One hundred fifty individual male and female blood samples were collected in two areas of eastern Slovakia, which are differently contaminated with PCBs, namely in the Micha-lovce district, where the commercial PCB mixtures were produced for a number of years, and in the Stropkov District, which was selected as a background area with low PCB contamination. The whole blood was collected from fasting subjects in S-Monovette Vacutainer tubes without anticoagulants, immediately centrifuged at 3,000 rpm for 15 minutes and stored in glass tubes at -70 degrees C. The samples of human male serum were extracted in the following manner: 5 ml serum was extracted with methanol and hexane/diethyl ether three times, and extracts were evaporated and dissolved in 1 ml dichloromethane. 0.5 ml extract was transferred to the vial, then evaporated and consequently dissolved in dimethylsulfoxide (DMSO). The second half of crude extract was placed on a sulfuric acid–activated silica column and eluted with 20 ml n-hexane:diethyl ether mixture, evaporated and redissolved in DMSO.

Chemical Analysis of POPs

Chemically derived TCDD equivalency factors (TEQs) were calculated from HPGC/MS data on blood concentrations of PCDD/Fs and non-ortho- and mono-ortho-chlorinated PCB. Concentrations of other prevalent (noncoplanar) PCB congeners and p,p'-DDE were determined by GC/MS method (Kočan et al. 2001; Kočan et al. 2004). The sum of PCBs used in the correlation and multivariate statistical analysis was based on the data on concentrations of both indicator PCBs and mono-ortho-chlorinated PCB congeners.

Determination of Effects Associated with AhR Activation

The CYP1A1 and CYP1B1 mRNA levels in human peripheral lymphocytes were determined by RNA extraction and a quantitative RT-PCR method using TaqMan technology (Canton et al. 2003). The *in vitro* potencies of POPs present in blood serum to activate the AhR were measured in sulfuric acid/silica treated extracts by a luciferase reporter gene assay (DR-CALUX, BioDetection Systems, Amsterdam, Netherlands) as described previously (Murk et al. 1996).

ER-mediated activity and determination of 17beta-estradiol (E2) in male blood samples

In 150 male serum samples, the ER-CALUX bioassay (BioDetection Systems, Amsterdam, Netherlands) was performed using human breast carcinoma T47D.Luc cell line (Legler et al. 1999; Machala et al. 2004). The ER-mediated activity was determined in the cells treated with either total serum hexane/diethyl ether extracts or with a fraction of POPs obtained by a consequent sulfuric acid/silica fractionation of total samples. Antiestrogenicity of POPs fractions was determined as a decrease in response to E2 in the cells cotreated with the POPs fraction. Concentrations of E2 were determined by a commercial ELISA kit (ADVIA Centaur

Estradiol-6 III assay, Bayer HealthCare, Tarrytown, New York) in 60 samples selected according to stratified PCB levels. Estrogenic and anti-estrogenic activities of 18 prevalent PCB congeners were determined by the ER-CALUX assay. Cytotoxicity of extracts or individual PCBs was determined by a neutral red uptake assay after 24 hours exposure.

Data Analyses

All calculations were performed with Microsoft Excel, SlideWrite 3.0 for Windows, or Statistica 6.1 for Windows. Nonparametric statistical analyses (Kruskal-Wallis analysis of variance and the Mann-Whitney U test) were used for data analysis. The relationships among biological and chemical data were determined by the correlation analysis and multivariate principal component analysis (PCA). The correlations among the compared parameters were assessed using nonparametric Spearman's rank coefficient (Rs). For the correlation and PCA analyses, all the data were normalized using the transformation log (X+1).

RESULTS

Estrogenic Activity in Male Blood and Its Association with Exposure to POPs

The estrogenic and anti-estrogenic activities *in vitro* were measured in both crude hexane/diethyl ether extracts of 150 human male serum samples and POP fractions. The samples were collected in two regions of eastern Slovakia, and they differed significantly in their overall PCB content. The total extracts of human male serum samples, containing both endogenous steroids and POPs, elicited significant estrogenic responses in the ER-CALUX assay, between 12.5 and 59.2 pg E2 equivalents (EEQs) per ml (table 13.1). A significant part of estrogenic activity might be attributed to the presence of E2, concentrations of which reached 43.5 pg/ml. Dioxin-like activities, as determined by the DR-CALUX bioassay, were in the range of 0.2–2.9 pg TCDD equivalents (TEQs) per ml (table 13.1).

This measurement was a part of a large study in which over 2000 human blood samples were evaluated for concentrations of PCBs, PCDD/Fs, DDT and metabolites of PCBs and DDT (Kočan et al. 2004; Hovander et al. 2004). The analytical data for the subset of 150 male samples, as summarized in table 13.1, were used for statistical evaluation of *in vitro* bioassay data. PCDDs did not contribute to higher levels of TEQs. High concentrations of p,p′-DDE were found in a majority of samples; however, only weak estrogenic and no dioxin-like activity of this compound was detected (see below). Therefore, biological effects of extracted organic contaminants could be attributed to PCBs. Additionally, data on induction of AhR-dependent expression of CYP1A1 and CYP1B1 mRNAs in blood lymphocytes, determined by real time-PCR (not presented here), were also included in the statistical analysis.

Table 13.1. Estrogenic and Dioxin-Like Activity (Expressed as EEQs and TEQs) and Concentrations of Estradiol and Major Persistent Organic Pollutants in Human Male Serum Samples.

	N	Concentration per ml Serum			Concentration per g Lipid		
		Range	Median	Mean	Range	Median	Mean
Estrogenic activity (pg EEQs)	150	12.5 59.2	28.2	29.2	1.3 11.6	4.1	4.3
E$_2$ (pg)	60	<1.0 43.5	15.5	15.8	0.1 5.4	2.0	2.1
Dioxin-like activity (pg TEQs)	144	0.2 2.9	0.6	0.7	11.9 434.0	83.6	92.0
Sum of PCBs/PCDD/Fs (pg TEQs)	100	0.05 0.5	0.1	0.2	7.5 57.9	18.2	20.8
Sum of 17 PCBs (ng)	150	2.0 175.5	7.8	14.7	345.8 32509.4	1123.6	2040.4
p,p′-DDE (ng)	150	1.7 116.5	11.9	17.1	268.9 11162.5	1800.0	2219.2

Note: Sum of PCDD/Fs and dioxin-like PCBs was calculated as TEQs according to WHO-TEF values (van den Berg et al. 1998). Data on total concentration of PCBs include sum of 17 congeners. N = number of samples.

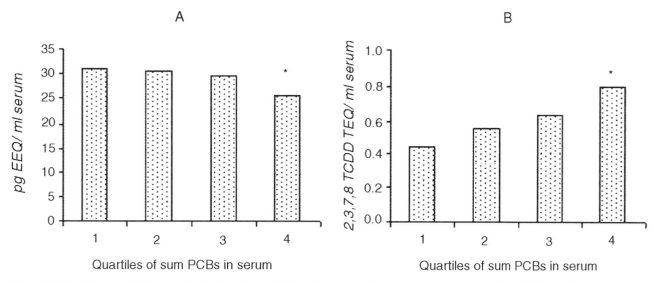

Figure 13.1.   Estrogenic (A) and dioxin-like (B) activities of extracts of human male serum samples. Median values of quartiles were stratified according to PCB concentrations. p = 0.02 (A) and p = 0.02 (B).

The estrogenic activity correlated with E2 concentrations (R = 0.510, p < 0.001), suggesting that E2 was responsible for a major part of the estrogenicity detected in male serum samples. The estrogenicity and concentrations of E2 were moderately decreased and AhR-mediated activity was increased in subjects with a high content of PCBs (fourth quartile). However, estrogenic and dioxin-like activities were modulated significantly only in samples with the highest PCB levels (fig. 13.1). Weak estrogenic and anti-estrogenic activities were also found in the fractions of POPs, but only in some

of the samples. Nevertheless, neither induction nor suppression of estrogenic activity reached more than 50 percent in these cases. The POP fractions from the less-polluted background area elicited ER-mediated activity with a higher incidence, while the anti-estrogenic activity was detected more frequently in the samples from the PCB-polluted region (data not shown).

Multivariate principal component analysis confirmed a close association between estrogenic activity of total extracts and concentrations of E2; PCA also revealed negative

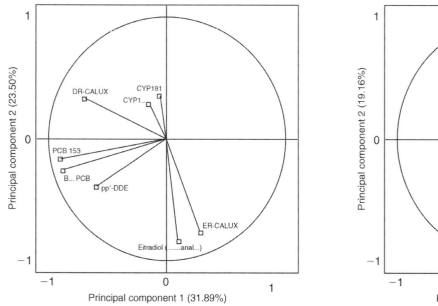

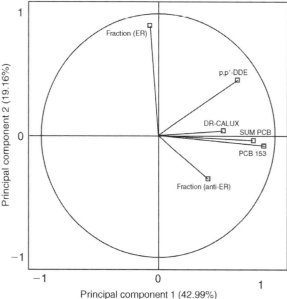

Figure 13.2.   Principal component analysis of the measured parameters of the serum samples: (A) ER-CALUX, overall estrogenic activity, DR-CALUX, dioxin-like activity; (B) fraction (ER), number of estrogenic samples of the fraction of POPs; fraction (anti-ER), number of anti-estrogenic samples of the fraction of POPs.

associations between estrogenicity and expression of CYP1A1 and CYP1B1 mRNAs, and between estrogenic and dioxin-like activities (fig. 13.2A). The total sum of PCBs, when samples were not discriminated into quartiles according to the PCB levels, did not correlate with a decrease in the ER-mediated activity; this statistical analysis confirmed that modulations of E2 levels and decreased estrogenic activity in overall extracts of male serum were significant only at the highest PCB exposure levels. On the other hand, anti-estrogenic activity of the POPs fractions was associated with concentration of PCBs, TEQ values obtained in the DR-CALUX assay and increased expression of CYP1A1 and CYP1B1 mRNA (fig. 13.2B).

Activities of PCBs, p,p′-DDE and Metabolites Found in Male Serum

In order to interpret the data on (anti)estrogenic activities of serum extracts, we determined (anti)estrogenic potencies of a large set of individual PCB congeners found to be prevalent in the human male blood samples, as well as additional "indicator" PCBs 28 and 52 that are routinely used in monitoring programs. Estrogenic and anti-estrogenic potencies of p,p′-DDE and major PCB and DDE metabolites were determined as well. Lower-molecular-weight PCBs No. 28, 52, 66, 74, 99, and 105 elicited ER-mediated activity at micromolar concentrations (table 13.2). The most prevalent PCB congeners No. 138, 153, 170, 180, and 187, as well as octachlorobiphenyls (PCBs 194, 199, and 203, found at high concentrations in human blood), did not induce ER-dependent luciferase activity in the ER-CALUX assay. Contrary to that, they all significantly decreased E2 response in this assay (table 13.2). Moreover, when a reconstituted mixture of the most prevalent PCBs was used, it had a higher anti-estrogenic activity, as compared to the potency of PCB 153 alone (fig. 13.3). Potent AhR agonists (PCBs 126, 118, 105, 156) did not significantly affect ER activation. The major hydroxy- and methylsulfonyl-PCB metabolites were found to be anti-estrogenic in the ER-CALUX assay. p,p′-DDE, which was present at high levels in the samples, as well as its major metabolite 3-methylsulfonyl-DDE, elicited only a weak estrogenic activity. The effects of PCB congeners, p,p′-DDE and their major metabolites are summarized in table 13.2, including the data on their EC50 values.

DISCUSSION

In vitro bioassays are a suitable tool for exposure assessment of dioxin-like and (anti)estrogenic compounds (van den Berg et al. 1998; Zacharewski 1997). Nevertheless, there are only limited data available on estrogenic and dioxin-like activities of complex samples of organic compounds collected from human blood. Sonnenschein et al. (1995) and Soto et al.

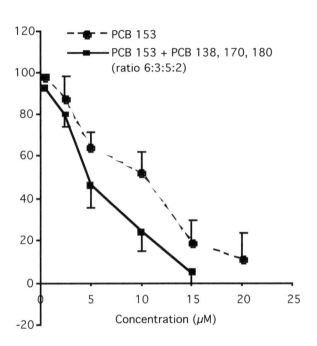

Figure 13.3.   Anti-estrogenic potencies of PCB 153 and artificial mixture of the most prevalent PCBs.

(1997) reported development of a serum extraction method for separation of POPs and endogenous steroids. Recently, an extraction and fractionation technique has been developed for combined chemical and *in vitro* assay analysis in human blood allowing for discrimination of effects of endogenous hormones and xenoestrogens (Fernandez et al. 2004). However, results of direct measurements of ER-mediated activity of serum extracts or total POPs fractions in a comprehensive set of human subjects have not been published yet. More information is available concerning *in vitro* bioassays of dioxin-like activity in human blood. The total TEQ values determined in human female serum and follicular fluid by the DR-CALUX assay have been reported to correlate well with the sum of four major PCB congeners 153, 138, 180, and 118 (Pauwels et al. 2000), in spite of the fact that only PCB 118 has a significant TEF. In the present study, suppression of E2 levels, a decrease of total estrogenic activity and increased dioxin-like activity were found in serum samples from human males chronically exposed to PCBs. However, as shown in figure 13.1, correlations with PCB concentrations were significant only in subjects with high exposure levels.

Both estrogenic and anti-estrogenic activities of PCBs have been reported, using various *in vitro* and *in vivo* models. Lower-molecular-weight PCBs have been reported to be estrogenic both *in vitro* and *in vivo*, with the exception of dioxin-like 3,3′,4,4′-tetrachlorobiphenyl (PCB 77), which elicited anti-estrogenicity *in vivo* and in some *in vitro* models (Ramamoorthy et al. 1999). 2,2′,6,6′-tetrachlorobiphenyl

(PCB 54), a fully ortho-substituted compound not occurring in the environment at significant levels (Hansen 1998), was estrogenic both in the MCF-7 cell focus assay and in the rat uterotropic assay (Arcaro et al. 1999). 3,3′,5,5′-tetrachlorobiphenyl (PCB 80), another model congener, was a weak ER agonist both *in vivo* and *in vitro,* while, surprisingly, PCB 52 was inactive in the same models (Nesaretnam and Darbre 1997). PCB 66 and 95 (2,2′,3,5′,6-pentachlorobiphenyl) have been reported to be estrogenic in BG1LucE2 cells at 10mM concentrations, while coplanar PCB 77 elicited no ER-mediated activity in this cellular model (Rogers and Denison 2000). PCB 52 and also PCB 77 caused a modest transient uterotrophic effect in weanling rats (Rose et al. 2002); however, PCB 77 attenuated the increase in uterine weight and cell proliferation in another report (Jansen et al. 1993). The uterotrophic effects after exposure to less persistent PCB congeners showed nonlinear dose responses and they decreased rapidly (Rose et al. 2002). However, all the above data have been obtained from various models and assays, and estrogenic/anti-estrogenic effects of both lower chlorinated and higher chlorinated environmentally significant PCBs have not yet been examined systematically in one assay.

In our study, PCBs 28, 52, 66, 74, 99 and 105, all found at significant levels in male serum samples, activated the ER at micromolar concentrations (table 13.2) suggesting that

Table 13.2. Concentration Levels in Blood and *in vitro* Anti-estrogenic Activities of Prevalent PCBs, p,p′-DDE, and Their Major Metabolites Determined in the ER-CALUX Bioassay.

IUPAC No. (Substitution)	Concentration Levels in Human Blood[a]	ER-Inducing Activity (EC$_{50}$)	Antiestrogenic Activity (Suppression of ER Activation by E$_2$) (EC$_{50}$)
28 (2,4,4′)	(+)	+ + (8 µM)	-
52 (2,2′,5,5′)	(+)	+ + (10 µM)	-
66 (2,3′,4,4′)	+ +	+ (24 µM)	-
74 (2,4,4′,5)	+ +	+ (17 µM)	-
99 (2,2′,4,4′,5)	+ +	(+)	-
105 (2,3,3′,4,4′)	+	(+)	-
118 (2,3′,4,4′,5)	+ +	-	-
126 (3,3′,4,4′,5)	(+)	-	++ [b]
138 (2,2′,3,4,4′,5′)	+ + +	-	+ + (10 µM)
153 (2,2′,4,4′,5,5′)	+ + +	-	+ + (6 µM)
156 (2,3,3′,4,4′,5)	+ +	-	(+)
170 (2,2′,3,3′,4,4′,5)	+ + +	-	+ (16 µM)
180 (2,2′,3,4,4′5,5′)	+ + +	-	+ + (9 µM)
187 (2,2′,3,4′,5,5′,6)	+ + +	-	+ + (7 µM)
194 (2,2′,3,3′,4,4′,5,5′)	+ +	-	+ (14 µM)
199 (2,2′,3,3′,4′,5,6,6′)	+ +	-	+ + (3 µM)
203 (2,2′,3,4,4′,5,5′,6)	+ +	-	+ + (3 µM)
4-OH-PCB 107	++	(+)	+ (cytotoxic)[c]
4-OH-PCB 146	++	-	+ (30 µM)[c]
4-OH-PCB 187	++	-	+ + (10 µM)[c]
4′-MeSO$_2$-CB 101	(+)	-	+ + (2.4-5 µM)[d,e]
4-MeSO$_2$-CB 149	(+)	-	+ + (3 µM)[d]
p,p′-DDE	+ + +	(+)[d,e]	-[d,e]
3-MeSO$_2$-4,4′-DDE	+	(+)[d]	-[d,e]

[a] In Kočan et al. 2004; Hovander et al. 2004.

[b] Total suppression of E2 response has been reported after cotreatment with 10 mM PCB 126 in the MCF- and T47D breast cancer cells transfected with reporter gene under control of creatine kinase B or cathepsin D gene promoters.

[c] In Machala et al. 2004.

[d] Plisková et al. (manuscript in preparation).

[e] In Letcher et al. 2002.

Note: Concentrations were expressed as follows: (+), quantified only in some samples; +, median value 0.2–10 ng/g lipid; ++, median value 10–100 ng/g lipid; +++, median value < 100 ng/g lipid.

ER activation could be one of the potential modes of action of low-molecular-weight PCBs. However, the decrease of total estrogenic activity and E2 concentrations observed in human serum samples of males exposed to high PCB levels (fig. 13.1), indicated that PCB mixtures elicit overall anti-estrogenic effects.

Antiestrogenicity might be caused by dioxin-like and prevalent high-molecular-weight PCBs. TCDD, a model toxicant for dioxin-like PCBs, exhibits potent anti-estrogenic activity (Buchanan et al. 2000; Buchanan et al. 2002; Cooke, Sato, and Buchanan 2001; Safe and Wormke 2003). TCDD has little effect on total ER levels (Gierthy et al. 1996), and no direct binding to ERs has been reported (ref. Safe and Wormke 2003). In this study, anti-estrogenicity was not elicited by coplanar PCB 126 (table 13.2) and TCDD showed only a weak anti-estrogenic potency (data not shown) in the T47D.Luc cells used in the ER-CALUX assay. It has been reported that TCDD or coplanar PCBs did not inhibit E2-induced activity of reporter construct containing the promoter insert from *creatine kinase* B in T47D cells, while dioxin-like compounds including PCB 77 and 126 prevented activation of other reporter constructs in both MCF-7 and T47D cells, although only at levels as high as 10 mM (Ramamoorthy et al. 1999). This suggests that the type of reporter construct used could affect detection of anti-estrogenic activity. One possible mechanism of anti-estrogenic activity of AhR ligands is the direct inhibition of E2-responsive genes through binding to inhibitory dioxin responsive elements (iDRE) in their promoter regions. Functional iDREs have been identified in promoter regions of *pS2, c-fos, Hsp27,* and *cathepsin D* genes (ref. in Safe and Wormke 2003). The lack of anti-estrogenic activity of coplanar PCBs observed in the T47D.Luc cell line might be explained by the missing iDREs in the reporter construct, which contains three tandem repeats of the consensus estrogen responsive elements (ERE) oligonucleotide (Legler et al. 1999).

Nevertheless, this cellular model allowed us to investigate a direct activation of ER and/or perturbation of the ER activation by E2. While low-molecular-weight PCBs elicited ER activation and ER-dependent gene expression, prevalent and more persistent high-molecular-weight PCB congeners were anti-estrogenic (table 13.2, fig. 13.3). It has been stated that pulses of exposure to more labile mixtures of lower chlorinated PCBs may contribute to transient endocrine disruption including increase in estrogenic activity (Hansen 1998; Rose et al. 2002). PCB 153 was estrogenic in the acute two-day immature rat uterine weight assay, albeit at very high concentrations (Li and Hansen 1994). Anti-estrogenic potencies of three of prevalent congeners (PCB 138, 153, 180) have been found both in a reporter gene and cell proliferation MCF-7 assays (Bonenfeld-Jorgensen et al. 2001).

This is in accordance with our data on anti-estrogenicity of high-molecular-weight PCBs in the ER-CALUX assay (table 13.2). Inhibition of ER activation by hexa-, hepta- and octachlorinated biphenyls and suppression of estrogenic signaling found in serum of males chronically exposed to PCBs (fig. 13.1) suggested that PCB 153 and other prevalent congeners could contribute to overall anti-estrogenic response. Because low-molecular-weight and high-molecular-weight PCBs elicited their effects on estrogenic activity at similar micromolar concentrations, it could be expected that the anti-estrogenic effects of the more prevalent higher chlorinated PCBs would prevail in human male blood. Nevertheless, (anti)estrogenic effects of PCBs appeared to be relatively weak when compared with the levels of E2 as the physiological estrogen and the ER-mediated activity detected in whole serum extracts. This holds true also for the effects of PCB metabolites, p,p′-DDE and its methylsulfonyl metabolite.

Our data suggested that exposure to high levels of PCBs might affect E2 blood levels. Currently, only limited information is available about a possible modulation of steroid hormone levels after exposure to PCBs. In a recently published Swedish study, a weak but significant negative correlation has been found between serum levels of the prevalent PCB 153 congener and testosterone in young men, and E2 concentrations (within a serum concentration range from 43 to 144 pM) were also slightly decreased in more exposed subjects (Richthoff et al. 2003). Concentrations of PCB 153 between 23 to 250 ng/g lipid have been found in these subjects—this is a much lower exposure than the PCB contamination observed both in this and in previous studies in eastern Slovakia (Kočan et al. 2001; Kočan et al. 2004). Our data are in accordance with the results of the Swedish study, and with reported lower testosterone and E2 serum levels and suppression of brain aromatase activity in rats exposed to a PCB mixture (Hany et al. 1999).

Decreased E2 concentrations could result from AhR activation by dioxin-like PCBs, which lead to enhanced CYP1A1/CYP1B1-catalyzed metabolism of E2 (Spink et al. 1990; Gierthy et al. 1996). Induced levels of CYP1A1 and CYP1B1 mRNAs in lymphocytes are considered to reflect increased exposure to dioxin-like compounds (Canton et al. 2003). Within this epidemiological study, increased levels of CYP1A1 and CYP1B1 mRNA were found in lymphocytes of males exposed to PCBs. This finding suggests that a physiologically significant AhR-dependent induction of E2-metabolizing CYP enzymes might occur in liver and other target tissues.

Besides CYP1A1, 1A2, and 1B1 isoenzymes, CYP3A4 has been suggested to play a major role in hydroxylation of E2 (Spink et al. 1990; Hayes et al. 1996; Pang et al. 1999; Yamazaki et al. 1998; Badawi, Cavalieri, and Rogan 2000;

Table 13.3. Possible Modes of Action of Dioxin-Like and Non-Dioxin-Like PCBs.

AhR-dependent induction of CYP1A1 / CYP1B1[a]	Dioxin-like PCBs	
Induction of CYP3A4[b]	Non-dioxin-like PCBs	
Antiestrogenicity via direct interactions of the AhR with iDRE, competition of AhR with ER for cofactors, induction of inhibitory factors or induction of ER degradation[c]	Dioxin-like PCBs	Suppression of ER-mediated gene expression
Antiestrogenicity mediated through direct suppression of ER activation[d]	Non-dioxin-like PCBs	Decrease in E2 response (at the micromolar concentrations only)
Effects on pituitary signaling? Modulation of steroidogenesis? Transport/bioavailability of estrogen hormones?	Dioxin-like/non-dioxin-like PCBs	Decrease in estrogen concentration in target tissues?

[a]See Spink et al. 1990; Hayes et al. 1996; Badawi, Cavalieri, and Rogan 2000.

[b]See Parkinson et al. 1982; Yamazaki et al. 1998; Gillette, Hansen, and Rose 2002.

[c]See Safe and Wormke 2003.

[d]This study.

Takemoto et al. 2004). While induction of CYP1A1 and 1B1 expression is dependent on AhR activation (Hayes et al. 1996; Pang et al. 1999), induction of CYP3A4 is a consequence of exposure to prevalent non-dioxin-like PCBs (Parkinson et al. 1982; Gillette, Hansen, and Rose 2002). Therefore, both coplanar and noncoplanar PCBs could increase E2 metabolism and reduce blood E2 concentrations. Both dioxin-like and non-dioxin-like PCBs might affect estrogen signaling by multiple mechanisms as summarized in table 13.3. Obviously, this list is not complete, as it could be speculated that PCBs might also disrupt the pathways associated with perturbation of hypothalamus-pituitary-gonadal axis hormone signaling and steroidogenesis, as another potential mechanism of E2 modulations by PCBs.

In summary, relationships between exposure to PCBs and effects on E2 levels and overall (anti)estrogenic activity in this study were significant only at high exposure levels. Although the prevalent noncoplanar PCBs elicited antiestrogenicity in the ER-CALUX assay, when tested as individual compounds or as a partially reconstituted mixture, a significant estrogenic activity was determined in whole serum extracts. Due to the presence of E2 in human male blood and its dominant role in total estrogenic activity of serum samples, reduction of E2 levels appears to be a more significant anti-estrogenic effect of high PCB exposure. Induction of CYP1A1, CYP1A2, CYP1B1 and CYP3A4 by PCBs, as enzymes responsible for metabolical inactivation of E2, could play a role in this effect. Since the data on estrogenicity and E2 concentrations were negatively correlated with CYP1A1 and CYP1B1 mRNA expression, modulation of blood concentrations of E2 by PCB-induced metabolism of E2 might perhaps play a more significant role in endocrine disruption than the direct effects on the ER-mediated activity. The significance of alternative mechanisms of depression of E2 levels by PCBs, such as perturbation of steroidogenesis and endocrine signaling preceding the biosynthesis of estrogens, remains to be examined, as well as the comparison of *in vivo* contribution of CYP isoenzymes, induced by AhR agonists and non-dioxin-like compounds, to reduction of E2 levels.

ACKNOWLEDGMENTS

This work was supported by European Union project No. QLK4-CT-2000-00488 and by the Czech Ministry of Agriculture (MZE 0002716201). The authors thank Marie Gajova (VRI, Brno) for her assistance with extraction and cleanup of serum samples and Dr. Stanislav Jursa (SHU, Bratislava) for preliminary selection of prevalent PCBs based on GC/MS data.

REFERENCES

Anderson, J. W., E. Y. Zeng, J. M. Jones. 1999. "Correlation between the Response of a Human Cell Line (P450RGS) and the Distribution of Sediment PAHs and PCBs on the Palos Verdes Shelf, California." *Environ. Toxicol. Chem.* 18: 1506–10.

Ankley, G. T., E. Mihaich, R. Stahl, D. Tillitt, T. Colborn, S. McMaster, R. Miller, J. Bantle, P. Campbell, N. Denslow, R. Dickerson, L. Folmar, M. Fry, J. P. Giesy, L. E. Gray, P. Guiney, T. Hutchinson, S. Kennedy, V. Kramer, G. LeBlanc, M. Mayes, A. Nimrod, R. Patino, R. Peterson, R. Purdy,

R. Ringer, P. Thomas, L. Touart, G. Van der Kraak, and T. Zacharewski. 1998. "Overview of a Workshop on Screening Methods for Detecting Potential (Anti-)Estrogenic/Androgenic Chemicals in Wildlife." *Environ. Toxicol. Chem.* 17:68–87.

Arcaro, K. F., L. Yi, R. F. Seegal, D. D. Vakharia, Y. Yang, D. C. Spink, K. Brosch, and J. F. Gierthy. 1999. "2,2′,6,6′-Tetrachlorobiphenyl Is Estrogenic in Vitro and in Vivo." *J. Cell. Biochem.* 72:94–102.

Badawi, A. F., E. L. Cavalieri, and E. G. Rogan. 2000. "Effect of Chlorinated Hydrocarbons on Expression of Cytochrome P450 1A1, 1A2 and 1B1 and 2- and 4-Hydroxylation of 17ß-Estradiol in Female Sprague–Dawley Rats." *Carcinogenesis* 21:1593–99.

Behnisch, P. A., K. Hosoe, and S-i. Sakai. 2001. "Bioanalytical Screening Methods for Dioxins and Dioxin-Like Compounds—A Review of Bioassay/Biomarker Technology." *Environ. Int.* 27:413–39.

Bonenfeld-Jorgensen, E. C., H. R. Andersen, T. H. Rasmussen, and A. M. Vinggaard. 2001. "Effect of Highly Bioaccumulated Polychlorinated Biphenyl Congeners on Estrogen and Androgen Receptor Activity." *Toxicology* 158:141–53.

Brack, W., H. Segner, M. Möder, and G. Schüürmann. 2000. "Fixed-Effect-Level Toxicity Equivalents—A Suitable Parameter for Assessing Ethoxyresorufin-*o*-Deethylase Induction Potency in Complex Environmental Samples." *Environ. Toxicol. Chem.* 19:2493–2501.

Brouwer, A., M. P. Longnecker, L. S. Birnbaum, J. Cogliano, P. Kostyniak, J. Moore, S. Schantz, and G. Winneke. 1999. "Characterization of Potential Endocrine-Related Health Effects at Low-Dose Levels of Exposure to PCBs." *Environ. Health Perspect.* 107 (Suppl. 4): 639–49.

Buchanan, D. L., T. Sato, R. E. Peterson, and P. S. Cooke. 2000. "Antiestrogenic Effects of 2,3,7,8-Tetrachlorodibenzo-*p*-Dioxin in Mouse Uterus: Critical Role of the Aryl Hydrocarbon Receptor in Stromal Tissue." *Toxicol. Sci.* 57:302–11.

Buchanan, D. L., S. Ohsako, C. Tohyama, P. S. Cooke, and T. Iguchi. 2002. "Dioxin Inhibition of Estrogen-Induced Mouse Uterine Epithelial Mitogenesis Involves Changes in Cyclin and Transforming Growth Factor-Beta Expression." *Toxicol. Sci.* 66:62–68.

Canton, R. F., H. T. Besselink, J. T. Sanderson, S. Botschuyver, B. Brouwer, and M. van den Berg. 2003. "Expression of CYP1A and 1B1 mRNA in Blood Lymphocytes from Two District Populations in Slovakia Compared to Total TEQs in Blood as Measured by the DRE-CALUX® Assay." *Organohalogen Compounds* 64:215–18.

Cooke, P. S., T. Sato, and D. L. Buchanan. 2001. "Disruption of Steroid Hormone Signaling by PCBs." In *PCBs: Recent Advances in Environmental Toxicology and Health Effects,* edited by L. W. Robertson and L. G. Hansen, 257–63. Lexington: University Press of Kentucky.

Daly, R. J., and P. D. Darbre. 1990. "Cellular and Molecular Events in Loss of Estrogen Sensitivity in ZR-75-1 and T-47-D Human Breast Cancer Cells. *Cancer Res.* 50:5868–75.

Daston, G. P., J. C. Cook, and R. J. Kavlock. 2003. "Uncertainties for Endocrine Disrupters: Our View on Progress." *Toxicol. Sci.* 74:245–52.

Fernandez, M. F., A. Rivas, F. Olea-Serrano, I. Cerrillo, J. M. Molina-Molina, P. Araque, J. L. Martinez-Vidal, and N. Olea. 2004. "Assessment of Total Effective Xenoestrogen Burden in Adipose Tissue and Identification of Chemicals Responsible for the Combined Estrogenic Effect." *Anal. Bioanal. Chem.* 379:163–70.

Gale, R. W., E. R. Long, T. R. Schwartz, and D. E. Tillitt. 2000. "Evaluation of Planar Halogenated and Polycyclic Aromatic Hydrocarbons in Estuarine Sediments Using Ethoxyresorufin-*o*-Deethylase Induction of H4IIE Cells." *Environ. Toxicol. Chem.* 19:1348–59.

Garrison, P. M., K. Tullis, J. M. Aarts, A. Brouwer, J. P. Giesy, and M. S. Denison. 1996. "Species-Specific Recombinant Cell Lines as Bioassay Systems for the Detection of 2,3,7,8-Tetrachlorodibenzo-*p*-Dioxin-Like Chemicals." *Fundam. Appl. Toxicol.* 30:194–203.

Gierthy, J., K. F. Arcaro, and M. Floyd. 1997. "Assessment of PCB Estrogenicity in a Human Breast Cancer Cell Line." *Chemosphere* 34:1495–1505.

Gierthy, J. F., B. C. Spink, H. L. Figge, B. T. Pentecost, and D. C. Spink. 1996. "Effects of 2,3,7,8-Tetrachlorodibenzo-*p*-Dioxin, 12-*O*-Tetradecanoylphorbol-13-Acetate and 17β-Estradiol on Estrogen Receptor Regulation in MCF-7 Human Breast Cancer Cells." *J. Cell. Biochem.* 60:173–84.

Gillesby, B. E., and T. R. Zacharewski. 1998. "Exoestrogens: Mechanisms of Action and Strategies for Identification and Assessment." *Environ. Toxicol. Chem.* 17:3–14.

Gillette, J. S., L. G. Hansen, and R. L. Rose. 2002. "Metabolic Effects of Episodic Polychlorinated Biphenyl (PCB) Congeners." *Reviews in Toxicology: Environmental Toxicology* 4: 129–59.

Gutendorf, B., and J. Westendorf. 2001. "Comparison of an Array of in Vitro Assays for the Assessment of the Estrogenic Potential of Natural and Synthetic Estrogens, Phytoestrogens and Xenoestrogens." *Toxicology* 166:79–89.

Hansen, L. G. 1998. "Stepping Backward to Improve Assessment of PCB Congener Toxicities." *Environ. Health Perspect.* 106: 171–89.

Hany, J., H. Lilienthal, A. Sarasin, A. Roth-Härer, A. Fastabend, L. Dunemann, W. Lichtensteiger, and G. Winneke. 1999. "Developmental Exposure of Rats to a Reconstituted PCB Mixture or Aroclor 1254: Effects on Organ Weights, Aromatase Activity, Sex Hormone Levels, and Sweet Preference Behavior." *Toxicol. Appl. Pharmacol.* 158:231–43.

Hayes, C. L., D. C. Spink, B. C. Spink, J. Q. Cao, N. J. Walker, and T. R. Sutter. 1996. "17β-Estradiol Hydroxylation Catalyzed by Human Cytochrome P450 1B1." *Proc. Natl. Acad. Sci. USA* 93:9776–81.

Hovander, L., L. Linderholm, M. Athanasiadou, I. Athanassiadis, T. Trnovec, and A. Kočan. 2004. "Analysis of PCB and PCB Metabolites in Humans from Eastern Slovakia." *Organohalogen Compounds* 66:3525–31.

Jansen, H. T., P. S. Cooke, J. Porcelli, T-C. Liu, and L. G. Hansen. 1993. "Estrogenic and Antiestrogenic Actions of PCBs in the Female Rat: In Vitro and in Vivo Studies." *Reprod. Toxicol.* 7:237–48.

Jobling, S., T. Reynolds, R. White, M. G. Parker, and J. P. Sumpter. 1995. "A Variety of Environmentally Persistent Chemicals, Including Some Phthalate Plasticizers, Are Weakly Estrogenic." *Environ. Health Perspect.* 103:582–87.

Kanno, J., L. Onyon, J. Haseman, P. Fenner-Crisp, J. Ashby, and W. Owens. 2001. "The OECD Program to Validate the Rat Uterotrophic Bioassay to Screen Compounds for *in Vivo* Estrogenic Responses: Phase 1." *Environ. Health Perspect.* 109:785–94.

Kavlock, R. J., G. P. Daston, C. DeRosa, P. Fenner-Crisp, L. E. Gray, S. Kaattari, G. Lucier, M. Luster, M. J. Mac, C. Maczka, R. Miller, J. Moore, R. Rolland, G. Scott, D. M. Sheehan, T. Sinks, and H. A. Tilson. 1996. "Research Needs for the Risk Assessment of Health and Environmental Effects of Endocrine Disruptors: A Report of the U.S. EPA-Sponsored Workshop." *Environ. Health Perspect.* 104 (Suppl. 4): 715–40.

Kočan, A., J. Petrík, S. Jursa, J. Chovancová, and B. Drobná. 2001. "Environmental Contamination with Polychlorinated Biphenyls in the Area of Their Former Manufacture in Slovakia." *Chemosphere* 43:595–600.

Kočan, A., B. Drobná, J. Petrík, S. Jursa, J. Chovancová, and K. Čonka et al. 2004. "Human Exposure to PCBs and Some Other Persistent Organochlorines in Eastern Slovakia as a Consequence of Former PCB Production." *Organohalogen Compounds* 66: 3539–46.

Legler, J., C. E. van den Brink, A. Brouwer, A. J. Murk, P. T. van der Saag, A. D. Vethaak, and B. van der Burg. 1999. "Development of a Stably Transfected Estrogen Receptor-Mediated Luciferase Reporter Gene Assay in the Human T47D Breast Cancer Cell Line." *Toxicol. Sci.* 48:55–66.

Letcher, R. J., I. van Holsteijn, H-J. Drenth, R. J. Norstrom, A. Bergman, S. Safe, R. Pieters, and M. van den Berg. 1999. "Cytotoxicity and Aromatase (CYP19) Activity Modulation by Organochlorines in Human Placental JEG-3 and JAR Choriocarcinoma Cells." *Toxicol. Appl. Pharmacol.* 160:10–20.

Letcher, R. J., J. G. Lemmen, B. van der Burg, A. Brouwer, A. Bergman, and J. P. Giesy et al. 2002. "*In Vitro* Antiestrogenic Effects of Aryl Methyl Sulfone Metabolites of Polychlorinated Biphenyls and 2,2-bis(4-chlorophenyl)-1,1-Dichloroethene on 17 Beta-Estradiol-Induced Gene Expression in Several Bioassay Systems." *Toxicol. Sci.* 69:362–72.

Li, M-H., and L. G. Hansen. 1996. "Responses of Prepubertal Female Rats to Environmental PCBs with High and Low Dioxin Equivalencies." *Fundam. Appl. Toxicol.* 33:282–93.

Li, M-H., Y-D. Zhao, and L. G. Hansen. 1994. "Multiple Dose Toxicokinetic Influence on the Estrogenicity of 2,2′,4,4′,5,5′-Hexachlorobiphenyl." *Bull. Environ. Contam. Toxicol.* 53: 583–90.

Machala, M., L. Bláha, J. Vondrácek, J. E. Trosko, J. Scott, and B. L. Upham. 2003. "Inhibition of Gap Junctional Intercellular Communication by Noncoplanar Polychlorinated Biphenyls: Inhibitory Potencies and Screening for Potential Mode(s) of Action." *Toxicol. Sci.* 76:102–11.

Machala, M., L. Bláha, H-J. Lehmler, M. Plísková, Z. Májková, P. Kapplová, I. Sovadinová, J. Vondrácek, T. Malmberg, and L. W. Robertson. 2004. "Toxicity of Hydroxylated and Quinoid PCB Metabolites: Inhibition of Gap Junctional Intercellular Communication and Activation of Aryl Hydrocarbon and Estrogen Receptors in Hepatic and Mammary Cells." *Chem. Res. Toxicol.* 17:340–47.

Markey, C. M., C. L. Michaelson, E. C. Veson, C. Sonnenschein, and A. M. Soto. 2001. "The Mouse Uterotrophic Assay: A Reevaluation of Its Validity in Assessing the Estrogenicity of Bisphenol A." *Environ. Health Perspect.* 109:55–60.

Moore, M., M. Mustain, K. Daniel, I. Chen, S. Safe, T. Zacharewski, et al. 1997. "Antiestrogenic Activity of Hydroxylated Polychlorinated Biphenyl Congeners Identified in Human Serum." *Toxicol. Appl. Pharmacol.* 142:160–168.

Murk, A. J., J. Legler, M. S. Denison, J. P. Giesy, C. van der Guchte, and A. Brouwer. 1996. "Chemical-Activated Luciferase Gene Expression (CALUX): A Novel *in Vitro* Bioassay for Ah Receptor Active Compounds in Sediments and Pore Water." *Fundam. Appl. Toxicol.* 33:149–60.

Nesaretnam, K., and P. Darbre. 1997. "3,5,3′,5′-Tetrachlorobiphenyl Is a Weak Oestrogen Agonist *in Vitro* and *in Vivo*." *J. Steroid Biochem. Mol. Biol.* 62:409–18.

Nilsen, B. M., K. Berg, A. Goksoyr. 1998. "Induction of Cytochrome P450 1A (CYP1A) in Fish: A Biomarker for Environmental Pollution." *Methods Mol. Biol.* 107:423–38.

Pang, S., J. Q. Cao, B. H. Katz, C. L. Hayes, T. R. Sutter, and D. C. Spink. 1999. "Inductive and Inhibitory Effects of Non-Ortho-Substituted Polychlorinated Biphenyls on Estrogen Metabolism and Human Cytochromes P450 1A1 and 1B1." *Biochem. Pharmacol.* 58:29–38.

Parkinson, A., L. Robertson, L. Safe, and S. Safe. 1982. "Polychlorinated Biphenyls as Inducers of Hepatic Microsomal Enzymes: Structure-Activity Rules." *Chem. Biol. Interact.* 30:271–85.

Pauwels, An, P. H. Cenijn, J. C. Paul, P. J. C. Schepens, and A. Brouwer. 2000. "Comparison of Chemical-Activated Luciferase Gene Expression Bioassay and Gas Chromatography for PCB Determination in Human Serum and Follicular Fluid." *Environ. Health. Perspect.* 108:553–57.

Pons, M., D. Gagne, J. C. Nicolas, and M. Mehtali. 1990. "A New Cellular-Model of Response to Estrogens: A Bioluminescent Test to Characterize (Anti)estrogen Molecules." *Biotechniques* 9:450–59.

Ramamoorthy, K., M. S. Gupta, G. Sun, A. McDougal, and S. H. Safe. 1999. "3,3′4,4′-Tetrachlorobiphenyl Exhibits Antiestrogenic and Antitumorigenic Activity in the Rodent Uterus and Mammary Cells and in Human Breast Cancer Cells." *Carcinogenesis* 20:115–23.

Richthoff, J., L. Rylander, Bo A. G. Jönsson, H. Åkesson, L. Hagmar, P. Nilsson-Ehle, M. Stridsberg, and A. Giwercman. 2003. "Serum Levels of 2,2′,4,4′,5,5′-Hexachlorobiphenyl (CB-153) in Relation to Markers of Reproductive Function in Young Males from the General Swedish Population." *Environ. Health Perspect.* 111:409–13.

Robertson, L. W., and L. G. Hansen. 2001. *PCBs: Recent Advances in Environmental Toxicology and Health Effects.* Lexington: University Press of Kentucky.

Rogers, J. M., and M. S. Denison. 2000. "Recombinant Cell Bioassays for Endocrine Disruptors: Development of a Stably Transfected Human Ovarian Cell Line for the Detection of Estrogenic and Anti-estrogenic Chemicals." *In Vitr. Mol. Toxicol.* 13:67–82.

Rose, R. L., M. A. Khan, M-H. Li, J. S. Gillette, and L. G. Hansen. 2002. "Endocrine Effects of Episodic Polychlorinated Biphenyl (PCB) Congeners." *Reviews in Toxicology: Environmental Toxicology* 2:1–18.

Safe, S. H. 1994. "Polychlorinated Biphenyls (PCBs): Environmental Impact, Biochemical and Toxic Responses, and Implications for Risk Assessment." *Crit. Rev. Toxicol.* 24:87–149.

Safe, S., and M. Wormke. 2003. "Inhibitory Aryl Hydrocarbon Receptor–Estrogen Receptor Alpha Cross-Talk and Mechanisms of Action." *Chem. Research Toxicol.* 16:807–16.

Seidel, S. D., V. Li, G. M. Winter, W. J. Rogers, E. I. Martinez, and M. S. Denison. 2000. "Ah Receptor-Based Chemical Screening Bioassays: Application and Limitations for the Detection of Ah Receptor Agonists." *Toxicol. Sci.* 55:107–15.

Sonnenschein, C., A. M. Soto, M. F. Fernandez, N. Olea, M. F. Olea-Serrano, and M. D. Ruiz-Lopez. 1995. "Development of a Marker of Estrogenic Exposure in Human Serum." *Clin. Chem.* 41:1888–95.

Soto, A. M., M. F. Fernandez, M. F. Luizzi, A. S. Oles Karasko, and C. Sonnenschein. 1997. "Developing a Marker of Exposure to Xenoestrogen Mixtures in Human Serum." *Environ. Health Perspect.* 105 (Suppl. 3): 647–54.

Spink, D. C., D. W. Lincoln II, H. W. Dickerman, and J. F. Gierthy. 1990. "2,3,7,8-Tetrachlorodibenzo-*p*-Dioxin Causes an Extensive Alteration of 17β-Estradiol Metabolism in MCF-7 Breast Tumor Cells." *Proc. Natl. Acad. Sci. USA.* 87:6917–21.

Takemoto, K., M. Nakajima, Y. Fujiki, M. Katoh, F. J. Gonzalez, and T. Yokoi. 2004. "Role of the Aryl Hydrocarbon Receptor and Cyp1b1 in the Antiestrogenic Activity of 2,3,7,8-Tetrachlorodibenzo-*p*-dioxin." *Arch. Toxicol.* 78:309–15.

van den Berg, M., L. Birnbaum, A. T. Bosveld, B. Brunstrom, P. Cook, M. Feeley, J. P. Giesy, A. Hanberg, R. Hasegawa, S. W. Kennedy, T. Kubiak, J. C. Larsen, F. X. van Leeuwen, A. K. Liem, C. Nolt, R. E. Peterson, L. Poellinger, S. Safe, D. Schrenk, D. Tillitt, M. Tysklind, M. Younes, F. Waern, and T. Zacharewski. 1998. "Toxic Equivalency Factors (TEFs) for PCBs, PCDDs, PCDFs for Humans and Wildlife." *Environ. Health Perspect.* 106:775–92.

van Duursen, M. B., J. T. Sanderson, M. van der Bruggen, J. van der Linden, and M. van den Berg. 2003. "Effects of Several Dioxin-Like Compounds on Estrogen Metabolism in the Malignant MCF-7 and Nontumorigenic MCF-10A Human Mammary Epithelial Cell Lines." *Toxicol. Appl. Pharmacol.* 190:241–50.

Wang, W. L., R. Smith III, and S. Safe. 1998. "Aryl Hydrocarbon Receptor-Mediated Antiestrogenicity in MCF-7 Cells: Modulation of Hormone-Induced Cell Cycle Enzymes." *Arch. Biochem. Biophys.* 356:239–48.

Yamasaki, K., M. Takeyoshi, Y. Yakabe, M. Sawaki, N. Imatanaka, and M. Takatsuki. 2002. "Comparison of Reporter Gene Assay and Immature Rat Uterotrophic Assay of Twenty-Three Chemicals." *Toxicology* 170:21–30.

Yamazaki, H., P. M. Shaw, F. P. Guengerich, and T. Shimada. 1998. "Roles of Cytochromes P450 1A2 and 3A4 in the Oxidation of Estradiol and Estrone in Human Liver Microsomes." *Chem. Res. Toxicol.* 11:659–65.

Zacharewski, T. 1997. "*In Vitro* Bioassay for Assessing Estrogenic Substances." *Environ. Sci. Technol.* 31:613–23.

Zacharewski, T. R., K. Berhane, and B. E. Gillesby. 1995. "Detection of Estrogen- and Dioxin-Like Activity in Pulp and Paper Mill Black Liquor Effluent Using *in Vitro* Recombinant Receptor/Reporter Gene Assays." *Environ. Sci. Technol.* 29: 2140–46.

14

Cross-Talk between the Aryl Hydrocarbon Receptor and Estrogen Receptor Signaling Pathways

Paul S. Cooke, *University of Illinois, Urbana-Champaign*
Motoko Mukai, *University of Illinois, Urbana-Champaign*
David L. Buchanan, *University of Illinois, Urbana-Champaign*

Estrogen receptor (ER) α and β are members of the steroid-thyroid hormone superfamily of nuclear receptors, and play critical roles in reproductive and nonreproductive effects of estrogens. A large and still evolving body of literature indicates that cross-talk between the liganded aryl hydrocarbon receptor (AhR) and ER produces anti-estrogenic effects, though the precise mechanism of this interaction is still debated. However, liganded AhR may not function simply and exclusively as an anti-estrogen. Other work has shown that exposure to the prototypical AhR ligand 2,3,7,8-tetrachlorodibenzo-*p*-dioxin is associated with increases in endometriosis and certain types of estrogen-dependent tumors, an outcome more consistent with an estrogenic effect. In addition, recent work suggests that some AhR ligands may be able to induce estrogenic effects by binding to AhR and then inducing activation of unliganded ER, ultimately producing phenotypic effects similar to that seen with estrogen exposure. Further complicating the elucidation of the cross-talk between these two signaling pathways, some AhR ligands, including the major ligand used in the studies indicating that liganded AhR can induce activation of unliganded ER, are themselves ER agonists. In addition, other recent work has failed to confirm the initial finding that liganded AhR can induce activation of unliganded ER. In conclusion, there is an extensive and complex interaction between the AhR and ER signaling pathways, but our knowledge of the mechanism of these effects, as well as the ultimate health consequences of these interactions, is incomplete.

INTRODUCTION

Estrogen receptors (ERs) α and β are members of the steroid-thyroid hormone superfamily of nuclear receptors. These receptors play critical roles in a wide variety of reproductive effects of estrogens in both the male and female reproductive tract. In addition, both ER α and β are widely distributed in nonreproductive tissues such as bone, adipose tissue, the brain, the heart, and the immune system, where they also function as critical regulators of both the development and function of these organs. The aryl hydrocarbon receptor (AhR) is a basic helix-loop-helix (bHLH) protein that is a member of the Per-ARNT-Sim (PAS) subfamily. Both AhR and ER are ligand-activated nuclear transcription proteins. ER binds its principal natural ligand, 17b-estradiol (E2), but also interacts with other endogenous estrogens as well as a large number of natural and man-made chemicals that have affinity for ER. In contrast, the natural ligand for AhR has not been definitively established. However, several endogenous ligands have recently been suggested, many of which are tryptophan metabolites (Heath-Pagliuso et al. 1998; Song and Pollenz 2002). The highly toxic environmental contaminant 2,3,7,8-tetrachlorodibenzo-*p*-dioxin (TCDD) and other aromatic hydrocarbons such as 3-methylcholanthrene also bind to AhR, and a number of polychlorinated biphenyls (PCBs) have affinity for AhR.

Due to the ubiquity of both ER and AhR, significant amounts of both are expressed in many reproductive and nonreproductive tissues. There is no obvious link between the biological function of ER and AhR, but a substantial body of literature developed over the past 20 years indicates that liganded AhR can affect signaling through ER. This chapter will give an overview of the literature indicating that liganded AhR can have anti-estrogenic actions and will concentrate on literature that has sought to establish a mechanistic basis for this effect. A surprising and potentially important development in the field of cross-talk between AhR and ER occurred approximately a year ago, when it was proposed that liganded AhR could interact with unliganded ER and initiate transcription (Ohtake et al. 2003). This finding has not yet been confirmed, and recent data has disputed the original conclusion (Pearce et al. 2004). Both the original evidence supporting the concept that liganded AhR could interact with unliganded ER and the recent evidence to the contrary are discussed here.

TCDD has anti-estrogenic effects *in vitro* and *in vivo*. TCDD and related chemicals produce cell-type and species specific changes in various endocrine responses (Barsotti, Abrahamson, and Allen 1979; Bruner-Tran et al. 1999), and a large body of literature dating back almost 20 years indicates that TCDD can inhibit estrogen signaling. For example, treatment with TCDD has been found to cause anti-estrogenic effects, including decreases in uterine wet weight in rodents (Gallo et al. 1986). TCDD also inhibited the E2-induced growth and function of the human breast tumor cell line MCF-7 *in vitro* (Gierthy et al. 1987). Furthermore, TCDD treatment inhibits mammary cell proliferation and gland development in pubertal rats, and this chemical also inhibited development and growth of mammary tumors in rodent models, indicating that the anti-estrogenic effects seen in MCF-7 cells may translate to similar effects in mammary tissue *in vivo* (Brown and Lamartiniere 1995; Gierthy et al. 1993; Holcomb and Safe 1994).

Anti-Estrogenic Effects of TCDD Are Mediated through AhR

The cellular and molecular events that occur in response to TCDD treatment have been studied extensively, and TCDD is known to produce effects through both AhR-dependent and AhR-independent pathways (Fernandez-Salguero et al. 1996). Conclusive data have shown that the anti-estrogenic effects of TCDD are mediated through AhR. For example, the degree of TCDD inhibition correlated with the amount of AhR, as MCF-7 cells with minimal AhR did not show the typical anti-estrogenic responses after TCDD exposure (Moore et al. 1994). The affinity of TCDD-like congeners for the AhR correlates with their anti-estrogenic potency, further suggesting that the anti-estrogenic effects of these compounds are mediated through AhR (Safe et al. 1991). In addition, strain differences in the anti-estrogenic effects of

TCDD in mice correlate with the degree of AhR expression (Umbreit et al. 1988). Finally, the anti-estrogenic effects of TCDD are partially blocked by the AhR antagonist alpha napthoflavone (Wang et al. 1996).

Mechanism of Anti-Estrogenic Effects by AhR Agonists

TCDD and related compounds have potent anti-estrogenic activities mediated through AhR. The obvious question then is how does binding of an AhR agonist to its receptor induce anti-estrogenic effects. No clear answer has been obtained here, although data from a variety of *in vivo* and *in vitro* systems suggests that this effect may involve alterations of ER signaling at various steps in the normal signaling cascade initiated by the binding of a ligand to ER. The majority of mechanistic insights as to how liganded AhR induces anti-estrogenic effects have come from *in vitro* systems with estrogen-responsive cell lines. Therefore, it is still unclear how relevant findings related to the mechanism of TCDD effects in cell lines *in vitro* may be for understanding the effects of TCDD or other AhR agonists in normal tissue *in vivo*.

The simplest mechanism for the anti-estrogenic effects of TCDD would involve TCDD binding to and occupying the ER in a manner analogous to clinically used anti-estrogens such as tamoxifen, and interfering with E2 signaling in this manner. However, early results (Poland and Knutson 1982) showed that TCDD did not bind to ER, indicating that its anti-estrogenic effects were more complicated than simply occupying ER and preventing binding of E2.

One of the classic effects of TCDD is the induction of CYP1A1 and CYP1B1, and the induction of these compounds can lead to increased metabolism of E2 and thus potentially reduce levels of the ER ligand and consequent ER signaling. However, this mechanism would appear to be unlikely to be the major explanation for the anti-estrogenic effects of TCDD. For example, some AhR agonists that do not induce CYP1A1 in MCF-7 cells still have anti-estrogenic actions, and some anti-estrogenic effects are seen prior to the increased expression of cytochrome P450 (reviewed in Safe and Wormke 2003). In addition, DeVito et al. (1992) did not observe changes in circulating E2 in response to a range of TCDD doses, though increases in cytochrome P450 levels were seen. In addition, TCDD appears to exert anti-estrogenic effects at doses lower than necessary for induction of cytochrome P450s (Wang et al. 1993).

Binding of liganded AhR to xenobiotic response elements (XREs) in MCF-7 cells can directly inhibit activation of E2-responsive genes. For example, liganded AhR complex inhibits the ability of liganded ER to bind to the estrogen response element (ERE) in the cathepsin D gene (Kharat and Saatcioglu 1996; Krishnan et al. 1995). Krishnan et al. (1995) showed that these EREs were near XREs in the pro-

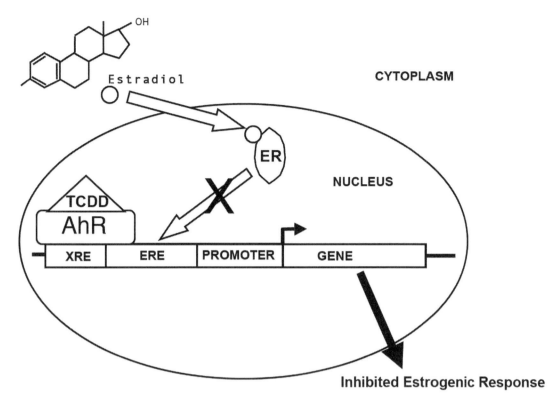

Figure 14.1.   Binding of liganded AhR could directly inhibit activation of E2-responsive genes. The binding of liganded AhR complex to xenobiotic response elements (XREs) in target genes such as cathepsin D could inhibit the ability of liganded ER to bind to neighboring estrogen response elements in the promoter region of these genes. Inhibitory effects of liganded AhR binding to XREs may also involve other mechanisms in addition to physically interfering with the interaction of E2/ER complexes with EREs. Liganded AhR could directly repress the activity of an estrogen-responsive target gene and could also act by other mechanisms such as interfering with the ability of AP-1 complexes, which are involved in the estrogen response of some genes, to bind and activate a promoter (Safe and Wormke 2003).

moter region of the cathepsin D gene, and binding of liganded AhR to the XRE could physically interfere with binding of E2/ER complexes to adjacent EREs (fig. 14.1). The proximity of XREs to EREs in other E2-responsive genes has also been shown, and inhibitory effects of liganded AhR binding to XREs may also involve other mechanisms in addition to physically inhibiting binding of E2/ER complexes to EREs (Gillesby et al. 1997).

Substantial evidence also exists to support the concept that TCDD or other AhR ligands could induce anti-estrogenic activity by decreasing the availability of ER (fig. 14.2). This mechanism appears to be a substantial factor in the overall anti-estrogenic effects of AhR agonists in cell lines, although its applicability to the *in vivo* situation is not as clear. The ability of AhR agonists to decrease ERα was first shown in MCF-7 cells (Harris, Zacharewski, and Safe 1990), although other data in MCF-7 cells have not corroborated this finding (Kharat and Saatcioglu 1996). Recent work by Wormke et al. (2003) has indicated that TCDD treatment (10 nM)

of ZR-75 cells decreased ERα by 60 percent. TCDD also decreases uterine ER levels in uteri of intact mice (DeVito et al. 1992). Initial results *in vivo* indicated that uterine ER levels in ovariectomized mice were not reduced by TCDD compared to intact untreated controls (DeVito et al. 1992). Newer findings (Wormke et al. 2003) have demonstrated that TCDD does decrease ER in the uterus of immature mice *in vivo,* which also have low E2 levels comparable to those seen in the ovariectomized mice used by DeVito et al. (1992). The effects of AhR agonists on ERα concentrations in cell lines do not appear to involve changes in expression of the ERα gene, but instead reflect increased proteosome degradation of ERα (Wormke et al. 2003).

AhR-ligand complexes could occupy ER, producing anti-estrogenic activity by preventing the more potent ligand E2 from binding. This hypothesis arose from observations that liganded AhR could activate ER in the absence of E2 (Ohtake et al. 2003), but these experiments also showed that in the presence of E2, liganded AhR had an anti-estrogenic effect

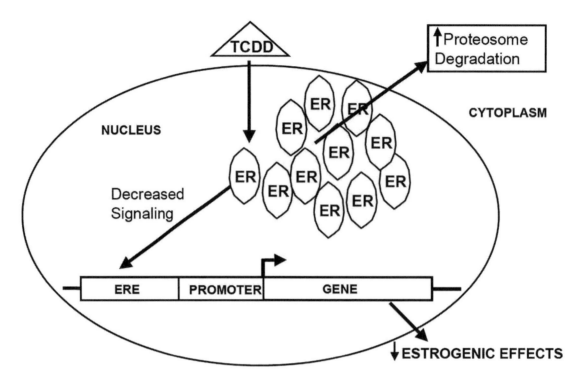

Figure 14.2.   TCDD and other AhR ligands may induce anti-estrogenic activity by decreasing the availability of ER. Anti-estrogenic effects of AhR agonists have been reported to decrease ERα concentrations *in vivo* and *in vitro* through proteosome degradation. Loss of ER would decrease the amount of ER available to bind ligands and subsequently interact with EREs in the promoters of estrogen-responsive genes and would result in an in estrogen responsiveness.

(Ohtake et al. 2003), suggesting that the interaction of the liganded AhR with ER might decrease binding of the more potent E2 and result in a net anti-estrogenic effect.

AhR and ERα signaling involve a complex interplay with a variety of nuclear co-regulatory proteins that function as coactivators or corepressors for steroid receptor signaling, but are also important for AhR signaling. Recently, Ricci et al. (1999) have shown that AhR and ERα can compete for common coregulatory proteins. This suggests that AhR signaling may interfere with ERα signaling through this squelching mechanism (fig. 14.3). Interestingly, although AhR inhibition of ERα signaling has received extensive attention, other reports have indicated that E2 can inhibit TCDD induction of CYP1A1 (Kharat and Saatcioglu 1996; Ricci et al. 1999), possibly by competing for common coregulatory proteins used by both receptors. However, here again seemingly conflicting results showing that estrogen potentiates cytochrome P450 induction by TCDD in rat liver have also been published (Sarkar et al. 2000). The potential modulation of TCDD signaling by E2 further complicates our models of AhR/ER interaction in that it suggests that ERα can modulate AhR signaling under some circumstances, in addition to the more extensively documented inhibition of ERα signaling by liganded AhR.

Other Interactions between AhR and Steroid Hormone Signaling Pathways

Initial conclusions that TCDD was anti-estrogenic were based on the reduction of uterine wet weight by TCDD *in vivo* (Gallo et al. 1986). One of the classic uterine responses to estrogenic stimulation is uterine epithelial proliferation, and we determined that uterine epithelial proliferation is dramatically decreased when E2-treated mice were pretreated with TCDD (Buchanan et al. 2000). Subsequent experiments with AhR knockout (AhRKO) mice established that AhR is obligatory for the anti-estrogenic effect of TCDD on both uterine epithelial proliferation and E2-induced uterine secretory protein production (Buchanan et al. 2000).

Uterine epithelial proliferation following E2 treatment is induced through stromal ER (Cooke et al. 1997). This suggests that AhR could potentially act to inhibit E2-induced uterine epithelial proliferation through AhR in the stroma, consistent with our demonstration that AhR was present in both uterine stroma and epithelium (Buchanan et al. 2000).

To determine if TCDD inhibited the E2-induced uterine epithelial proliferative response through AhR in the epithelia, stroma or both, tissue recombinants using uterine tissues from wild-type and AhRKO mice were prepared and grafted

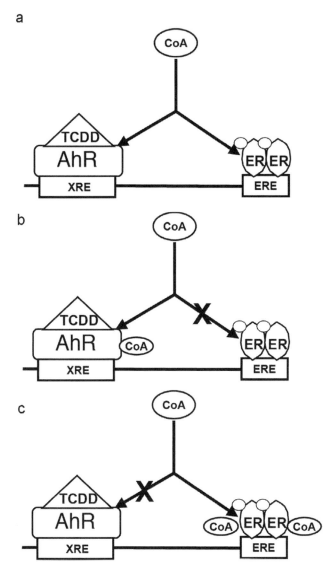

a

b

c

Figure 14.3. Anti-estrogenic effects of AhR agonists may involve competition with ERα for common nuclear coregulatory proteins. Some proteins act as coactivators (CoA) or core-pressors for both steroid receptor and AhR signaling (a), and AhR signaling may interfere with ERα signaling by decreasing levels of these coregulators below the level that is needed to allow maximal responses through ERα (b). This possible mechanism for AhR inhibition of ERα signaling implies that E2 would be able to inhibit signaling through AhR through a similar mechanism (c), and there are also data suggesting that this can occur.

in vivo. Our results indicated that AhR expression was necessary in the stroma, but not the epithelium, of tissue recombinants in order for TCDD to have inhibitory effects on E2-induced uterine epithelial proliferation (fig. 14.4; Buchanan et al. 2000). Thus TCDD acts through stromal AhR to inhibit E2-stimulated uterine epithelial proliferation, while

epithelial AhR is not involved. A likely explanation for these results is that liganded stromal AhR inhibits signaling through stromal ER, which is critical for mediation of the E2-stimulated response on epithelial proliferation.

Uterine epithelial DNA synthesis (S-phase) begins 8.5 hours after E2 treatment in the mouse (Martin, Finn and Trinder 1973b). Earlier effects of E2 stimulation in mouse uterus include changes in expression of delayed early genes (e.g., TGFβs and cyclins; Geum et al. 1997; Martin, Pollard, and Fagg 1976; Takahashi et al. 1994), which are critical for normal progression of the uterotrophic response. The steroid-inducible cytokine TGFβ may be an important mediator of stromal-epithelial interactions (Thompson et al. 1989). TGFβ is anti-mitogenic in uterine and other epithelia but typically promotes growth of stromal tissue (Thompson et al. 1989). E2 and P alter uterine TGFβ levels, and TGFβ mRNA is expressed in mouse uterine myometrium, stroma and epithelium, but mRNA levels for all TGFβ isoforms (TGFβ1, TGFβ2, and TGFβ3) were transiently increased only in epithelium after E2 treatment (Takahashi et al. 1994). By six hours after E2, uterine mRNA levels for the three isoforms

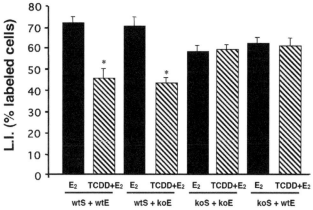

Figure 14.4. Labeling Index (L.I.) of uterine tissue recombinant luminal epithelia. Data are shown as mean percentage of total labeled cells ± SEM. Uterine stroma from either wild-type (wtS) or AhR knockout (koS) mice was recombined with epithelia from wild-type (wtE) or AhR knockout (koE) mice and grafted to ovariectomized host mice that were later treated with either oil, E_2, or 5 mg/kg TCDD + E_2. L.I. in tissues containing wtS and either wtE or koE removed from host animals given TCDD before E_2 was reduced by 34 and 37 percent, respectively, compared to the same tissue recombinant type in hosts given E_2 alone. In contrast, L.I. in tissue recombinants containing koS was similar between E_2- and TCDD + E_2-treated groups. Epithelial L.I. in tissue recombinants from oil-treated hosts was low (not shown). Asterisks indicate significant differences ($p < 0.05$) between treatment groups within each tissue recombinant type.

were suppressed to or below those of oil controls (Buchanan et al. 2002; Takahashi et al. 1994). Sustained increases in TGFβ proteins occur specifically in epithelium after E2 treatment (Takahashi et al. 1994). Importantly, TGFβ activity has been closely associated with down-regulation of G1 phase cyclins (Takahashi et al. 1994), which could contribute to inhibition of epithelial mitogenesis.

Other delayed early genes in the uterus are cyclins. Cyclins are regulatory subunits of cyclin-dependent kinases and they control transitions through different specific stages of the cell cycle. E2-regulation of cyclin expression has been demonstrated (Geum et al. 1997). In mouse uterine epithelia, E2-induced cyclin E protein levels are two- to five-fold lower than those of cyclin A2 throughout the G1 and S phases of the cell cycle (Tong and Pollard 1999), pointing to A2 as critical for initiation of uterine epithelial DNA synthesis.

When mice were given E2 with or without TCDD, we found that the anti-estrogenic TCDD effect on epithelial proliferation involved alterations in uterine TGFβ and cyclin gene expression (Buchanan et al. 2002). Low uterine TGFβ levels and increased cyclin A2 expression were observed after E2 treatment. These effects were blocked by TCDD pretreatment, similar to the TCDD effects others have reported on cell cycle regulatory proteins (Weber et al. 1996). This deregulation of E2-induced uterine TGFβ and cyclin expression by TCDD may be a critical aspect of the mechanism by which liganded AhR inhibits E2-induced epithelial proliferation (Buchanan et al. 2002). Interestingly, progesterone (P) increases TGFβ, which may contribute to the inhibition of uterine epithelial proliferation by P. This raises the question of possible cross-talk or convergent mechanistic pathways between the liganded AhR and the progesterone receptor. Consistent with this idea, interactions and similarities between progesterone and TCDD signaling pathways have been described (Kuil et al. 1998; Romkes and Safe 1988).

Although E2 is a critical regulator of uterine cell growth, proliferation and protein synthesis, P also plays an essential role in many critical uterine processes. While E2 induces mitogenesis in uterine epithelium, P antagonizes E2-stimulated effects on proliferation and induces epithelial differentiation in preparation for embryo implantation (Paria et al. 1998). In addition, P regulates mitogenic and differentiative responses in uterine stroma, presumably in preparation for decidualization (Martin, Finn, and Trinder 1973a). P+E2 cotreatment of ovariectomized mice stimulates stromal cells to progress through G1 and enter S-phase but drastically reduces epithelial proliferation. When given alone, E2 treatment does not produce stromal mitogenesis, as uterine stromal cell proliferation requires P. Cellular responses induced by P+E2

include increases in uterine proteins such as cyclins and cyclin-dependent kinases, which are critical for the onset and normal progression of stromal cell G1 activity and DNA synthesis.

Preliminary data from our recent *in vivo* studies indicate that liganded AhR may inhibit the production and activation (phosphorylation) of cyclins and cyclin-dependent kinases, and thus may inhibit the action of P on uterine stroma. We recently have determined that TCDD inhibits uterine stromal responses stimulated by P+E2 cotreatment. P+E2 significantly increased uterine weight and uterine/body weight ratio at time points encompassing onset and maximal stromal mitogenic activity, compared to mice given oil. Absolute and relative uterine weights after TCDD treatment were similar to those of mice given P+E2 alone. The percentage of BrdU-labeled cells, indicative of DNA synthesis and mitogenic activity, in peri-luminal stroma of uteri from P+E2-treated animals significantly increased compared to oil controls. In contrast, when TCDD was given prior to hormone treatment, peri-luminal stromal DNA synthesis was significantly reduced by 15 and 18 hours after final hormone treatments. Uterine cyclin A2, which is critical for the G1/S transition, showed a marked increase by 12 hours after P+E2, compared to oil, but this increase was not seen at 15 and 18 hours after hormone treatments. This decline in cyclin A2 in the hormone-treated mice at 15 and 18 hours was accompanied by significant increases in phosphorylated cyclin A2, consistent with normal stromal cell cycle progression. However, when animals were given TCDD prior to P+E2, cyclin A2 levels decreased at 12 hours and increased at 15 and 18 hours relative to mice receiving hormones but no TCDD. These TCDD-induced changes in cyclin A2 were accompanied by increased phosphorylated cyclin A2 at 12 hours and decreases at 15 and 18 hours compared to P+E2 alone. Thus TCDD inhibited the normal changes in cyclin A2 and phosphorylated cyclin A2 induced by P+E2. Cyclin-dependent kinase 2 (cdk2), which phosphorylates cyclin A2, was increased by P+E2 at all time points compared to oil. The P+E2-induced increase in cdk2 was augmented by TCDD at 12 h, corresponding to the observed increase in phosphorylated cyclin A2 and the decrease in cyclin A2 after TCDD at this time point. Importantly, TCDD decreased P+E2-induced cdk2 levels at 15 and 18 hours when cyclin A2 was increased and phosphorylated cyclin A2 was decreased by TCDD compared to P+E2 alone. These findings indicate for the first time that TCDD inhibits steroid-induced uterine stromal cell mitogenesis but does not alter early P+E2-induced increases in uterine weight and suggest TCDD inhibits P+E2-induced progression of the uterine stromal cell cycle sometime after initiation, most likely during G1.

In summary, stromal proliferation in response to P+E2 cotreatment is inhibited by TCDD and this involves uterine stromal cyclin A2 production and activation. These initial data indicating that TCDD blocks P+E2-induced stromal DNA synthesis also indicate that the TCDD effect occurs after the stimulation of cyclin A2 protein synthesis that occurs during late G1. The fact that P+E2-induced increases in uterine weights were not altered by TCDD suggest the inhibitory effect of liganded AhR on P+E2-induced stromal proliferation was specific. Progesterone inhibits E2-induced cyclins (including cyclin A2) and other G1 phase proteins in mouse uterine epithelium (Tong and Pollard 1999). Our results are the first indication that TCDD inhibition of steroid-induced DNA synthesis in uterine stroma is accompanied by deregulation of P+E2-controlled cyclin A2 activation and cdk2 protein levels, which are essential for normal stromal cell S-phase activity (Huet et al. 1996; Resnitzky, Hengst, and Reed 1995). Whether liganded AhR acts to specifically inhibit P-induced events or works through anti-estrogenic mechanisms is unclear, as the inhibition of the uterine response could occur through ER- and/or PR-mediated pathways.

Can Liganded AhR Have Estrogenic Effects?

Ohtake et al. (2003) have postulated that the binding of AhR to an agonist, in this case 3-methylcholanthrene (3-MC), causes the liganded AhR to interact with both ER α and β, and activate the ER in the absence of the endogenous ligand E2 in MCF-7 cells (fig. 14.5). This resulted in expression of estrogen target genes and activation of phenotypic responses typical of estrogen, such as mitogenesis in glandular epithelium, and increased uterine weight. These effects of 3-MC were absent in cells lacking AhR or ERα, indicating the dependence of the response on both receptors. In the presence of E2, 3-MC had an anti-estrogenic effect, also possibly through the 3-MC/AhR complex interacting with ERα, inhibiting binding of E2 and thus producing an anti-estrogenic effect.

The unexpected results of Ohtake et al. suggested a need to reassess our understanding of AhR-ER crosstalk. However,

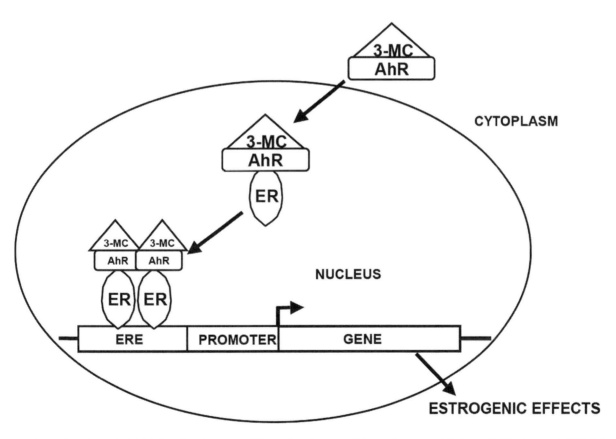

Figure 14.5.   Can liganded AhR bind and activate ER in the absence of the endogenous ER ligand? Ohtake et al (2003) have presented data suggesting that the AhR agonist 3-MC may bind AhR and that this liganded AhR complex can interact with ER and induce the expression of estrogen target genes in the absence of the endogenous ER ligand.

there has been no confirmation of these results. Subsequent work by Pearce et al. (2004) was not consistent with the original findings of Ohtake et al., and the possibility that liganded AhR activates ER independent of ligand binding must be regarded as unclear as of the writing of this review.

The work of Ohtake et al. used the AhR agonist 3-MC in the majority of their experiments. However, more recent work has suggested that 3-MC is an ER agonist (Pearce et al. 2004). This complicates determination of which effects may be mediated by 3-MC/AhR complexes interacting with ER and which may result from direct effects of 3-MC on ER. Ohtake et al. also demonstrated that TCDD could activate transcriptional activity of an estrogen response element linked to a reporter gene. This experiment would seem to indicate that it was the liganded AhR that was having the effect on the ERE, rather than a direct effect of 3-MC on ER, since TCDD was assumed to be working solely through AhR. However, Abdelrahim, Smith, and Safe (2003) recently showed that TCDD had estrogenic activity in MCF-7 cells in which the AhR had been silenced with siRNA. These authors suggested that the lack of the normal anti-estrogenic effect of TCDD resulting from the lack of AhR allowed the modest TCDD estrogenic effects to be manifested.

Another critical piece of evidence from Ohtake et al. supporting the hypothesis that AhR-ligand complexes were acting to stimulate activity of ER was that the estrogenic effects of their AhR agonist were not seen in AhR knockout mice. Pearce et al. (2004) have recently reported that another chemical that had been considered to be an AhR agonist, 6-methyl-1,3,8-trichlorodibenzofuran (MCDF), also stimulated the growth of MCF-7 cells and activated an estrogen response element–reporter gene construct. In contrast to the results obtained by Ohtake et al. (2003), Pearce et al. (2004) reported that AhR was not required for the estrogenic effects of MCDF, which were still seen when AhR was suppressed by siRNA. This finding indicates that estrogenicity may be a more common property of the AhR agonists than previously realized. Although this result would be consistent with the idea that estrogenic effects of MCDF do not involve AhR, it is not possible to conclude this definitively based on this result, because despite the knockdown of AhR, it is difficult to exclude the possibility that residual AhR still expressed in the cell might be sufficient to bind MCDF and allow the MCDF-AhR complexes to activate unliganded ER. In addition, earlier work by Hoffer, Chang, and Puga (1996) that used cells lacking AhR to demonstrate that TCDD can induce transcription of E2-responsive genes by Ah receptor-independent pathways must also be considered as a possible explanation for these results. Thus, at present the recent concept that liganded AhR can interact with and induce activation of ERα (Ohtake et al. 2003) has not been supported by subsequent results (Pearce et al. 2004), and the use of what has subsequently been suggested to be an estrogenic agonist for the original studies also makes the experiments of Ohtake et al. more complicated. Further work will be required to clarify the present inconsistencies in the literature, and clearly the idea that liganded AhR can activate ERα is still under debate.

SUMMARY AND CONCLUSION

Extensive evidence from both *in vivo* and *in vitro* studies indicates that liganded AhR can have anti-estrogenic effects, although the effects of TCDD and the liganded AhR are age, cell-type and species specific. A number of different mechanisms to account for the anti-estrogenic effects of liganded AhR have been proposed; these are not mutually exclusive, and it appears most likely that liganded AhR may act by several mechanisms to produce its anti-estrogenic effects. Recent results suggesting that liganded AhR could interact with ER and initiate transcription of E2-responsive genes in the absence of the normal E2 ligand have not been confirmed, and these initial results were complicated by the use of an AhR ligand that also had affinity for ER. More recent reports dispute the basic hypothesis that liganded AhR can interact with ER and induce estrogenic effects. The initial results suggesting liganded AhR can activate ER must therefore be viewed with caution, although additional studies in the area will probably be needed before a clear consensus on this idea emerges.

ACKNOWLEDGMENTS

The authors gratefully acknowledge the support of this work by NIH grants AG15500 and ES01332 and by the Thanis A. Field Endowment. The work from the laboratory of Paul S. Cooke described here was conducted in a facility constructed with support from Research Facilities Improvement Program Grant No. C06 RR16515 from the National Center for Research Resources, National Institutes of Health.

REFERENCES

Abdelrahim, M., R. Smith III, and S. Safe. 2003. "Aryl Hydrocarbon Receptor Gene Silencing with Small Inhibitory RNA Differentially Modulates Ah-Responsiveness in MCF-7 and HepG2 Cancer Cells." *Mol. Pharmacol.* 63 (6): 1373–81.

Barsotti, D. A., L. J. Abrahamson, and J. R. Allen. 1979. "Hormonal Alterations in Female Rhesus Monkeys Fed a Diet Containing 2,3,7,8-Tetrachlorodibenzo-*p*-Dioxin." *Bull. Environ. Contam. Toxicol.* 21 (4–5): 463–69.

Brown, N. M., and C. A. Lamartiniere. 1995. "Xenoestrogens Alter Mammary Gland Differentiation and Cell Proliferation in the Rat." *Environ. Health Perspect.* 103 (7–8): 708–13.

Bruner-Tran, K. L., S. E. Rier, E. Eisenberg, and K. G. Osteen. 1999. "The Potential Role of Environmental Toxins in the Pathophysiology of Endometriosis." *Gynecol. Obstet. Invest.* 48 (Suppl. 1): 45–56.

Buchanan, D. L., T. Sato, R. E. Peterson, and P. S. Cooke. 2000. "Antiestrogenic Effects of 2,3,7,8-Tetrachlorodibenzo-*p*-dioxin in Mouse Uterus: Critical Role of the Aryl Hydrocarbon Receptor in Stromal Tissue." *Toxicol. Sci.* 57 (2): 302–11.

Buchanan, D. L., S. Ohsako, C. Tohyama, P. S. Cooke, and T. Iguchi. 2002. "Dioxin Inhibition of Estrogen-Induced Mouse Uterine Epithelial Mitogenesis Involves Changes in Cyclin and Transforming Growth Factor-Beta Expression." *Toxicol. Sci.* 66 (1): 62–68.

Cooke, P. S., D. L. Buchanan, P. Young, T. Setiawan, J. Brody, K. S. Korach, J. Taylor, D. B. Lubahn, and G. R. Cunha. 1997. "Stromal Estrogen Receptors Mediate Mitogenic Effects of Estradiol on Uterine Epithelium." *Proc. Natl. Acad. Sci. USA* 94 (12): 6535–40.

DeVito, M. J., T. Thomas, E. Martin, T. H. Umbreit, and M. A. Gallo. 1992. "Antiestrogenic Action of 2,3,7,8-Tetrachlorodibenzo-*p*-dioxin: Tissue-Specific Regulation of Estrogen Receptor in CD1 Mice." *Toxicol. Appl. Pharmacol.* 113 (2): 284–92.

Fernandez-Salguero, P. M., D. M. Hilbert, S. Rudikoff, J. M. Ward, and F. J. Gonzalez. 1996. "Aryl-Hydrocarbon Receptor-Deficient Mice Are Resistant to 2,3,7,8-Tetrachlorodibenzo-*p*-Dioxin-Induced Toxicity." *Toxicol. Appl. Pharmacol.* 140 (1): 173–79.

Gallo, M. A., E. J. Hesse, G. J. Macdonald, and T. H. Umbreit. 1986. "Interactive Effects of Estradiol and 2,3,7,8-Tetrachlorodibenzo-*p*-Dioxin on Hepatic Cytochrome P-450 and Mouse Uterus." *Toxicol. Lett.* 32 (1–2): 123–32.

Geum, D., W. Sun, S. K. Paik, C. C. Lee, and K. Kim. 1997. "Estrogen-Induced Cyclin D1 and D3 Gene Expressions during Mouse Uterine Cell Proliferation in Vivo: Differential Induction Mechanism of Cyclin D1 and D3." *Mol. Reprod. Dev.* 46 (4): 450–58.

Gierthy, J. F., D. W. Lincoln, M. B. Gillespie, J. I. Seeger, H. L. Martinez, H. W. Dickerman, and S. A. Kumar. 1987. "Suppression of Estrogen-Regulated Extracellular Tissue Plasminogen Activator Activity of MCF-7 Cells by 2,3,7,8-Tetrachlorodibenzo-*p*-dioxin." *Cancer Res.* 47 (23): 6198–6203.

Gierthy, J. F., J. A. Bennett, L. M. Bradley, and D. S. Cutler. 1993. "Correlation of in Vitro and in Vivo Growth Suppression of MCF-7 Human Breast Cancer by 2,3,7,8-Tetrachlorodibenzo-*p*-Dioxin." *Cancer Res.* 53 (13): 3149–53.

Gillesby, B. E., M. Stanostefano, W. Porter, S. Safe, Z. F. Wu, and T. R. Zacharewski. 1997. "Identification of a Motif within the 5′ Regulatory Region of pS2 Which Is Responsible for AP-1 Binding and TCDD-Mediated Suppression." *Biochemistry* 36 (20): 6080–89.

Harris, M., T. Zacharewski, and S. Safe. 1990. "Effects of 2,3,7,8-Tetrachlorodibenzo-*p*-Dioxin and Related Compounds on the Occupied Nuclear Estrogen Receptor in MCF-7 Human Breast Cancer Cells." *Cancer Res.* 50 (12): 3579–84.

Heath-Pagliuso, S., W. J. Rogers, K. Tullis, S. D. Seidel, P. H. Cenijn, A. Brouwer, and M. S. Denison. 1998. "Activation of the Ah Receptor by Tryptophan and Tryptophan Metabolites." *Biochemistry* 37 (33): 11508–15.

Hoffer, A., C. Y. Chang, and A. Puga. 1996. "Dioxin Induces Transcription of Fos and Jun Genes by Ah Receptor-Dependent and -Independent Pathways." *Toxicol. Appl. Pharmacol.* 141 (1): 238–47.

Holcomb, M., and S. Safe. 1994. "Inhibition of 7,12-Dimethyl-benzanthracene-Induced Rat Mammary Tumor Growth by 2,3,7,8-Tetrachlorodibenzo-*p*-Dioxin." *Cancer Lett.* 82 (1): 43–47.

Huet, X., J. Rech, A. Plet, A. Vie, and J. M. Blanchard. 1996. "Cyclin A Expression Is Under Negative Transcriptional Control during the Cell Cycle." *Mol. Cell. Biol.* 16 (7): 3789–98.

Kharat, I., and F. Saatcioglu. 1996. "Antiestrogenic Effects of 2,3,7,8-Tetrachlorodibenzo-*p*-Dioxin Are Mediated by Direct Transcriptional Interference with the Liganded Estrogen Receptor: Cross-Talk between Aryl Hydrocarbon- and Estrogen-Mediated Signaling." *J. Biol. Chem.* 271 (18): 10533–37.

Krishnan, V., W. Porter, M. Santostefano, X. Wang, and S. Safe. 1995. "Molecular Mechanism of Inhibition of Estrogen-Induced Cathepsin D Gene Expression by 2,3,7,8-Tetrachlorodibenzo-*p*-Dioxin (TCDD) in MCF-7 Cells." *Mol. Cell. Biol.* 15 (12): 6710–19.

Kuil, C. W., A. Brouwer, P. T. van der Saag, and B. van der Burg. 1998. "Interference between Progesterone and Dioxin Signal Transduction Pathways: Different Mechanisms Are Involved in Repression by the Progesterone Receptor A and B Isoforms." *J. Biol. Chem.* 273 (15): 8829–34.

Martin, L., C. A. Finn, and G. Trinder. 1973a. "DNA Synthesis in the Endometrium of Progesterone-Treated Mice." *J. Endocrinol.* 56 (2): 303–7.

———. 1973b. "Hypertrophy and Hyperplasia in the Mouse Uterus after Oestrogen Treatment: An Autoradiographic Study." *J. Endocrinol.* 56 (1): 133–44.

Martin, L., J. W. Pollard, and B. Fagg. 1976. "Oestriol, Oestradiol-17beta and the Proliferation and Death of Uterine Cells." *J. Endocrinol.* 69 (1): 103–15.

Moore, M., X. Wang, Y. F. Lu, M. Wormke, A. Craig, J. H. Gerlach, R. Burghardt, R. Barhoumi, and S. Safe. 1994. "Benzo[a]pyrene-Resistant MCF-7 Human Breast Cancer Cells: A Unique Aryl Hydrocarbon-Nonresponsive Clone." *J. Biol. Chem.* 269 (16): 11751–59.

Ohtake, F., K. Takeyama, T. Matsumoto, H. Kitagawa, Y. Yamamoto, K. Nohara, C. Tohyama, A. Krust, J. Mimura, P. Chambon, J. Yanagisawa, Y. Fujii-Kuriyama, and S. Kato. 2003. "Modulation of Oestrogen Receptor Signaling by Association with the Activated Dioxin Receptor." *Nature* 423 (6939): 545–50.

Paria, B. C., N. Das, S. K. Das, X. Zhao, K. N. Dileepan, and S. K. Dey. 1998. "Histidine Decarboxylase Gene in the Mouse Uterus Is Regulated by Progesterone and Correlates with Uterine Differentiation for Blastocyst Implantation." *Endocrinology* 139 (9): 3958–66.

Pearce, S. T., H. Liu, I. Radhakrishnan, M. Abdelrahim, S. Safe, and V. C. Jordan. 2004. "Interaction of the Aryl Hydrocarbon Receptor Ligand 6-Methyl-1,3,8-Trichlorodibenzofuran with Estrogen Receptor Alpha." *Cancer Res.* 64 (8): 2889–97.

Poland, A., and J. C. Knutson. 1982. "2,3,7,8-Tetrachlorodibenzo-p-Dioxin and Related Halogenated Aromatic Hydrocarbons: Examination of the Mechanism of Toxicity." *Annu. Rev. Pharmacol. Toxicol.* 22:517–54.

Resnitzky, D., L. Hengst, and S. I. Reed. 1995. "Cyclin A-Associated Kinase Activity Is Rate Limiting for Entrance into S Phase and Is Negatively Regulated in G1 by p27Kip1." *Mol. Cell. Biol.* 15 (8): 4347–52.

Ricci, M. S., D. G. Toscano, C. J. Mattingly, and W. A. Toscano Jr. 1999. "Estrogen Receptor Reduces CYP1A1 Induction in Cultured Human Endometrial Cells." *J. Biol. Chem.* 274 (6): 3430–38.

Romkes, M., and S. Safe. 1988. "Comparative Activities of 2,3,7,8-Tetrachlorodibenzo-p-Dioxin and Progesterone as Antiestrogens in the Female Rat Uterus." *Toxicol. Appl. Pharmacol.* 92 (3): 368–80.

Safe, S., and M. Wormke. 2003. "Inhibitory Aryl Hydrocarbon Receptor-Estrogen Receptor Alpha Cross-Talk and Mechanisms of Action." *Chem. Res. Toxicol.* 16 (7): 807–16.

Safe, S., B. Astroff, M. Harris, T. Zacharewski, R. Dickerson, M. Romkes, and M. Biegel. 1991. "2,3,7,8-Tetrachlorodibenzo-p-Dioxin (TCDD) and Related Compounds as Antioestrogens: Characterization and Mechanism of Action." *Pharmacol. Toxicol.* 69 (6): 400–409.

Sarkar, S., N. R. Jana, J. Yonemoto, C. Tohyama, and H. Sone. 2000. "Estrogen Enhances Induction of Cytochrome P-4501A1 by 2,3,7, 8-Tetrachlorodibenzo-p-Dioxin in Liver of Female Long-Evans Rats." *Int. J. Oncol.* 16 (1): 141–47.

Song, Z., and R. S. Pollenz. 2002. "Ligand-Dependent and Independent Modulation of Aryl Hydrocarbon Receptor Localization, Degradation, and Gene Regulation." *Mol. Pharmacol.* 62 (4): 806–16.

Takahashi, T., B. Eitzman, N. L. Bossert, D. Walmer, K. Sparrow, K. C. Flanders, J. McLachlan, and K. G. Nelson. 1994. "Transforming Growth Factors Beta 1, Beta 2, and Beta 3 Messenger RNA and Protein Expression in Mouse Uterus and Vagina during Estrogen-Induced Growth: A Comparison to Other Estrogen-Regulated Genes." *Cell Growth Differ.* 5 (9): 919–35.

Thompson, N. L., K. C. Flanders, J. M. Smith, L. R. Ellingsworth, A. B. Roberts, and M. B. Sporn. 1989. "Expression of Transforming Growth Factor-Beta 1 in Specific Cells and Tissues of Adult and Neonatal Mice." *J. Cell. Biol.* 108 (2): 661–69.

Tong, W., and J. W. Pollard. 1999. "Progesterone Inhibits Estrogen-Induced Cyclin D1 and cdk4 Nuclear Translocation, Cyclin E- and Cyclin A-cdk2 Kinase Activation, and Cell Proliferation in Uterine Epithelial Cells in Mice." *Mol. Cell. Biol.* 19 (3): 2251–64.

Umbreit, T. H., E. J. Hesse, G. J. Macdonald, and M. A. Gallo. 1988. "Effects of TCDD-Estradiol Interactions in Three Strains of Mice." *Toxicol. Lett.* 40 (1): 1–9.

Wang, W. L., J. S. Thomsen, W. Porter, M. Moore, and S. Safe. 1996. "Effect of Transient Expression of the Oestrogen Receptor on Constitutive and Inducible CYP1A1 in Hs578T Human Breast Cancer Cells." *Br. J. Cancer.* 73 (3): 316–22.

Wang, X., W. Porter, V. Krishnan, T. R. Narasimhan, and S. Safe. 1993. "Mechanism of 2,3,7,8-Tetrachlorodibenzo-p-Dioxin (TCDD)-Mediated Decrease of the Nuclear Estrogen Receptor in MCF-7 Human Breast Cancer Cells." *Mol. Cell Endocrinol.* 96 (1–2): 159–66.

Weber, T. J., R. S. Chapkin, L. A. Davidson, and K. S. Ramos. 1996. "Modulation of Protein Kinase C-Related Signal Transduction by 2,3,7,8-Tetrachlorodibenzo-p-Dioxin Exhibits Cell Cycle Dependence." *Arch. Biochem. Biophys.* 328 (2): 227–32.

Wormke, M., M. Stoner, B. Saville, K. Walker, M. Abdelrahim, R. Burghardt, and S. Safe. 2003. "The Aryl Hydrocarbon Receptor Mediates Degradation of Estrogen Receptor Alpha through Activation of Proteasomes." *Mol. Cell Biol.* 23 (6): 1843–55.

15

Polychlorinated Biphenyls as Disruptors of Thyroid Hormone Action

Kelly J. Gauger, *University of Massachusetts, Amherst*
David S. Sharlin, *University of Massachusetts, Amherst*
R. Thomas Zoeller, *University of Massachusetts, Amherst*

Polychlorinated biphenyls (PCBs) are ubiquitous environmental contaminants routinely found in human and animal tissues. Developmental exposure to PCBs is associated with neuropsychological deficits, which may be related to effects on thyroid hormone (TH) signaling in the developing brain. However, PCBs may interfere with TH signaling solely by reducing circulating levels of thyroid hormone or they may exert direct effects on thyroid hormone receptors (TRS). In this chapter, we review the evidence that PCB exposure interferes with TH signaling and the potential consequences of this action.

INTRODUCTION

Polychlorinated biphenyls are neurotoxins, as evidenced by a number of epidemiological studies that have indicated that children developmentally exposed to PCBs exhibit a variety of measurable cognitive deficits (Ayotte et al. 2003; Huisman et al. 1995; Jackson et al. 1997; Korrick and Altshul 1998; Osius et al. 1999; Walkowiak et al. 2001). These studies show that incidental exposure to PCBs is associated with shorter gestation, lower birth weight, and smaller head circumference (Fein et al. 1984), as well as deficits in gross motor performance and visual recognition memory (Jacobson et al. 1985; Longnecker, Rogan, and Lucier 1997; Rogan and Gladen 1992). The specific neuropsychological domains affected by developmental exposure to PCBs overlap with those affected by maternal thyroid hormone (TH) insufficiency, including lower IQ, visual memory deficits, and motor function and attention deficits (Haddow et al. 1999; Morreale de Escobar, Obregon, and Escobar del Rey 2000; Pop et al. 1999).

Considering these observations, several investigators have hypothesized that PCBs may impact brain development by interfering with TH signaling (Colborn 2004; McKinney and Waller 1998; Porterfield 2000; Porterfield and Hendry 1998). This hypothesis has been explored in the human population by testing whether PCB body burden is associated with measures of thyroid function; several major studies have reported that the concentration of PCBs, or of specific PCB congeners, in maternal and/or cord blood is associated with lower TH levels in both the mother and infant (Koopman-Esseboom et al. 1994; Schantz, Widholm, and Rice 2003). Moreover, studies using experimental animals have consistently shown that PCBs cause a decrease in circulating levels of thyroxine (T_4) (Bastomsky 1974; Bastomsky, Murthy, and Banovac 1976; Brouwer et al. 1998).

Thus the current prevailing presumption is that PCBs produce adverse effects on the developing nervous system, at least in part by producing a state of TH insufficiency. However, several observations are not consistent with this presumption. Several epidemiological studies have failed to identify an association between various measures of thyroid function and PCB body burden (Hagmar et al. 2001; Longnecker et al. 2000; Matsuura and Konishi 1990; Sala et al. 2001; Steuerwald et al. 2000), and not all effects of PCBs in experimental animals are consistent with the effects of experimentally produced hypothyroidism (Goldey et al. 1995a; Hood and Klaassen 2000a; Kolaja and Klaassen 1998; Zoeller, Dowling, and Vas 2000). In fact, PCBs can exert a TH-like effect on the expression of several TH-responsive genes in the developing brain despite a simultaneous and severe reduction in serum TH levels (Bansal et al. 2003; Gauger et al. 2004; Zoeller, Dowling, and Vas 2000).

Therefore, the hypothesis that PCB exposure causes cognitive deficits by producing hypothyroidism does not account for all of these observations. The goal of this chapter is to describe the complexities of TH signaling during brain development, to argue that the effects of PCBs on neurodevelopment cannot be fully explained by a reduction in circulating levels of TH, and to propose a unifying hypothesis to account for the apparent paradox described above.

REGULATION OF THE THYROID HORMONE SYSTEM

The hypothesis that PCBs reduce circulating levels of thyroid hormones, thereby producing adverse effects on neurodevelopment, requires an understanding of both the complexities of the regulation of TH signaling and the effect of TH insufficiency on the developing brain. In the section below, we present a definition of "the regulation of TH signaling" that includes both the regulation of serum hormone levels and the regulation of TH action at the receptor.

Regulating Hormone Levels in Blood

Hormone levels in serum represent a balance between the rate of new hormone secretion from its glandular source and the removal of hormone from circulation. Thyroid hormone secretion is controlled by the pituitary gland, which in turn is controlled by the hypothalamus and by the negative feedback action of TH. Thus the hypothalamic-pituitary-thyroid (HPT) axis is a classic neuroendocrine system, the activity of which will maintain TH levels within narrow limits (Andersen et al. 2002). The role of the hypothalamus in the regulation of the HPT axis is mediated largely by thyrotropin-releasing hormone (TRH), a tripeptide released from nerve terminals in the median eminence whose cell bodies reside in the hypothalamic paraventricular nucleus (PVN). TRH promotes the synthesis and secretion of the pituitary glycoprotein hormone, thyrotropin (TSH) (Wondisford, Magner, and Weintraub 1996), which then regulates the activity of the thyroid gland including the synthesis and secretion of THs (Rapoport and Spaulding 1986). The hormones secreted from the thyroid gland, thyroxine (T_4) and triiodothyronine (T_3), are synthesized by coupling two iodinated tyrosyl residues that make up the larger hormone precursor, thyroglobulin (Tg).

Thus TSH primarily controls the rate of TH secretion from the thyroid gland. However, the regulation of TH clearance is more complicated. Once T_4 and T_3 are released into the bloodstream, they become bound to thyroxine-binding proteins including thyroxine-binding globulin (TBG) and transthyretin (TTR) (Schussler 2000). Serum binding proteins likely have a large impact on total serum hormone

level both within and between species (Palha 2002). However, "free" hormone, that fraction of total hormone that is not bound to protein, is very similar in different species that exhibit differences in total hormone level (Refetoff, Robin, and Fang 1970; Stockigt 2000). In addition, genetic defects in TBG or TTR in humans lead to altered levels of total hormone, but there are no associated symptoms of hypo- or hyperthyroidism (Refetoff, Dwulet, and Benson 1986; Rosen et al. 1993). Likewise, rodents (rat and mouse) have both TBG and TTR, though TBG exhibits a developmental profile. Specifically, TBG is two to three times higher in mouse fetuses than in dams, then further increases after birth, reaching maximum values between three and five days that are seven to eight times higher than the adult (Vranckx et al. 1990). This pattern is not correlated with the ontogenesis of TTR. In a follow-up study in rats, this group found that the mRNA encoding rat TBG in liver, cloned by Tani et al. (Tani et al. 1994), exhibits a similar developmental pattern. Both serum TBG and hepatic TBG mRNA are nearly undetectable at eight weeks of age following a transient rise after birth. Interestingly, TBG expression was induced by thyroidectomy in the eight-week-old male rat and T_3 replacement suppressed it.

These observations indicate that serum binding proteins in humans and in rodents play a major role in determining total TH levels in the blood but that the free fraction is an important correlate of hormone action in tissues. In addition, free hormone appears to be more closely associated with tissue levels. The practical implication is that total T_4 will be a valid measure of thyroid function if and only if serum thyroxine binding proteins are equivalent in the experimental groups being compared. This is a fact that is seldom considered in toxicological studies. However, there appear to be tissue-specific mechanisms that contribute to the control of tissue levels of TH and control TH action; thus the control of serum total and free T_4 may not be the only mechanism by which thyroid hormone action in tissues is controlled.

For example, once in the blood, thyroid hormones are selectively taken up into tissues where they act. Thyroid hormones are lipophilic and are generally thought to diffuse passively across the plasma and nuclear membranes. However, there is some evidence for facilitated transport across plasma membranes and high-affinity TH binding sites in the plasma membranes of different cells (Ekins et al. 1994; Friesema et al. 1999; Moreau, Lejeune, and Jeanningros 1999). There is also evidence for a stereo-specific transporter of T_3 into the nucleus considering that L-T_3 is concentrated in the nucleus at a 58-fold higher concentration than D-T_3 (Nishii et al. 1993). One potential transporter may be the multidrug resistance P glycoprotein that can modulate TH concentration when overexpressed in cells (Neves et al. 2002).

Another family of transporters appears to be the organic anion transporter proteins that have been shown to import TH into hepatocytes (Friesema et al. 1999). Finally, recent studies show that severe neurological symptoms are associated with a defect in the gene coding for a T_3 transporter, MCT8 (Dumitrescu et al. 2004; Friesema et al. 2004). If the regulation of TH uptake into specific tissues or cells within tissues is an important point of physiological regulation, then chemicals that interfere with this uptake may produce tissue-specific thyroid disease which would be difficult to identify.

Once in the tissues, iodothyronine deiodinases catalyze the removal of an iodine atom in the 5′ position of the T_4 molecule (outer ring deiodinases, D1 and D2), resulting in the generation of T_3 (St. Germain and Galton 1997). D1 is largely responsible for maintaining serum levels of T_3; it has been estimated that 80 percent of T_3 is derived from peripheral deiodination of T_4 (Chopra and Sabatino 2000). Therefore, in addition to controlling the *de novo* production of T_3 in tissues, deiodinases contribute to the majority of T_3 to the circulation, and in both cases represent an important regulator of TH signaling that may be disrupted by environmental chemicals.

The regulation of TH clearance from blood is controlled largely by metabolism in the liver; largely by glucuronidation and sulfation of the phenolic hydroxyl group, and by oxidative deamination of the alanine side chain by UDP-glucuronyltransferases and sulfotransferases, respectively (Siegrist-Kaiser and Burger 1994). Iodothyronine glucuronides are rapidly excreted in the bile, as glucuronidation is an important reaction for the metabolic clearance of T_3 and T_4. In contrast, iodothyronine sulfates rarely appear in the bile because they are rapidly deiodinated in the liver, a mechanism that leads to the irreversible inactivation of TH (Otten et al. 1983; Visser et al. 1984).

Thus the regulation of circulating levels of TH levels within a "normal" range is accomplished by a variety of mechanisms, including the negative feedback system within the HPT axis, by the uptake and deiodination in tissues, and by TH clearance by the liver. It is likely that these mechanisms interact both to maintain blood concentrations of hormone within a limited range, and to protect tissues from small fluctuations in serum TH.

MECHANISMS OF THYROID HORMONE ACTION

Once thyroid hormone enters the cell, the majority of known biological actions of TH are mediated by their receptors (TRs)—nuclear proteins that interact mainly with T_3 (Hu and Lazar 2000; Wu, Xu, and Koenig 2001). But the availability of hormone to the TR—by changes in serum hormone levels, by differential uptake into the cells, or by differences in deiodination—are not the only mechanisms by which TH signaling is regulated. Different TR isoforms likely control regional and temporal differences in TH action. These actions are likely modified further by the kinds of accessory proteins that interact with TRs. These accessory proteins include cofactors that mediate repression or activation functions of the TRs, and dimerization partners that are believed to play a role in targeting specific genes for regulation (Yen 2001). Moreover, these actions may be conditioned by the specific TH response element itself.

Thyroid Hormone Receptor Subtypes

TRs are members of the steroid/thyroid superfamily of ligand-dependent transcription factors (Lazar 1993, 1994; Mangelsdorf and Evans 1995) and are encoded by two genes, designated alpha- and beta- c-*erbA* (Sap et al. 1986; Weinberger et al. 1986). These two genes produce at least four functional TRs: TRa1, TRb1, TRb2, and TRb3 (Hodin et al. 1989; Izumo and Mahdavi 1988; Koenig et al. 1988; Murray et al. 1988; Thompson et al. 1987). Although the binding affinity for T_3 and for T_4 is not different among the various TR isoforms (Oppenheimer 1983; Schwartz et al. 1992), the TRs exhibit a 50-fold greater affinity for T_3 than for T_4, making T_3 the physiologically important regulator of TR action.

The ability of TRs to affect gene transcription requires them to interact with nuclear cofactors, which are requisite mediators of ligand-dependent transcriptional activation or repression of hormone-responsive genes (Glass and Rosenfeld 2000; Hermanson, Glass, and Rosenfeld 2002; Mckenna and O'Malley 2002; Rosenfeld and Glass 2001). There are two categories of nuclear receptor cofactors: corepressors and coactivators (Glass and Rosenfeld 2000; Leo and Chen 2000). In the absence of TH, TRs are believed to bind to regulatory regions of genes and repress basal transcription by recruiting corepressors such as Silencing Mediator of Retinoic acid and Thyroid hormone receptor (SMRT) or Nuclear Co-Repressor (NcoR) (Horlein et al. 1995; Koenig 1998). When T_3 binds to the receptor, the TR changes shape sufficiently to cause the release of the bound corepressor and recruit a coactivator complex that can include steroid receptor coactivator 1 (SRC-1) (Koenig 1998; Onate et al. 1995). Cofactors remodel local chromatin structure enabling nuclear receptors to activate or repress gene transcription. The specific recruitment of a cofactor complex with histone acetyltransferase activity, such as SRC-1, appears to play a regulatory role in activating gene transcription, whereas the recruitment of a cofactor complex with histone deacetylase activity, such as NcoR and SMRT, appears to play a regulatory role in gene repression (Struhl 1998).

These molecular studies make it clear that the regulation of hormone delivery to a cell's nucleus is only one part of the regulation of TH signaling. The specific effect of TH within a cell will clearly depend on the cell type and the developmental stage of the cell, and these pleiotropic effects of TH are controlled by the variety of proteins associating with the TRs to mediate their function. Any of these points of regulation may be targets of PCB action, but these possibilities have not presently been investigated.

THYROID HORMONE AND BRAIN DEVELOPMENT

If developmental exposure to PCBs can affect the course of brain development by causing a reduction in circulating levels of thyroid hormone, then it is important to identify the effects of TH insufficiency. We know that TH is essential for brain development during the neonatal period in humans, especially as revealed in the disorder known as congenital hypothyroidism (CH) (Calvo et al. 1990; Delange 1997; Dussault and Ruel 1987; Fisher et al. 1979; Foley 1996; Kooistra et al. 1994; Krude et al. 1977; Miculan, Turner, and Paes 1993; Rovet 2000; van Vliet 1999). CH is a disorder of newborns that affects about 1 in 3,500 newborns and was once a leading cause of mental retardation in the United States. However, since the advent of newborn screening programs, children are now being diagnosed and treated early in infancy, before the appearance of associated symptomatology. As a consequence, mental retardation due to CH has been virtually eradicated (Klein 1980; Klein and Mitchell 1996). Nevertheless, affected children still experience reduced IQ levels by about 6 points on average (Derksen-Lubsen and Verkerk 1996) as well as mild to moderate impairments in visuospatial, motor, language, memory, and attention abilities (Fuggle et al. 1991; Gottschalk, Richman, and Lewandowski 1994; Kooistra et al. 1994; Kooistra et al. 1996; Rovet, Ehrlich, and Sorbara 1992; Rovet et al. 1996). About 20 percent of cases also have a mild sensorineural hearing loss (François et al. 1993; Rovet et al. 1996), which contributes to difficulties in initially learning to read (Rovet et al. 1996). Among children with CH, there exists a wide degree of variability in the severity of their symptoms, which reflects factors associated with the severity of their disease and its management (LaFranchi 1999).

Experimental studies have revealed some of the developmental effects of TH on brain development. These include effects on neuronal proliferation, differentiation, migration and synaptogenesis (Dussault and Ruel 1987; Koibuchi and Chin 2000; Muñoz and Bernal 1997; Thompson 1999; Thompson and Potter 2000). Much of this work has focused on the postnatal development of the cerebellum. However, recent work is beginning to focus on the effects of TH insufficiency on brain development during fetal development. These studies include identification of TH-responsive genes in the fetal cortex (Bansal et al. 2003; Dowling et al. 2001; Dowling et al. 2000; Dowling and Zoeller 2000), and quantification of effects of mild TH insufficiency in neuronal migration in the early cortex (Auso et al. 2004; Lavado-Autric et al. 2003). These findings will lay the groundwork for studies focused on evaluating effects of PCBs on the fetal brain.

DEVELOPMENTAL EXPOSURE TO PCBS PRODUCES COGNITIVE DEFICITS

Humans

Several independent epidemiological studies, based in a number of countries around the world, have evaluated the neuropsychological impacts of inadvertent PCB exposure (Schantz, Widholm, and Rice 2003). Many of these studies have found an association between PCB body burden and measures of intrauterine growth, including lower birth weight and smaller head circumference (Fein et al. 1984). Studies in Michigan (Jacobson and Jacobson 1996; Jacobson, Jacobson, and Humphrey 1990a; Jacobson, Jacobson, and Humphrey 1990b; Jacobson et al. 1985), New York (Darvill et al. 2000), the Netherlands (Patandin et al. 1997), Germany (Walkowiak et al. 2001; Winneke et al. 1998), and the Faroe Islands (Grandjean et al. 2001) have all identified a negative association between PCB exposure and various aspects of neurocognitive function in children. Interestingly, comparing these studies reveals that the neuropsychological domains associated with PCB exposure appear to differ at the age of testing (Schantz, Widholm, and Rice 2003). This is important because the same statement can be said for the association between serum T_4 levels during development and the neuropsychological domains that reveal the greatest sensitivity to TH insufficiency (Oerbeck et al. 2003; Zoeller and Rovet 2004).

Descriptions of the health effects of accidental PCB exposure of humans also support the concept that these chemicals disrupt neurocognitive development. The first accidental human population exposure occurred in Japan in 1968 resulting in Yusho (oil disease), so called because commercially distributed cooking oil had become contaminated with PCBs and their thermal degradation products. Adults exposed to the resulting high levels of PCBs exhibited epidermal abnormalities, behavioral deficits and hypothyroxinemia (Kashimoto et al. 1981). Children born to mothers who consumed this oil were exposed to PCBs through the placenta and by breast feeding (Masuda et al. 1978; Nishimura et al. 1977) and exhibited a number of physical and behavioral deficits, including apathy, inactivity, hypothyroidism, and generally lower IQ scores. A similar incident occurred in Taiwan 10 years later, resulting in Yu-Cheng. Follow-up studies showed that children born up to 12 years after their mothers' exposure exhibited delays in several neurological measures of development (Guo et al. 1994; Rogan et al. 1988; Yu et al. 1991).

Although these unfortunate accidents provide support that high-dose exposure to PCBs may affect TH signaling and brain development, studies of the incidental exposure to PCBs through a contaminated environment also provide important support for this hypothesis. For example, lower IQ, visual memory deficits, motor function and attention deficits are symptoms of TH insufficiency (Haddow et al. 1999; Morreale de Escobar, Obregon, and Escobar del Rey 2000; Pop et al. 1999) that overlap with effects of incidental PCB exposure described above. Moreover, the concentrations of PCBs, or of specific PCB congeners, in maternal and cord blood have been associated in some studies with lower TH levels in both the mother and infant (Koopman-Esseboom et al. 1994; Schantz, Widholm, and Rice 2003). These PCB levels are also negatively correlated with birth size and early growth rate (Patandin et al. 1998) as well as with various neurological measures (Koopman-Esseboom et al. 1997). This is important because it demonstrates that incidental exposure of pregnant women to PCBs may reduce levels of TH, which may impact fetal development.

Animals

PCB research in animals clearly supports the hypothesis that these environmental contaminants may interfere with TH signaling during brain development, and may further explain why simple associations between PCB body burden and point-estimates of thyroid function are not always correlated. Developmental exposure to PCBs also produces deficits in brain development in laboratory animals. For example, perinatal exposure to PCBs diminishes muscarinic receptor binding in the brain (Eriksson 1988) as well as choline acetyltransferase activity in the cerebral cortex (Ku et al. 1994). Goldey et al. demonstrated that developmental exposure to PCBs produced deficits in hearing (Goldey et al. 1995a) that are similar to those affected by hypothyroidism (Goldey and Crofton 1998). Several studies have evaluated the effects of PCB exposure on various behaviors in rats (Ku et al. 1994; Schantz et al. 1990; Seo et al. 1995; Weinand-Harer et al. 1997), and although many of these behavior disturbances are similar to those produced by perinatal hypothyroidism, most of these reports were not designed

to provide information about the mechanism by which the behavioral deficits were produced.

Ulbrich and Stahlmann (Ulbrich and Stahlmann 2004) recently published an excellent review of the literature providing information about the effects of PCBs on neurodevelopment performed in experimental animals. The reader is referred to this review for a comprehensive review; in this chapter, we will focus on a brief summary of the observed effects of PCB exposure in experimental animals. First, experimental studies consistently find that developmental exposure to a mixture of PCBs, Aroclor 1254, decreases circulating levels of T_4 in rats (Bastomsky 1974; Bastomsky, Murthy, and Banovac 1976; Brouwer et al. 1998). Moreover, congener-specific studies demonstrate that both coplanar and noncoplanar PCB congeners can reduce circulating levels of TH. For example, coplanar PCB congeners (PCBs 28, 77, 126 or dioxin) cause a significant reduction in serum total T_4 at weaning when the animals were exposed prenatally (Ness et al. 1993; Seo et al. 1995). Likewise, noncoplanar PCBs such as PCB 153 also cause a reduction in serum total T_4 (Hussain et al. 2000; Ness et al. 1993). There also is evidence that PCB exposure can reduce tissue levels of thyroid hormone, in addition to effects on serum total hormone. For example, maternal exposure to 4-OH-CB107, one of the major PCB metabolites detected in human blood, reduces TH levels in fetal rat brain (Meerts et al. 2002).

The mechanisms by which PCBs can cause a reduction in serum TH levels have been the focus of intense investigation. At least three independent, potentially interacting, mechanisms may account for the ability of PCBs to reduce circulating levels of TH (Brouwer et al. 1998). These include direct effects on thyroid function (Collins and Capen 1980; Collins et al. 1977; Kasza et al. 1978), the ability of some PCB congeners to displace T_4 from TTR (Brouwer et al. 1998; Chauhan et al. 2000), and the induction of UDP-glucuronosyltransferases (Hood and Klaassen 2000b; Visser et al. 1993) leading to increased biliary secretion of T_4 (Bastomsky, Murthy, and Banovac 1976). Debate remains concerning the relative importance of these mechanisms in the PCB-induced reduction in serum total T_4 (Kato et al. 2004); however, establishing whether the observed reduction in circulating total T_4 is related to adverse effects on brain development is more important to the central theme of this chapter.

ARE THE EFFECTS OF PCBS ON NEURODEVELOPMENT CAUSED BY REDUCING TH LEVELS?

Despite the finding that PCBs uniformly reduce circulating levels of TH in experimental animals, the subsequent effects are not fully consistent with experimentally produced hypothyroidism using goitrogens such as propylthiouracil (PTU). For example, developmental exposure to PCBs in experimental animals induces a hearing loss (Crofton et al. 2000a; Crofton et al. 2000b; Goldey et al. 1995b), a reduction in choline acetyltransferase in the cerebral cortex (Ku et al. 1994), and an increase in testicular growth, all consistent to some degree with effects produced by PTU. However, T_4 replacement only partially ameliorates these effects (Goldey and Crofton 1998; Ku et al. 1994). In addition, developmental hypothyroidism induced by PTU causes a significant increase in serum TSH levels (Connors and Hedge 1981), reduced body and brain weight and brain size of rat pups (Schwartz, 1983), and a delay in eye opening and tooth eruption (Varma et al. 1978). In contrast, PCB exposure at doses that lower serum TH do not always produce these effects (Goldey et al. 1995b; Hood and Klaassen 2000b; Kolaja and Klaassen 1998; Zoeller, Dowling, and Vas 2000). Therefore, there is a significant discrepancy between the ability of PCBs to reduce circulating levels of TH and their ability to produce symptoms of hypothyroidism.

It is difficult to demonstrate that PCBs affect brain development by interfering with TH action if the only measures obtained are circulating levels of the hormone. Rather, to determine whether PCBs affect TH action, it is essential to evaluate the effects of PCBs on direct measures of TH action, such as on specific TH-responsive genes or on TH-responsive developmental events.

Considering that the effect of PCBs on circulating levels of TH does not accurately reflect their effect on symptoms of hypothyroidism, our laboratory tested the hypothesis that PCBs interfere with TH action in the developing rodent brain. Initially, we evaluated the effect of PCB exposure (Aroclor 1254) on circulating levels of TH and on the expression of TH-responsive genes in the developing brain (Zoeller, Dowling, and Vas 2000). Despite reducing circulating T_4 to below detectable levels in the pups, PCB exposure increased the expression of mRNAs encoding RC3/Neurogranin and MBP in the brain. Three features of these findings indicate that PCBs directly alter RC3/Neurogranin gene expression. First, TH regulates RC3/Neurogranin expression on postnatal day 15, but not on postnatal days 5 and 30 (Ibarrola and Rodriguez-Pena 1997) and we found the same temporal change in sensitivity to PCBs. Second, RC3 expression is not sensitive to TH regulation in all brain areas (Guadano-Ferraz et al. 1997), and we found that PCBs affected RC3/Neurogranin expression only in those brain areas known to be affected by TH. Finally, RC3/Neurogranin expression is transcriptionally regulated by TH (Arrieta et al. 1999), and we demonstrated that cellular levels of RC3/Neurogranin mRNA were elevated by PCB exposure, consistent with a transcriptional effect. Thus PCB exposure can have TH-like effects in the developing brain independent of its effects on serum TH.

We have also catalogued several additional examples of PCBs exerting TH-like effects in the developing rat brain. An important observation is the induction of oligodendrocyte differentiation by PCB exposure. Although we have shown that PCB exposure increases MBP expression in the postnatal rat brain (Zoeller, Dowling, and Vas 2000), we were aware that our results could be explained by an increase in the expression of the gene coding for MBP or by an increase in the number of oligodendrocytes expressing MBP. We employed myelin-associated glycoprotein (MAG) as a probe to discriminate between these two possibilities. We found that perinatal exposure to PCBs increased the number of oligodendrocytes in the internal capsule (manuscript in preparation). Considering that previous work has demonstrated that TH can induce oligodendrocyte differentiation *in vitro* (Johe et al. 1996) and *in vivo* (Baas et al. 1994; Baas et al. 1997; Bury et al. 2002), these results suggest that PCBs alter oligodendrocyte development in a manner that is consistent with the known effects of TH.

Interestingly, the numbers of oligodendrocytes in the internal capsule decline on P30. This seemingly contradictory finding may result from TH-like effects on early fate specification of neural progenitors in the fetal rat brain. Specifically, PCBs may enhance gliogenesis at the expense of neurogenesis, leaving the animal with fewer neurons in the adult. Moreover, oligodendrocytes that fail to wrap an axon undergo apoptosis (Calver et al. 1998) because they do not receive the trophic support required for survival (Fruttiger, Calver, and Richardson 2000; Fruttiger et al. 1999; van Heyningen et al. 2001). Thus PCBs mimicking TH early in development may decrease neuronal number, while PCBs mimicking TH later in development increase oligodendrocyte number. However, ultimately, oligodendrocyte number will reflect neuronal number due to programmed cell death resulting from their inability to wrap available axons.

In addition, we have found that A1254 significantly reduced circulating levels of T_3 and T_4 in pregnant rats, but fetuses derived from A1254-treated dams exhibited a significant increase in RC3/Neurogranin and Oct-1 expression in the brain (Gauger et al. 2004). These data provide strong evidence that PCBs can produce TH-like effects in the fetal brain because maternal TH increases the expression of these genes in the fetal brain (Dowling et al., 2000; Dowling and Zoeller 2000).

The simplest explanation for our findings is that individual or classes of PCB congeners in A1254, or classes of congeners, can directly activate TRs either as parent congeners or following either hydroxylation or methylation. To test this, we employed a mixture of 6 PCB congeners based on their *ortho* substitution pattern, including PCBs 77, 126 (non-*ortho*), PCBs 105 and 118 (mono-*ortho*), and PCBs 138 and 153 (di-*ortho*) to identify potential TR agonists. This mixture, prepared in ratios present in Aroclor 1254, significantly reduced serum TH levels in pregnant rats on gestational day (G) 16 but simultaneously upregulated the expression of a well-known TH-regulated gene, malic enzyme (ME), in the maternal liver. This mixture also activated a TH response element (TRE) in a transient transfection system using rat pituitary GH3 cells, indicating that at least one of these congeners was activating the TR. However, none of the individual PCB congeners comprising this mixture were active in this system. Rather, we found that a minimal mixture of PCB 126, 105, and 118 was required to activate Luciferase from a TRE. Using the AhR antagonist a-naphthoflavone, or the CYP1A1 antagonist ellipticine, we showed that the effect of the mixture on the TRE required AhR and CYP1A1. Thus we propose that PCB 126 induces CYP1A1 through the AhR, and that CYP1A1 activates PCB 105 and/or 118 to a form that acts as a TR agonist. These data suggest that some tissues may be especially vulnerable to PCBs interfering directly with TH signaling due to their capacity to mount a CYP1A1 response to coplanar PCBs (or other dioxin-like molecules) if sufficient mono-*ortho* PCBs are present. This may account for why the effect of PCB exposure is different in white matter compared to the regions of neurons we had previously investigated (e.g., hippocampus).

Therefore, we tested a number of PCB congeners and PCB metabolites for their ability to bind to TRs using a well-established binding assay (DeGroot and Torresani 1975). We found that neither the parent PCB congeners nor the hydroxylated or methylsulfonyl metabolites could significantly displace T_3 from rat hepatic nuclei (Gauger et al. 2004). It is not likely that these observations are false-negatives because the observed K_i for several control ligands including T_3, T_4, Tetrac, and Triac were all within the published range (Goslings et al. 1976; Ichikawa and DeGroot 1987; Weinberger et al. 1986). Thus, our findings indicate that neither individual PCB congeners nor their metabolites interact with the TR in a competitive manner.

ALTERNATE MECHANISMS BY WHICH PCBS MAY AFFECT TH SIGNALING

If PCBs can produce effects on the expression of TH-responsive genes in the developing brain but not bind to the ligand-binding site of the TR, there must be an alternate mechanism by which PCBs exert this effect. We propose that there are two fundamental mechanisms that could account

for these observations. First, PCBs may affect TR activity, either directly or indirectly, in response to T_3. Second, PCBs may increase the availability of T_3 to the receptor. In both cases, it is possible that these effects could be tissue- or developmental stage–specific. The following discussion illustrates these concepts.

Novel Effects of PCBs on TR Activity

Several recent reports have begun to characterize the effects of PCBs on TR action. Iwasaki et al. (Iwasaki et al., 2002) showed that the minor metabolite 4′-OH-PCB106 suppressed T_3-induced activation of TRb1 in several cell lines. They employed a canonical TRE for this work (DR4), which is highly responsive to T_3 in the presence of SRC-1, and found that 10^{-10} M 4′-OH-PCB106 significantly reduced the ability of T_3 to activate the DR4 by acting on the TRb1. This appeared to be specific to the TR because it did not suppress glucocorticoid receptor-mediated transactivation. Although this study supports the concept that PCBs can exert a direct action on the TRb1, it does not help us explain the ability of PCBs to exert TH-like effects.

A further study by Yamada-Okabe et al. (2004) reported similar effects of the same hydroxylated PCB used by Iwasaki et al. described above on the ability of T_3 to induce the activity of a DR4 promoter. However, they also examined the effect of 4′-OH-PCB106 on the expression of a number of genes expressed in HeLa cells that were stably transfected with the human TRb1. Interestingly, they found that 4′-OH-PCB106 could suppress the T_3-activation of some genes (fmfc, PSG7, and TRAF1), augmented the T_3-activation of others (BMP-6), and had no effect on still others (4-1BB, PSCA, RANTES, TRAF2). Thus, in a single cell line (HeLa), a single hydroxylated PCB congener has different effects on T_3-mediated gene expression depending on the target gene.

Bogazzi et al. also reported that the mixture Aroclor 1254 can suppress the ability of T_3 to activate a canonical malic enzyme TRE in COS-7 cells (Bogazzi et al. 2003). Interestingly, they also observed that Aroclor could displace T_3 binding to the TR. However, Aroclor altered the sensitivity of the TR to protease digestion, indicating that they interacted directly with the TR, changing its shape. Finally, Miyazaki et al. (Miyazaki et al. 2004) report that 4′-OH-PCB106 can cause TR:RXR heterodimers to partially dissociate from the TRE using electrophoretic mobility shift assays.

These studies provide strong evidence that PCBs can exert a direct effect on the TR (or TRs). However, there is no clear picture emerging about the mechanism(s) by which PCBs can influence T_3 signaling through the TR or the consequences of these effects on TR function. Generally, PCBs appear to uniformly inhibit T_3-mediated gene expression in assays employing canonical TREs. In contrast, the effect of PCB exposure on native genes (in HeLa cells) depends on the gene in question and can include inhibitory and stimulatory effects on T_3-mediated activation. Finally, the mixture of Aroclor 1254 appears to affect T_3-mediated gene transcription in a manner that does not affect its ability to interact with DNA, but 4'-OH-PCB106 appears to cause the TR to dissociate from the canonical TRE. A number of possibilities exist to reconcile these apparent differences including differences in the regulation of TR-mediated gene expression in episomes compared to native genes in the nucleus. However, formal studies need to provide some empirical clarification before speculation is useful.

The nuclear cofactors involved in TH-responsive gene expression remodel local chromatin structure, which results in the activation or repression of transcription. If PCBs interfere with these specific regulatory processes, they may alter gene expression in the absence of TR binding. Specifically, PCBs may increase histone acetyltransferase activity, an intrinsic property of the nuclear cofactor SRC-1, which would upregulate gene transcription. Alternatively, the nuclear corepressors, NcoR and SMRT, recruit histone deactylases and if PCBs repress histone deacetylase activity, transcriptional repression would be alleviated, causing an increase in TH-mediated gene expression. Recently, the activation of gene expression by a xenobiotic via the mechanism described above has been described. Jansen et al. (Jansen et al. 2004) demonstrated that methoxyacetic acid (MAA) increases ER-mediated gene expression but does not compete with 17-b-estradiol for ER binding. Interestingly, MAA can inhibit endogenous histone deacetylase activity and increase histone acetylation *in vitro* and *in vivo*. Therefore, it is plausible, though speculative, that PCBs increase TR-mediated gene expression in a similar manner.

Several lines of evidence implicate TR phosphorylation in the regulation of TH-responsive gene expression (Chen et al. 2003; Goldberg et al. 1988; Jones et al. 1994; Li et al. 2002; Lin et al. 1992). Moreover, phosphorylation of SMRT by mitogen-activated protein kinase (MAPK) inhibits the ability of this nuclear coregulator to mediate transcriptional repression (Hong and Privalsky 2000; Hong et al. 1998). Thus if PCBs interfere with phosphorylation signal transduction cascades in such a way that increases the activity of protein kinases, there would be an increase in TH-mediated gene expression either by direct phosphorylation of the TR or by phosphorylation induced dissociation of SMRT from the TR. Interestingly, it has been shown that PCBs can activate MAPKs *in vitro* (Fischer, Wagner, and Madhukar 1999; Olivero and Ganey 2000). Taken together, the PCB induced increase in TH-responsive gene expression that we observe *in vivo* may be a result of either TR or SMRT phosphorylation.

Effects of PCBs on T_3 Availability

Several reports describe effects of PCBs on TH metabolism by tissue deiodinases, thereby potentially changing the amount of hormone available to the TR. PCB exposure can increase type 2 deiodinase activity in the adult (Hood and Klaassen 2000b) and fetal (Meerts et al., 2002; Morse et al., 1996) rat brain. These changes in deiodinase expression and activity may represent an important tissue-autonomous mechanism controlling the sensitivity of target tissues to TH fluctuations (Kohrle 1999), especially the developing brain (Kester et al. 2004), and could protect tissues from small fluctuations in serum TH. Consequently, the PCB induced reductions in serum TH levels are likely to have direct effects on the expression and activity of tissue specific deiodinases. If PCB exposure increases D2 activity by reducing serum T_4, a state of overcompensation may occur, which could alter the cellular concentration of TH and perhaps generate more T_3 to upregulate TH-responsive gene expression.

Recently, a nicotinamide adenine dinucleotide phosphate (NADPH)-dependent cytosolic T_3-binding protein (CTBP) has been implicated in the intracellular storage and translocation of T_3 *in vitro* (Hashizume et al. 1989a; Hashizume et al. 1989b; Hashizume et al. 1989c; Hashizume et al. 1989d; Hashizume et al. 1991). Furthermore, CTBP increases cytoplasmic as well as nuclear content of T_3 but paradoxically suppresses T_3-responsive gene expression in CTBP-expressing stable cell lines (Mori et al. 2002). These data indicate that the expression of CTBP is likely to play a fundamental role in stabilizing T_3-mediated transcription when extracellular T_3 concentration is altered either physiologically or pathologically. Therefore, if PCBs alter the expression and/or activity of CTBP, it is possible that the intracellular distribution of T_3 as well as the expression of TH-responsive genes will be altered.

FUTURE DIRECTIONS

Ultimately, it will be essential to identify how specific PCB congeners, or classes of congeners, alter thyroid hormone action because the toxic effects of PCBs depend on their degree of chlorination and pattern of chlorine substitution. Non-*ortho* substituted PCBs can adopt a planar structure similar to that of dioxin and can bind to and activate the aryl hydrocarbon receptor (AhR) (Tilson and Kodavanti 1997). In contrast, *ortho*-substituted PCBs adopt a non-coplanar conformation that does not act through the AhR, but nevertheless produce neurotoxic effects (Fisher 1998; Seegal and Shain 1992). Moreover, dioxin-like and non-dioxin-like PCBs have differential effects on thyroid hormone homeostasis (Desaulniers et al. 1999; Khan and Hansen

2003; Khan et al. 2002; Seo et al. 1995). However, the pattern and duration of PCB exposure and the timing of T_4 measurements, both relative to PCB exposure and timing during development, are quite different among these studies, which makes it difficult to draw general conclusions concerning the ability of different classes of PCB congeners to affect different aspects of TH signaling.

The effect of high molecular weight *ortho*-substituted PCBs on TH-responsive endpoints will be particularly informative because studies that have examined PCB levels in human serum, milk and adipose samples have demonstrated that *ortho*-substituted PCB congeners are the most prevalent form in tissues (Nims et al. 1994; Seegal and Shain 1992; Shain et al. 1986). Additionally, PCBs that contain more chlorine substitutions (penta- and hexachlorobiphenyls) are preferentially represented in tissues. For example, Kodavanti et al. (Kodavanti et al. 1998) measured individual congeners present in Aroclor 1254 and congeners present in various rat tissues following continuous exposure to this PCB mixture. The *ortho*-substituted congeners 105, 118, 132, 138, 153, and 163 represented 26 percent of the Aroclor mixture and 54 percent of the congeners present in the brain. In contrast, other *ortho*-substituted congeners represented over 30 percent of the Aroclor mixture but less than 4 percent of the congeners extracted from brain. Furthermore, several studies have also evaluated the specific PCB congeners concentrated in fetal and neonatal brain. Morse et al. (Morse et al. 1996) found that the major metabolite 2,3,3′,4′,5-pentachloro-4-biphenylol (4-04 PCB 107) and 2,2′,4,4′,5,5′-hexachlorobiphenyl (PCB 153) are the most abundant PCBs in fetal tissue following maternal exposure to Aroclor 1254 (Morse et al. 1996). Considering that *ortho*-substituted congeners are selectively retained in tissues, and their effects on TSH are not predicted by reduced levels of TH, it will be essential to investigate the mechanism by which noncoplanar PCBs, or metabolites, interfere with TH signaling.

CONCLUSION

PCB exposure is associated with cognitive deficits in humans and behavioral abnormalities in animals. These effects may be related to their ability to reduce circulating levels of TH in both humans and in animals, but work in experimental animals does not fully support this hypothesis since the symptoms of developmental exposure to PCBs are not always consistent with the symptoms of developmental hypothyroidism. Furthermore, mechanistic studies in animals have recently provided important evidence that PCBs may be disrupting TH signaling in the developing brain in novel ways and that different PCB congeners, or classes of congeners, affect TH signaling in different ways. Therefore, it will

be important to ascertain the precise molecular and cellular mechanisms by which PCBs interfere with TH signaling and to ultimately develop biomarkers of TH signaling events in humans to aid in epidemiological studies.

REFERENCES

Andersen, S., K. M. Pedersen, N. H. Bruun, and P. Laurberg. 2002. "Narrow Individual Variations in Serum T(4) and T(3) in Normal Subjects: A Clue to the Understanding of Subclinical Thyroid Disease." *J. Clin. Endocrinol. Metab.* 87 (3): 1068–72.

Arrieta, CMd., B. Morte, A. Coloma, and J. Bernal. 1999. "The Human RC3 Gene Homolog, NRGN Contains a Thyroid Hormone-Responsive Element Located in the First Intron." *Endocrinology* 140:335–43.

Auso, E., R. Lavado-Autric, E. Cuevas, F. Escobar Del Rey, G. Morreale De Escobar, and P. Berbel. 2004. "A Moderate and Transient Deficiency of Maternal Thyroid Function at the Beginning of Fetal Neocorticogenesis Alters Neuronal Migration." *Endocrinology* 145:4037–47.

Ayotte, P., G. Muckle, J. L. Jacobson, S. W. Jacobson, and E. Dewailly. 2003. "Assessment of Pre- and Postnatal Exposure to Polychlorinated Biphenyls: Lessons from the Inuit Cohort Study." *Environ. Health Perspect.* 111 (9): 1253–58.

Baas, D., D. Bourbeau, J. L. Carre, L. L. Sarlieve, J. H. Dussault, and J. Puymirat. 1994. "Expression of Alpha and Beta Thyroid Receptors During Oligodendrocyte Differentiation." *Neuroreport* 5 (14): 1805–8.

Baas, D., D. Bourbeau, L. L. Sarlieve, M. E. Ittel, J. H. Dussault, and J. Puymirat. 1997. "Oligodendrocyte Maturation and Progenitor Cell Proliferation Are Independently Regulated by Thyroid Hormone." *Glia* 19 (4): 324–32.

Bansal, R., C. T. A. Herzig, E. A. Iannacone, R. T. Zoeller, and S. L. Petersen. 2003. "PCBs Exert a Thyroid Hormone-Like Effect on the Expression of Glial Determination Genes in the Early Cortex." *Society for Neuroscience Abstract* 455.

Bastomsky, C. H. 1974. "Effects of a Polychlorinated Biphenyl Mixture (Aroclor 1254) and DDT on Biliary Thyroxine Excretion in Rats." *Endocrinol.* 95:1150–55.

Bastomsky, C. H., P. V. N. Murthy, and K. Banovac. 1976. "Alterations in Thyroxine Metabolism Produced by Cutaneous Application of Microscope Immersion Oil: Effects Due to Polychlorinated Biphenyls." *Endocrinol.* 98:1309–14.

Bogazzi, F., F. Raggi, F. Ultimieri, D. Russo, A. Campomori, J. D. McKinney, A. Pinchera, L. Bartalena, and E. Martino. 2003. "Effects of a Mixture of Polychlorinated Biphenyls (Aroclor 1254) on the Transcriptional Activity of Thyroid Hormone Receptor." *J. Endocrinol. Invest.* 26 (10): 972–78.

Brouwer, A., D. C. Morse, M. C. Lans, A. G. Schuur, A. J. Murk, E. Klasson-Wehler, A. Bergman, and T. J. Visser. 1998. "Interactions of Persistent Environmental Organohalides with the Thyroid Hormone System: Mechanisms and Possible Consequences for Animal and Human Health." *Toxicol. Ind. Health* 14 (1/2): 59–84.

Bury, F., J. L. Carre, S. Vega, M. S. Ghandour, A. Rodriguez-Pena, K. Langley, and L. L. Sarlieve. 2002. "Coexpression of Thyroid Hormone Receptor Isoforms in Mouse Oligodendrocytes." *J. Neurosci. Res.* 67 (1): 106–13.

Calver, A. R., A. C. Hall, W. P. Yu, F. S. Walsh, J. K. Heath, C. Betsholtz, and W. D. Richardson. 1998. "Oligodendrocyte Population Dynamics and the Role of PDGF in Vivo." *Neuron* 20 (5): 869–82.

Calvo, R., M. J. Obregon, C. Ruiz de Ona, F. Escobar del Rey, and G. Morreale de Escobar. 1990. "Congenital Hypothyroidism, as Studied in Rats." *J. Clin. Invest.* 86:889–99.

Chauhan, K. R., P. R. Kodavanti, and J. D. McKinney. 2000. "Assessing the Role of Ortho-Substitution on Polychlorinated Biphenyl Binding to Transthyretin, a Thyroxine Transport Protein." *Toxicol. Appl. Pharmacol.* 162 (1): 10–21.

Chen, S. L., Y. J. Chang, Y. H. Wu, and K. H. Lin. 2003. "Mitogen-Activated Protein Kinases Potentiate Thyroid Hormone Receptor Transcriptional Activity by Stabilizing Its Protein." *Endocrinology* 144 (4): 1407–19.

Chopra, I. J., and L. Sabatino. (2000). "Nature and Sources of Circulating Thyroid Hormones." *The Thyroid: A Fundamental and Clinical Text,* edited by L. E. Braverman and R. D. Utiger, 136–73. Philadelphia: Lippincott-Raven.

Colborn, T. 2004. "Neurodevelopment and Endocrine Disruption." *Environ. Health Perspect.* 112 (9): 944–49.

Collins, W. T., Jr., and C. C. Capen. 1980. "Fine Structural Lesions and Hormonal Alterations in Thyroid Glands of Perinatal Rats Exposed in Utero and by the Milk to Polychlorinated Biphenyls." *Am. J. Pathol.* 99 (1): 125–42.

Collins, W. T., C. C. Capen, L. Kasza, C. Carter, and R. E. Dailey. 1977. "Effect of Polychlorinated Biphenyl (PCB) on the Thyroid Gland of Rats. Ultrastructural and biochemical investigations." *Am. J. Pathol.* 89:119–30.

Connors, J. M., and G. A. Hedge. 1981. "Feedback Regulation of Thyrotropin by Thyroxine under Physiological Conditions." *Am. J. Physiol.* 240 (3): E308–13.

Crofton, K. M., P. R. Kodavanti, E. C. Derr-Yellin, A. C. Casey, and L. S. Kehn. 2000a. "PCBs, Thyroid Hormones, and Ototoxicity in Rats: Cross-Fostering Experiments Demonstrate the Impact of Postnatal Lactation Exposure." *Toxicol. Sci.* 57 (1): 131–40.

Crofton, K. M., D. Ding, R. Padich, M. Taylor, and D. Henderson. 2000b. "Hearing Loss Following Exposure during Development to Polychlorinated Biphenyls: A Cochlear Site of Action." *Hear. Res.* 144 (1–2): 196–204.

Darvill, T., E. Lonky, J. Reihman, P. Stewart, and J. Pagano. 2000. "Prenatal Exposure to PCBs and Infant Performance on the Fagan Test of Infant Intelligence." *Neurotoxicology* 21 (6): 1029–38.

DeGroot, L. J., and J. Torresani. 1975. "Triiodothyronine Binding to Isolated Liver Cell Nuclei." *Endocrinology* 96:357–69.

Delange, F. 1997. "Neonatal Screening for Congenital Hypothyroidism: Results and Perspectives." *Horm. Res.* 48:51–61.

Derksen-Lubsen, G., and P. H. Verkerk. 1996. "Neuropsychologic Development in Early-Treated Congenital Hypothyroidism: Analysis of Literature Data." *Pediatric Research* 39:561–66.

Desaulniers, D., K. Leingartner, M. Wade, E. Fintelman, A. Yagminas, and W. G. Foster. 1999. "Effects of Acute Exposure to PCBs 126 and 153 on Anterior Pituitary and Thyroid Hormones and FSH Isoforms in Adult Sprague Dawley Male Rats." *Toxicol. Sci.* 47 (2): 158–69.

Dowling, A. L. S., and R. T. Zoeller. 2000. "Thyroid Hormone of Maternal Origin Regulates the Expression of RC3/Neurogranin mRNA in the Fetal Rat Brain." *Brain Res.* 82:126–32.

Dowling, A. L. S., E. A. Iannacone, and R. T. Zoeller. 2001. "Maternal Hypothyroidism Selectively Affects the Expression of Neuroendocrine-Specific Protein-A Messenger Ribonucleic Acid in the Proliferative Zone of the Fetal Rat Brain Cortex." *Endocrinol.* 142:390–99.

Dowling, A. L. S., G. U. Martz, J. L. Leonard, and R. T. Zoeller. 2000. "Acute Changes in Maternal Thyroid Hormone Induce Rapid and Transient Changes in Specific Gene Expression in Fetal Rat Brain." *J. Neurosci.* 20:2255–65.

Dumitrescu, A. M., X. H. Liao, T. B. Best, K. Brockmann, and S. Refetoff. 2004. "A Novel Syndrome Combining Thyroid and Neurological Abnormalities is Associated with Mutations in a Monocarboxylate Transporter Gene." *Am. J. Hum. Genet.* 74(1):168–75. Erratum in: Am J Hum Genet. 2004. Mar; 74(3):598.

Dussault, J. H., and J. Ruel. 1987. "Thyroid Hormones and Brain Development." *Annu. Rev. Physiol.* 49:321–34.

Ekins, R. P., A. K. Sinha, M. R. Pickard, I. M. Evans, and F. al Yatama. 1994. "Transport of Thyroid Hormones to Target Tissues." *Acta Med. Austriaca* 21 (2): 26–34.

Eriksson, P. 1988. "Effects of 3,3′,4,4′-Tetrachlorobiphenyl in the Brain of the Neonatal Mouse." *Toxicol.* 49:43–48.

Fein, G. G., J. L. Jacobson, S. W. Jacobson, P. M. Schwartz, and J. K. Dowler. 1984. "Prenatal Exposure to Polychlorinated Biphenyls: Effects on Birth Size and Gestational Age." *J. Pediatr.* 105:315–20.

Fischer, L. J., M. A. Wagner, and B. V. Madhukar. 1999. "Potential Involvement of Calcium, CaM kinase II, and MAP Kinases in PCB-Stimulated Insulin Release from RINm5F Cells." *Toxicol. Appl. Pharmacol.* 159 (3): 194–203.

Fisher, D. A. 1998. "Thyroid Function in Premature Infants: The Hypothyroxinemia of Prematurity." *Clin. Perinatol.* 25 (4): 999–1014, viii.

Fisher, D. A., J. H. Dussault, T. P. Foley, A. H. Klein, S. LaFranchi, P. R. Larsen, M. L. Mitchell, W. H. Murphey, and P. G. Walfish. 1979. "Screening for Congenital Hypothyroidism: Results of Screening One Million North American Infants." *J. Pediatr.* 94:700–705.

Foley, T. P. (1996). "Congenital Hypothyroidism." In *The Thyroid: A Fundamental and Clinical Text,* edited by L. E. Braverman and R. D. Utiger, 988–94. Philadelphia: Lippincott-Raven.

François, M., P. Bonfils, J. Leger, P. Czernichow, and P. J. Narcy. 1993. "Role of Congenital Hypothyroidism in Hearing Loss in Children." *Pediatrics* 123:444–46.

Friesema, E. C. H., R. Docter, E. P. C. M. Moerings, B. Stieger, B. Hagenbuch, P. J. Meier, E. P. Krenning, G. Hennemann, and T. J. Visser. 1999. "Identification of Thyroid Hormone Transporters." *Biochem. Biophys. Res. Commun.* 254:497–501.

Friesema, E. C., A. Grueters, H. Biebermann, H. Krude, A. von Moers, M. Reeser, T. G. Barrett, E. E. Mancilla, J. Svensson, M. H. Kester, G. G. Kuiper, S. Balkassmi, A. G. Uitterlinden, J. Koehrle, P. Rodien, A. P. Halestrap, and T. J. Visser. 2004. "Association between Mutations in a Thyroid Hormone Transporter and Severe X-linked Psychomotor Retardation." *Lancet.* 364(9443):1435–7.

Fruttiger, M., A. R. Calver, and W. D. Richardson. 2000. "Platelet-Derived Growth Factor Is Constitutively Secreted from Neuronal Cell Bodies but Not from Axons." *Curr. Biol.* 10 (20): 1283–86.

Fruttiger, M., L. Karlsson, A. C. Hall, A. Abramsson, A. R. Calver, H. Bostrom, K. Willetts, C. H. Bertold, J. K. Heath, C. Betsholtz, and W. D. Richardson. 1999. "Defective Oligodendrocyte Development and Severe Hypomyelination in PDGF-A Knockout Mice." *Development* 126 (3): 457–67.

Fuggle, P. W., D. B. Grant, I. Smith, and G. Murphy. 1991. "Intelligence, Motor Skills and Behaviour at 5 Years in Early-Treated Congenital Hypothyroidism." *Eur. J. Pediatr.* 150 (8): 570–74.

Gauger, K. J., Y. Kato, K. Haraguchi, H. J. Lehmler, L. W. Robertson, R. Bansal, and R. T. Zoeller. 2004. "Polychlorinated Biphenyls (PCBs) Exert Thyroid Hormone-like Effects in the Fetal Rat Brain but Do Not Bind to Thyroid Hormone Receptors." *Environ. Health Perspect.* 112 (5): 516–23.

Glass, C. K., and M. G. Rosenfeld. 2000. "The Coregulator Exchange in Transcriptional Functions of Nuclear Receptors." *Genes and Development* 14:121–41.

Goldberg, Y., C. Glineur, J. C. Gesquiere, A. Ricouart, J. Sap, B. Vennstrom, and J. Ghysdael. 1988. "Activation of Protein Kinase C or cAMP-Dependent Protein Kinase Increases Phosphorylation of the c-erbA-Encoded Thyroid Hormone Receptor and of the v-erbA-Encoded Protein." *Embo. J.* 7 (8): 2425–33.

Goldey, E. S., and K. M. Crofton. 1998. "Thyroxine Replacement Attenuates Hypothyroxinemia, Hearing Loss, and Motor Deficits Following Developmental Exposure to Aroclor 1254 in Rats." *Toxicol. Sci.* 45 (1): 94–105.

Goldey, E. S., L. S. Kehn, G. L. Rehnberg, and K. M. Crofton. 1995a. "Effects of Developmental Hypothyroidism on Auditory and Motor Function in the Rat." *Toxicol. Appl. Pharmacol.* 135:67–76.

Goldey, E. S., L. S. Kehn, C. Lau, G. L. Rehnberg, and K. M. Crofton. 1995b. "Developmental Exposure to Polychlorinated Biphenyls (Aroclor 1254) Reduces Circulating Thyroid Hormone Concentrations and Causes Hearing Deficits in Rats." *Toxicol. Appl. Pharmacol.* 135 (1): 77–88.

Goslings, B., H. L. Schwartz, W. Dillmann, M. I. Surks, and J. H. Oppenheimer. 1976. "Comparison of the Metabolism and Distribution of L-Triiodothyronine and Triiodothyroacetic Acid in the Rat: A Possible Explanation of Differential Hormonal Potency." *Endocrinology* 98:666–75.

Gottschalk, B., R. Richman, and L. Lewandowski. 1994. "Subtle Speech and Motor Deficits of Children with Congenital Hypothyroidism Treated Early." *Developmental Medicine Child Neurology* 36:216–20.

Grandjean, P., P. Weihe, V. W. Burse, L. L. Needham, E. Storr-Hansen, B. Heinzow, F. Debes, K. Murata, H. Simonsen, P. Ellefsen, E. Budtz-Jorgensen, N. Keiding, and R. F. White. 2001. "Neurobehavioral Deficits Associated with PCB in 7-Year-Old Children Prenatally Exposed to Seafood Neurotoxicants." *Neurotoxicol. Teratol.* 23 (4): 305–17.

Guadano-Ferraz, A., M. J. Escamez, B. Morte, P. Vargiu, and J. Bernal. 1997. "Transcriptional Induction of RC3/Neurogranin by Thyroid Hormone: Differential Neuronal Sensitivity Is Not Correlated with Thyroid Hormone Receptor Distribution in the Brain." *Molecular Brain Research* 49:37–44.

Guo, Y. L., Y. C. Chen, M. L. Yu, and C. C. Hsu. 1994. "Early Development of Yu-Cheng Children Born Seven to Twelve Years after the Taiwan PCB Outbreak." *Chemosphere* 29: 2395–2404.

Haddow, J. E., G. E. Palomaki, W. C. Allan, J. R. Williams, G. J. Knight, J. Gagnon, C. E. O'Heir, M. L. Mitchell, R. J. Hermos, S. E. Waisbren, J. D. Faix, and R. Z. Klein. 1999. "Maternal Thyroid Deficiency during Pregnancy and Subsequent Neuropsychological Development of the Child." *N. Engl. J. Med.* 341 (8): 549–55.

Hagmar, L., L. Rylander, E. Dyremark, E. Klasson-Wehler, and E. M. Erfurth. 2001. "Plasma Concentrations of Persistent Organochlorines in Relation to Thyrotropin and Thyroid Hormone Levels in Women." *Int. Arch. Occup. Environ. Health* 74 (3): 184–88.

Hashizume, K., T. Miyamoto, M. Kobayashi, S. Suzuki, K. Ichikawa, K. Yamauchi, H. Ohtsuka, and T. Takeda. 1989a. "Cytosolic 3,5,3′-Triiodo-L-Thyronine (T3)-Binding Protein (CTBP) Regulation of Nuclear T3 Binding: Evidence for the Presence of T3-CTBP Complex-Binding Sites in Nuclei." *Endocrinology* 124 (6): 2851–56.

Hashizume, K., T. Miyamoto, K. Yamauchi, K. Ichikawa, M. Kobayashi, H. Ohtsuka, A. Sakurai, S. Suzuki, and T. Yamada. 1989b. "Counterregulation of Nuclear 3,5,3′-Triiodo-L-Thyronine (T3) Binding by Oxidized and Reduced-Nicotinamide Adenine Dinucleotide Phosphates in the Presence of Cytosolic T3-Binding Protein in Vitro." *Endocrinology* 124 (4): 1678–83.

Hashizume, K., T. Miyamoto, K. Ichikawa, K. Yamauchi, A. Sakurai, H. Ohtsuka, M. Kobayashi, Y. Nishii, and T. Yamada. 1989c. "Evidence for the Presence of Two Active Forms of Cytosolic 3,5,3′-Triiodo-L-Thyronine (T3)-Binding Protein (CTBP) in Rat Kidney: Specialized Functions of Two CTBPs in Intracellular T3 Translocation." *J. Biol. Chem.* 264 (9): 4864–71.

Hashizume, K., T. Miyamoto, K. Ichikawa, K. Yamauchi, M. Kobayashi, A. Sakurai, H. Ohtsuka, Y. Nishii, and T. Yamada. 1989d. "Purification and Characterization of NADPH-Dependent Cytosolic 3,5,3′-Triiodo-L-Thyronine Binding Protein in Rat Kidney." *J. Biol. Chem.* 264 (9): 4857–63.

Hashizume, K., S. Suzuki, K. Ichikawa, and T. Takeda. 1991. "Purification of Cytosolic 3,5,3′-Triiodo-L-Thyronine(T3)-Binding Protein (CTBP) Which Regulates Nuclear T3 Translocation." *Biochem. Biophys. Res. Commun.* 174 (3): 1084–89.

Hermanson, O., C. K. Glass, and M. G. Rosenfeld. 2002. "Nuclear Receptor Coregulators: Multiple Modes of Modification." *Trends Endocrinol. Metab.* 13 (2): 55–60.

Hodin, R. A., M. A. Lazar, B. I. Wintman, D. S. Darling, and W. W. Chin. 1989. "Identification of a Thyroid Hormone Receptor That Is Pituitary-Specific." *Science* 244:76–79.

Hong, S. H., and M. L. Privalsky. 2000. "The SMRT Corepressor Is Regulated by a MEK-1 Kinase Pathway: Inhibition of Co-repressor Function Is Associated with SMRT Phosphorylation and Nuclear Export." *Mol. Cell. Biol.* 20 (17): 6612–25.

Hong, S. H., C. W. Wong, and M. L. Privalsky. 1998. "Signaling by Tyrosine Kinases Negatively Regulates the Interaction between Transcription Factors and SMRT (Silencing Mediator of Retinoic Acid and Thyroid Hormone Receptor) Corepressor." *Molec. Endocrinol.* 12:1161–71.

Hood, A., and C. D. Klaassen. 2000a. "Differential Effects of Microsomal Enzyme Inducers on in Vitro Thyroxine (T(4)) and Triiodothyronine (T(3)) Glucuronidation." *Toxicol. Sci.* 55 (1): 78–84.

———. 2000b. "Effects of Microsomal Enzyme Inducers on Outer-Ring Deiodinase Activity Toward Thyroid Hormones in Various Rat Tissues." *Toxicol. Appl. Pharmacol.* 163 (3): 240–48.

Horlein, A. J., A. M. Naar, T. Heinzel, J. Torchia, B. Gloss, R. Kurowkawa, A. Ryan, Y. Kamei, M. Soderstrom, C. K. Glass, and M. G. Rosenfeld. 1995. "Ligand-Independent Repression by the Thyroid Hormone Receptor Mediated by a Nuclear Co-Repressor." *Nature* 377:397–404.

Hu, X., and M. A. Lazar. 2000. "Transcriptional Repression by Nuclear Hormone Receptors." *Trends Endocrinol. Metab.* 11 (1): 6–10.

Huisman, M., C. Koopman-Esseboom, C. I. Lanting, C. G. van der Paauw, L. G. Tuinstra, V. Fidler, N. Weisglas-Kuperus, P. J. Sauer, E. R. Boersma, and B. C. Touwen. 1995. "Neurological Condition in 18-Month-Old Children Perinatally Exposed to Polychlorinated Biphenyls and Dioxins." *Early Hum. Dev.* 43:165–76.

Hussain, R. J., J. Gyori, A. P. DeCaprio, and D. O. Carpenter. 2000. "In Vivo and in Vitro Exposure to PCB 153 Reduces Long-Term Potentiation." *Environ. Health Perspect.* 108 (9): 827–31.

Ibarrola, N., and A. Rodriguez-Pena. 1997. "Hypothyroidism Coordinately and Transiently Affects Myelin Protein Gene Expression in Most Rat Brain Regions during Postnatal Development." *Brain Res.* 752 (1–2): 285–93.

Ichikawa, K., and L. J. DeGroot. 1987. "Purification and Characterization of Rat Liver Nuclear Thyroid Hormone Receptors." *Proc. Natl. Acad. Sci. USA* 84 (10): 3420–24.

Iwasaki, T., W. Miyazaki, A. Takeshita, Y. Kuroda, and N. Koibuchi. 2002. "Polychlorinated Biphenyls Suppress Thyroid Hormone-Induced Transactivation." *Biochem. Biophys. Res. Commun.* 299 (3): 384–88.

Izumo, S., and V. Mahdavi. 1988. "Thyroid Hormone Receptor Alpha Isoforms Generated by Alternative Splicing Differentially Activate Myosin HC Gene Transcription." *Nature* 334:539–42.

Jackson, T. A., J. K. Richer, D. L. Bain, G. S. Takimoto, L. Tung, and K. B. Horwitz. 1997. "The Partial Agonist Activity of Antagonist-Occupied Steroid Receptors Is Controlled by a Novel Hinge Domain-Binding Coactivator L7/SPA and the Corepressors N-CoR or SMRT." *Mol. Endocrinol.* 11 (6): 693–705.

Jacobson, J. L., and S. W. Jacobson. 1996. "Intellectual Impairment in Children Exposed to Polychlorinated Biphenyls in Utero." *New. Engl. J. Med.* 335:783–89.

Jacobson, J. L., S. W. Jacobson, and H. E. B. Humphrey. 1990a. "Effects of Exposure to PCBs and Related Compounds on Growth and Activity in Children." *Neurotoxicol. Teratol.* 12: 319–326.

———. 1990b. "Effects of in Utero Exposure to Polychlorinated Biphenyls and Related Contaminants on Cognitive Functioning in Young Children." *J. Pediatr.* 116:38–45.

Jacobson, W. W., G. G. Fein, J. L. Jacobson, P. M. Schwartz, and J. K. Dowler. 1985. "The Effect of Intrauterine PCB Exposure on Visual Recognition Memory." *Child Development* 56: 853–60.

Jansen, M. S., S. C. Nagel, P. J. Miranda, E. K. Lobenhofer, C. A. Afshari, and D. P. McDonnell. 2004. "Short-Chain Fatty Acids Enhance Nuclear Receptor Activity through Mitogen-Activated Protein Kinase Activation and Histone Deacetylase Inhibition." *Proc. Natl. Acad. Sci. USA* 101 (18): 7199–7204.

Johe, K. K., T. G. Hazel, T. Muller, M. M. Dugich-Djordjevic, and R. D. McKay. 1996. "Single Factors Direct the Differentiation of Stem Cells from the Fetal and Adult Central Nervous System." *Genes Dev.* 10 (24): 3129–40.

Jones, K. E., J. H. Brubaker, and W. W. Chin. 1994. "Evidence That Phosphorylation Events Participate in Thyroid Hormone Action." *Endocrinology* 134 (2): 543–48.

Kashimoto, T., H. Miyata, S. Kunita, T. Tung, T-C. Hsu, K-J. Chang, S-Y. Tang, G. Oni, J. Nakagawa, and S-I. Yamamoto. 1981. "Role of Polychlorinated Dibenzofuran in Yusho (PCB Poisoning)." *Arch. Environ. Health* 36:321–25.

Kasza, L., W. T. Collins, C. C. Capen, L. H. Garthoff, and L. Friedman. 1978. "Comparative Toxicity of Polychlorinated Biphenyl and Polybrominated Biphenyl in the Rat Thyroid Gland: Light and Electron Microscopic Alterations after Subacute Dietary Exposure." *J. Environ. Pathol. Toxicol.* 1:587–99.

Kato, Y., S. Ikushiro, K. Haraguchi, T. Yamazaki, Y. Ito, H. Suzuki, R. Kimura, S. Yamada, T. Inoue, and M. Degawa. 2004. "A Possible Mechanism for Decrease in Serum Thyroxine Level by Polychlorinated Biphenyls in Wistar and Gunn Rats." *Toxicol Sci.* 81:309–15.

Kester, M. H., R. Martinez de Mena, M. J. Obregon, D. Marinkovic, A. Howatson, T. J. Visser, R. Hume, and G. Morreale de Escobar. 2004. "Iodothyronine Levels in the Human Developing Brain: Major Regulatory Roles of Iodothyronine Deiodinases in Different Areas." *J. Clin. Endocrinol. Metab.* 89 (7): 3117–28.

Khan, M. A., and L. G. Hansen. 2003. "Ortho-Substituted Polychlorinated Biphenyl (PCB) Congeners (95 or 101) Decrease

Pituitary Response to Thyrotropin Releasing Hormone." *Toxicol. Lett.* 144 (2): 173–82.

Khan, M. A., C. A. Lichtensteiger, O. Faroon, M. Mumtaz, D. J. Schaeffer, and L. G. Hansen. 2002. "The Hypothalamo-Pituitary-Thyroid (HPT) Axis: A Target of Nonpersistent Ortho-Substituted PCB Congeners." *Toxicol. Sci.* 65 (1): 52–61.

Klein, R. 1980. "History of Congenital Hypothyroidism." In *Neonatal Thyroid Screening,* edited by G. N. Burrow and J. H. Dussault, 51–59. New York: Raven Press.

Klein, R. Z., and M. L. Mitchell. 1996. "Neonatal Screening for Hypothyroidism." In *The Thyroid: A Fundamental and Clinical Text,* edited by L. E. Braverman and R. D. Utiger, 984–88. Philadelphia: Lipponcott-Raven.

Kodavanti, P. R. S., T. R. Ward, E. C. Derr-Yellin, W. R. Mundy, A. C. Casey, B. Bush, and H. A. Tilson. 1998. "Congener-Specific Distribution of Polychlorinated Biphenyls in Brain Regions, Blood, Liver, and Fat of Adult Rats Following Repeated Exposure to Aroclor 1254." *Toxicol. Appl. Pharmacol.* 153:199–210.

Koenig, R. J. 1998. "Thyroid Hormone Receptor Coactivators and Corepressors." *Thyroid* 8:703–13.

Koenig, R. J., R. L. Warne, G. A. Brent, and J. W. Harney. 1988. "Isolation of a cDNA Clone Encoding a Biologically Active Thyroid Hormone Receptor." *Proc. Natl. Acad. Sci. USA* 85: 5031–35.

Kohrle, J. 1999. "Local Activation and Inactivation of Thyroid Hormones: The Deiodinase Family." *Mol. Cell. Endocrinol.* 151 (1–2): 103–19.

Koibuchi, N., and W. W. Chin. 2000. "Thyroid Hormone Action and Brain Development." *Trends Endocrinol. Metab.* 11 (4): 123–28.

Kolaja, K. L., and C. D. Klaassen. 1998. "Dose-Response Examination of UDP-Glucuronosyltransferase Inducers and Their Ability to Increase Both TGF-Beta Expression and Thyroid Follicular Cell Apoptosis." *Toxicol. Sci.* 46 (1): 31–37.

Kooistra, L., C. Laane, T. Vulsma, J. M. H. Schellekens, J. J. van der Meere, and A. F. Kalverboer. 1994. "Motor and Cognitive Development in Children with Congenital Hypothyroidism." *J. Pediatr.* 124:903–9.

Kooistra, L., J. J. van der Meere, T. Vulsma, and A. F. Kalverboer. 1996. "Sustained Attention Problems in Children with Early Treated Congenital Hypothyroidism." *Acta Paediatr.* 85 (4): 425–29.

Koopman-Esseboom, C., D. C. Morse, N. Weisglas-Kuperus, I. J. Lutkeschiphost, C. B. van der Paauw, L. G. M. T. Tuinstra, A. Brouwer, and P. J. J. Sauer. 1994. "Effects of Dioxins and Polychlorinated Biphenyls on Thyroid Hormone Status of Pregnant Women and Their Infants." *Pediatr. Res.* 36:468–73.

Koopman-Esseboom, C., M. Huisman, B. C. Touwen, E. R. Boersma, A. Brouwer, P. J. Sauer, and N. Weisglas-Kuperus. 1997. "Newborn Infants Diagnosed as Neurologically Abnormal with Relation to PCB and Dioxin Exposure and Their Thyroid-Hormone Status." *Dev. Med. Child. Neurol.* 39:785.

Korrick, S. A., and L. Altshul. 1998. "High Breast Milk Levels of Polychlorinated Biphenyls (PCBs) among Four Women Living Adjacent to a PCB-Contaminated Waste Site." *Environ. Health Perspect.* 106 (8): 513–18.

Krude, H., H. Biebermann, H. P. Krohn, and A. Gruters. 1977. "Congenital Hyperthyroidism." *Exp. Clin. Endocrinol. Diabetes* 105:6–11.

Ku, L. M. Jd., M. Sharma-Stokkermans, and L. A. Meserve. 1994. "Thyroxine Normalizes Polychlorinated Biphenyl (PCB) Dose-Related Depression of Choline Acetyltransferase (ChAT) Activity in Hippocampus and Basal Forebrain of 15-Day-Old Rats." *Toxicol.* 94:19–30.

LaFranchi, S. 1999. "Congenital Hypothyroidism: Etiologies, Diagnosis, and Management." *Thyroid* 9:735–40.

Lavado-Autric, R., E. Auso, J. V. Garcia-Velasco, C. Arufe Mdel, F. Escobar del Rey, P. Berbel, and G. Morreale de Escobar. 2003. "Early Maternal Hypothyroxinemia Alters Histogenesis and Cerebral Cortex Cytoarchitecture of the Progeny." *J. Clin. Invest.* 111 (7): 1073–82.

Lazar, M. A. 1993. "Thyroid Hormone Receptors: Multiple Forms, Multiple Possibilities." *Endocr. Rev.* 14:184–93.

———. 1994. "Thyroid Hormone Receptors: Update 1994." *Endocr. Rev. Monogr.* 3:280–83.

Leo, C., and J. D. Chen. 2000. "The SRC Family of Nuclear Receptor Coactivators." *Gene* 245:1–11.

Li, J., Q. Lin, H. G. Yoon, Z. Q. Huang, B. D. Strahl, C. D. Allis, and J. Wong. 2002. "Involvement of Histone Methylation and Phosphorylation in Regulation of Transcription by Thyroid Hormone Receptor." *Mol. Cell Biol.* 22 (16): 5688–97.

Lin, K. H., K. Ashizawa, and S. Y. Cheng. 1992. "Phosphorylation Stimulates the Transcriptional Activity of the Human Beta 1 Thyroid Hormone Nuclear Receptor." *Proc. Natl. Acad. Sci. USA* 89 (16): 7737–41.

Longnecker, M. P., W. J. Rogan, and G. Lucier. 1997. "The Human Health Effects of DDT (Dichlorodiphenyl-Trichloroethane) and PCBs (Polychlorinated Biphenyls) and an Overview of Organochlorines in Public Health." *Annu. Rev. Public Health* 18:211–44.

Longnecker, M. P., B. C. Gladen, D. G. Patterson Jr., and W. J. Rogan. 2000. "Polychlorinated Biphenyl (PCB) Exposure in Relation to Thyroid Hormone Levels in Neonates." *Epidemiology* 11 (3): 249–54.

Mangelsdorf, D. J., and R. M. Evans. 1995. "The RXR Heterodimers and Orphan Receptors." *Cell* 83:841–50.

Masuda, Y., R. Kagawa, I. Kuroki, I. Taki, M. Kusuda, F. Yamashita, M. Hayashi, M. Kuratsune, and T. Yoshimura. 1978. "Transfer of Polychlorinated Biphenyls from Mothers to Foetuses and Infants." *Fd. Cosmet. Toxicol.* 16:543–46.

Matsuura, N., and J. Konishi. 1990. "Transient Hypothyroidism in Infants Born to Mothers with Chronic Thyroiditis—A Nationwide Study of Twenty-three Cases." The Transient Hypothyroidism Study Group. *Endocrinol. Jpn.* 37 (3): 369–79.

Mckenna, N. J., and B. W. O'Malley. 2002. "Minireview: Nuclear Receptor Coactivators—An Update." *Endocrinol.* 143 (7): 2461–65.

McKinney, J. D., and C. L. Waller. 1998. "Molecular Determinants of Hormone Mimicry: Halogenated Aromatic Hydrocarbon Environmental Agents." *J. Toxicol. Env. Health*. Pt. B, 1:27–58.

Meerts, I. A., Y. Assink, P. H. Cenijn, J. H. Van Den Berg, B. M. Weijers, A. Bergman, J. H. Koeman, and A. Brouwer. 2002. "Placental Transfer of a Hydroxylated Polychlorinated Biphenyl and Effects on Fetal and Maternal Thyroid Hormone Homeostasis in the Rat." *Toxicol. Sci.* 68 (2): 361–71.

Miculan, J., S. Turner, and B. A. Paes. 1993. "Congenital Hypothyroidism: Diagnosis and Management." *Neonatal. Netw.* 12 (6): 25–34; quiz 34–38.

Miyazaki, W., T. Iwasaki, A. Takeshita, Y. Kuroda, and N. Koibuchi. 2004. "Polychlorinated Biphenyls Suppress Thyroid Hormone Receptor-Mediated Transcription through a Novel Mechanism." *J. Biol. Chem.* 279 (18): 18195–202.

Moreau, X., P. J. Lejeune, and R. Jeanningros. 1999. "Kinetics of Red Blood Cell T3 Uptake in Hypothyroidism with or without Hormonal Replacement, in the Rat." *J. Endocrinol. Invest.* 22:257–61.

Mori, J., S. Suzuki, M. Kobayashi, T. Inagaki, A. Komatsu, T. Takeda, T. Miyamoto, K. Ichikawa, and K. Hashizume. 2002. "Nicotinamide Adenine Dinucleotide Phosphate-Dependent Cytosolic T(3) Binding Protein as a Regulator for T(3)-Mediated Transactivation." *Endocrinology* 143 (4): 1538–44.

Morreale de Escobar, G., M. J. Obregon, and F. Escobar del Rey. 2000. "Is Neuropsychological Development Related to Maternal Hypothyroidism or to Maternal Hypothyroxinemia?" *J. Clin. Endocrinol. Metab.* 85 (11): 3975–87.

Morse, D. C., E. K. Wehler, W. Wesseling, J. H. Koeman, and A. Brouwer. 1996. "Alterations in Rat Brain Thyroid Hormone Status Following Pre- and Postnatal Exposure to Polychlorinated Biphenyls (Aroclor 1254)." *Toxicol. Appl. Pharmacol.* 136:269–79.

Muñoz, A., and J. Bernal. 1997. "Biological Activities of Thyroid Hormone Receptors." *Eur. J. Endocrinol.* 137:433–45.

Murray, M. B., N. D. Zilz, N. L. McCreary, M. J. MacDonald, and H. C. Towle. 1988. "Isolation and Characterization of Rat cDNA Clones for Two Distinct Thyroid Hormone Receptors." *J. Biol. Chem.* 263:12770–77.

Ness, D. K., S. L. Schantz, J. Moshtaghian, and L. G. Hansen. 1993. "Effects of Perinatal Exposure to Specific PCB Congeners on Thyroid Hormone Concentrations and Thyroid Histology in the Rat." *Toxicol. Lett.* 68:311–23.

Neves, F. A., R. R. Cavalieri, L. A. Simeoni, D. G. Gardner, J. D. Baxter, B. F. Scharschmidt, N. Lomri, and R. C. Ribeiro. 2002. "Thyroid Hormone Export Varies among Primary Cells and Appears to Differ from Hormone Uptake." *Endocrinology* 143 (2): 476–83.

Nims, R. W., S. D. Fox, H. J. Issaq, and R. A. Lubet. 1994. "Accumulation and Persistence of Individual Polychlorinated Biphenyl Congeners in Liver, Blood, and Adipose Tissue of Rats Following Dietary Exposure to Aroclor 1254." *Arch. Environ. Contam. Toxicol.* 27:513–20.

Nishii, Y., K. Hashizume, K. Ichikawa, T. Takeda, M. Kobayashi, T. Nagasawa, M. Katai, H. Kobayashi, and A. Sakurai. 1993. "Induction of Cytosolic Triiodo-L-Thyronine (T3) Binding Protein (CTBP) by T3 in Primary Cultured Rat Hepatocytes." *Endocr. J.* 40 (4): 399–404.

Nishimura, H., K. Shiota, T. Tanimura, T. Mizutani, M. Matsumoto, and M. Ueda. 1977. "Levels of Polychlorinated Biphenyls and Organochlorine Insecticides in Human Embryos and Fetuses." *Pediatrics* 6:45–47.

Oerbeck, B., K. Sundet, B. F. Kase, and S. Heyerdahl. 2003. "Congenital Hypothyroidism: Influence of Disease Severity and L-Thyroxine Treatment on Intellectual, Motor, and School-Associated Outcomes in Young Adults." *Pediatrics* 112 (4): 923–30.

Olivero, J., and P. E. Ganey. 2000. "Role of Protein Phosphorylation in Activation of Phospholipase A2 by the Polychlorinated Biphenyl Mixture Aroclor 1242." *Toxicol. Appl. Pharmacol.* 163 (1): 9–16.

Onate, S. A., S. Y. Tsai, M-J. Tsai, and B. W. O'Malley. 1995. "Sequence and Characterization of a Coactivator for the Steroid Hormone Receptor Superfamily." *Science* 270:1354–57.

Oppenheimer, J. H. 1983. "The Nuclear Receptor-Triiodothyronine Complex: Relationship to Thyroid Hormone Distribution, Metabolism, and Biological Action." In *Molecular Basis of Thyroid Hormone Action*, edited by J. H. Oppenheimer and H. H. Samuels, 1–35. New York, Academic Press.

Osius, N., W. Karmaus, H. Kruse, and J. Witten. 1999. "Exposure to Polychlorinated Biphenyls and Levels of Thyroid Hormones in Children." *Environ. Health Perspect.* 107:843–89.

Otten, M. H., J. A. Mol, and T. J. Visser. 1983. "Sulfation Preceding Deiodination of Iodothyronines in Rat Hepatocytes." *Science* 221 (4605): 81–83.

Palha, J. A. 2002. "Transthyretin as a Thyroid Hormone Carrier: Function Revisited." *Clin. Chem. Lab. Med.* 40 (12): 1292–1300.

Patandin, S., N. Weisglas-Kuperus, M. A. de Ridder, C. Koopman-Esseboom, W. A. van Staveren, C. G. van der Paauw, and P. J. Sauer. 1997. "Plasma Polychlorinated Biphenyl Levels in Dutch Preschool Children Either Breast-Fed or Formula-Fed during Infancy." *Am. J. Public Health* 87:1711–14.

Patandin, S., C. Koopman-Esseboom, M. A. de Ridder, N. Weisglas-Kuperus, and P. J. Sauer. 1998. "Effects of Environmental Exposure to Polychlorinated Biphenyls and Dioxins on Birth Size and Growth in Dutch Children." *Pediatr. Res.*, 538–45.

Pop, V. J., J. L. Kuijpens, A. L. van Baar, G. Verkerk, M. M. van Son, J. J. de Vijlder, T. Vulsma, W. M. Wiersinga, H. A. Drexhage, and H. L. Vader. 1999. "Low Maternal Free Thyroxine Concentrations during Early Pregnancy Are Associated with Impaired Psychomotor Development in Infancy." *Clin. Endocrinol.* 50 (2): 149–55.

Porterfield, S. P. 2000. "Thyroidal Dysfunction and Environmental Chemicals—Potential Impact on Brain Development." *Environ. Health Perspect.* 108 (Suppl. 3): 433–38. Available on line at http://www.porterfield/438porterfield/abstract.html.

Porterfield, S. P., and L. B. Hendry. 1998. "Impact of PCBs on Thyroid Hormone Directed Brain Development." *Toxicol. Ind. Health* 14:103–20.

Rapoport, B., and S. W. Spaulding. 1986. "Mechanism of Action of Thyrotropin and Other Thyroid Growth Factors." In *Werner and Ingbar's The Thyroid: A Fundamental and Clinical Text,* edited by L. E. Braverman and R. D. Utiger, 207–19. Philadelphia: Lippincott-Raven.

Refetoff, S., F. E. Dwulet, and M. D. Benson. 1986. "Reduced Affinity for Thyroxine in Two of Three Structural Thyroxine-Binding Prealbumin Variants Associated with Familial Amyloidotic Polyneuropathy." *J. Clin. Endocrinol. Metab.* 63 (6): 1432–37.

Refetoff, S., N. I. Robin, and V. S. Fang. 1970. "Parameters of Thyroid Function in Serum of 16 Selected Vertebrate Species: A Study of PBI, Serum T4, Free T4, and the Pattern of T4 and T3 Binding to Serum Proteins." *Endocrinology* 86 (4): 793–805.

Rogan, W. J., and B. C. Gladen. 1992. "Neurotoxicology of PCBs and Related Compounds." *Neurotoxicol.* 13:27–35.

Rogan, W. J., B. C. Gladen, K. L. Hung, S. L. Koong, L. Y. Shih, J. S. Taylor, Y. C. Wu, D. Yang, N. B. Ragan, and C. C. Hsu. 1988. "Congenital Poisoning by Polychorinated Biphenyls and Their Contaminants in Taiwan." *Science* 241:334–36.

Rosen, H. N., A. C. Moses, J. R. Murrell, J. J. Liepnieks, and M. D. Benson. 1993. "Thyroxine Interactions with Transthyretin: A Comparison of 10 Different Naturally Occurring Human Transthyretin Variants." *J. Clin. Endocrinol. Metab.* 77 (2): 370–74.

Rosenfeld, M. G., and C. K. Glass. 2001. "Coregulator Codes of Transcriptional Regulation by Nuclear Receptors." *J. Biol. Chem.* 276 (40): 36865–68.

Rovet, J. F., ed. 2000. *Neurobehavioral Consequences of Congenital Hypothyroidism Identified by Newborn Screening: Therapeutic Outcome of Endocrine Disorders. Efficacy, Innovation, and Quality of Life.* New York: Springer-Verlag.

Rovet, J. F., R. Ehrlich, and D. Sorbara. 1992. "Neurodevelopment in Infants and Preschool Children with Congenital Hypothyroidism: Etiological and Treatment Factors Affecting Outcome." *J. Pediatr. Psychol.* 17:187–213.

Rovet, J., W. Walker, B. Bliss, L. Buchanan, and R. Ehrlich. 1996. "Long-Term Sequelae of Hearing Impairment in Congenital Hypothyroidism." *J. Pediatr.* 128:776–83.

Sala, M., J. Sunyer, C. Herrero, J. To-Figueras, and J. Grimalt. 2001. "Association between Serum Concentrations of Hexachlorobenzene and Polychlorobiphenyls with Thyroid Hormone and Liver Enzymes in a Sample of the General Population." *Occup. Environ. Med.* 58 (3): 172–77.

Sap, J., A. Munoz, K. Damm, Y. Goldberg, J. Ghysdael, A. Lentz, H. Beug, and B. Vennstrom. 1986. "The c-erbA Protein Is a High Affinity Receptor for Thyroid Hormone." *Nature* 324: 635–40.

Schantz, S. L., J. J. Widholm, and D. C. Rice. 2003. "Effects of PCB Exposure on Neuropsychological Function in Children." *Environmental Health Perspectives* 111 (3): 357–76.

Schantz, S. L., J. L. Jacobson, S. W. Jacobson, and H. E. B. Humphrey. 1990. "Behavioral Correlates of Polychlorinated Biphenyl (PCB) Body Burden in School-Aged Children." *Toxicologist* 10:303.

Schussler, G. C. 2000. "The Thyroxine-Binding Proteins." *Thyroid* 10 (2): 141–49.

Schwartz, H. L. 1983. "Effect of Thyroid Hormone on Growth and Development." In *Molecular Basis of Thyroid Hormone Action,* edited by J. H. Oppenheimer and H. H. Samuels, 413–44. New York: Academic Press.

Schwartz, H. L., K. A. Strait, N. C. Ling, and J. H. Oppenheimer. 1992. "Quantitation of Rat Tissue Thyroid Hormone Binding Receptor Isoforms by Immunoprecipitation of Nuclear Triiodothyronine Binding Capacity." *J. Biol. Chem.* 267: 11794–99.

Seegal, R. F., and W. Shain. 1992. "Neurotoxicity of Polychlorinated Biphenyls: The Role of Ortho-Substituted Congeners in Altering Neurochemical Function." In *The Vulnerable Brain and Environmental Risks,* edited by R. L. Isaacson and K. F. Jensen, 2. New York: Plenum Press.

Seo, B-W., M-H. Li, L. G. Hansen, R. W. Moore, R. E. Peterson, and S. L. Schantz. 1995. "Effects of Gestational and Lactational Exposure to Coplanar Polychlorinated Biphenyl (PCB) Congeners or 2,3,7,8-Tetrachlorodibenzo-*p*-Dioxin (TCDD) on Thyroid Hormone Concentrations in Weanling Rats." *Toxicol. Lett.* 78:253–62.

Shain, W., S. R. Overmann, L. R. Wilson, J. Kostas, and B. Bush. 1986. "A Congener Analysis of Polychlorinated Biphenyls Accumulating in Rat Pups after Perinatal Exposure." *Arch. Environ. Contam. Toxicol.* 15:687–707.

Siegrist-Kaiser, C. A., and A. G. Burger. 1994. "Modification of the Side Chain of Thyroid Hormone." In *Thyroid Hormone Metabolism: Molecular Biology and Alternate Pathways,* edited by T. J. Visser, chap. 10. Boca Raton, Fla.: CRC Press.

Steuerwald, U., P. Weihe, P. J. Jorgensen, K. Bjerve, J. Brock, B. Heinzow, E. Budtz-Jorgensen, and P. Grandjean. 2000. "Maternal Seafood Diet, Methylmercury Exposure, and Neonatal Neurologic Function." *J. Pediatr.* 136 (5): 599–605.

St. Germain, D. L., and V. A. Galton. 1997. "The Deiodinase Family of Selenoproteins." *Thyroid* 7 (4): 655–68.

Stockigt, J. R. 2000. "Serum Thyrotropin and Thyroid Hormone Measurements and Assessment of Thyroid Hormone Transport." In *The Thyroid: A Fundamental and Clinical Text,* edited by L. E. Braverman and R. D. Utiger, 376–92. Philadelphia: Lippincott-Raven.

Struhl, K. 1998. "Histone Acetylation and Transcriptional Regulatory Mechanisms." *Genes and Development* 12:599–606.

Tani, Y., Y. Mori, Y. Miura, H. Okamoto, A. Inagaki, H. Saito, and Y. Oiso. 1994. "Molecular Cloning of the Rat Thyroxine-Binding Globulin Gene and Analysis of Its Promoter Activity." *Endocrinology* 135 (6): 2731–36.

Thompson, C. 1999. "Molecular Mechanisms of Thyroid Hormone Action in Neural Development." *Dev. Neuropsychol.* 16:365–67.

Thompson, C. C., and G. B. Potter. 2000. "Thyroid Hormone Action in Neural Development." *Cereb. Cortex.* 10:939–45.

Thompson, C. C., C. Weinberger, R. Lebo, and R. M. Evans. 1987. "Identification of a Novel Thyroid Hormone Receptor Expressed in the Mammalian Central Nervous System." *Science* 237:1610–14.

Tilson, H. A., and P. R. S. Kodavanti. 1997. "Neurochemical Effects of Polychlorinated Biphenyls: An Overview and Identification of Research Needs." *NeuroToxicology* 13:727–44.

Ulbrich, B., and R. Stahlmann. 2004. "Developmental Toxicity of Polychlorinated Biphenyls (PCBs): A Systematic Review of Experimental Data." *Arch. Toxicol.* 78 (5): 252–68.

van Heyningen, P., A. R. Calver, and W. D. Richardson. 2001. "Control of Progenitor Cell Number by Mitogen Supply and Demand." *Curr. Biol.* 11 (4): 232–41.

van Vliet, G. 1999. "Neonatal Hypothyroidism: Treatment and Outcome." *Thyroid* 9:79–84.

Varma, S. K., R. Murray, and J. B. Stanbury. 1978. "Effect of Maternal Hypothyroidism and Triiodothyronine on the Fetus and Newborn in Rats." *Endocrinology* 102 (1): 24–30.

Visser, T. J., M. H. Otten, J. A. Mol, R. Docter, and G. Hennemann. 1984. "Sulfation Facilitates Hepatic Deiodination of Iodothyronines." *Horm. Metab. Res. Suppl.* 14:35–41.

Visser, T. J., E. Kaptein, H. van Toor, J. A. G. M. van Raay, K. J. van den Berg, C. J. T. Joe, J. G. M. van Engelen, and A. Brouwer. 1993. "Glucuronidation of Thyroid Hormone in Rat Liver: Effects of in Vivo Treatment with Microsomal Enzyme Inducers and in Vitro Assay Conditions." *Endocrinology* 133: 2177–86.

Vranckx, R., L. Savu, M. Maya, and E. A. Nunez. 1990. "Characterization of a Major Development-Regulated Serum Thyroxine-Binding Globulin in the Euthyroid Mouse." *Biochem. J.* 271 (2): 373–79.

Walkowiak, J., J. A. Wiener, A. Fastabend, B. Heinzow, U. Kramer, E. Schmidt, H. J. Steingruber, S. Wundram, and G. Winneke. 2001. "Environmental Exposure to Polychlorinated Biphenyls and Quality of the Home Environment: Effects on Psychodevelopment in Early Childhood." *Lancet* 358 (9293): 1602–7.

Weinand-Harer, A., H. Lilienthal, K-A. Bucholski, and G. Winneke. 1997. "Behavioral Effects of Maternal Exposure to an Ortho-Chlorinated or a Coplanar PCB Congener in Rats." *Env. Tox. Pharmacol.* 3:97–103.

Weinberger, C., C. C. Thompson, E. S. Ong, R. Lebo, D. J. Gruol, and R. M. Evans. 1986. "The c-erbA Gene Encodes a Thyroid Hormone Receptor." *Nature* 324:641–46.

Winneke, G., A. Bucholski, B. Heinzow, U. Kramer, E. Schmidt, J. Walkowiak, J. A. Wiener, and H. J. Steingruber. 1998. "Developmental Neurotoxicity of Polychlorinated Biphenyls (PCBS): Cognitive and Psychomotor Functions in 7-Month Old Children." *Toxicol. Lett.* 102–3:423–38.

Wondisford, F. E., J. A. Magner, and B. D. Weintraub. 1996. "Thyrotropin." In *The Thyroid: A Fundamental and Clinical Text,* edited by L. E. Braverman and R. D. Utiger, 190–206. Philadelphia: Lippincott-Raven.

Wu, Y., B. Xu, and R. J. Koenig. 2001. "Thyroid Hormone Response Element Sequence and the Recruitment of Retinoid X Receptors for Thyroid Hormone Responsiveness." *J. Biol. Chem.* 276 (6): 3929–36.

Yamada-Okabe, T., T. Aono, H. Sakai, Y. Kashima, and H. Yamada-Okabe. 2004. "2,3,7,8-Tetrachlorodibenzo-p-Dioxin Augments the Modulation of Gene Expression Mediated by the Thyroid Hormone Receptor." *Toxicol. Appl. Pharmacol.* 194 (3): 201–10.

Yen, P. M. 2001. "Physiological and Molecular Basis of Thyroid Hormone Action." *Physiol. Rev.* 81 (3): 1097–1142.

Yu, M. L., C. C. Hsu, B. C. Gladen, and W. J. Rogan. 1991. "In Utero PCB/PCDF Exposure: Relation of Developmental Delay to Dysmorphology and Dose." *Neurotoxicol. Teratol.* 13:195–202.

Zoeller, R. T., A. L. Dowling, and A. A. Vas. 2000. "Developmental Exposure to Polychlorinated Biphenyls Exerts Thyroid Hormone-like Effects on the Expression of RC3/Neurogranin and Myelin Basic Protein Messenger Ribonucleic Acids in the Developing Rat Brain." *Endocrinol.* 141 (1): 181–89.

Zoeller, R. T., and J. Rovet. 2004. "Timing of Thyroid Hormone Action in the Developing Brain—Clinical Observations and Experimental Findings." *Journal of Neuroendocrinology* 16(10): 809–18.

16

Nutrition Modulates PCB Toxicity: Implications in Atherosclerosis

Bernhard Hennig, *University of Kentucky*
Michal Toborek, *University of Kentucky*
Gudrun Reiterer, *University of Kentucky*
Zuzana Majkova, *University of Kentucky*
Elizabeth Oesterling, *University of Kentucky*
Purushothaman Meerarani, *Mount Sinai School of Medicine*
Larry W. Robertson, *University of Iowa*

We hypothesize that nutrition can modulate the toxicity of environmental pollutants and thus modulate health and disease outcome associated with chemical insult. There is now increasing evidence that exposure to polychlorinated biphenyls (PCBs) can contribute to the development of inflammatory diseases such as atherosclerosis. Activation, chronic inflammation. and dysfunction of the vascular endothelium are critical events in the initiation and acceleration of atherosclerotic lesion formation. Our studies indicate that an increase in cellular oxidative stress and an imbalance in antioxidant status are critical events in PCB-mediated induction of inflammatory genes and endothelial cell dysfunction. Furthermore, we found that specific dietary fats (e.g., linoleic acid, the parent unsaturated fatty acid of the omega-6 family) can further increase endothelial dysfunction induced by selected PCBs, probably by contributing to oxidative stress and the production of toxic lipid metabolites such as leukotoxins. We also observed that antioxidant nutrients (such as vitamin E and dietary flavonoids) can protect against endothelial cell damage mediated by PCBs or polyunsaturated dietary fats. Our recent data suggest that membrane lipid rafts such as caveolae may play a major role in the regulation of PCB-induced inflammatory signaling in endothelial cells. In addition, PCB and lipid-induced inflammation can be down-regulated by ligands of antiatherogenic peroxisome proliferator activated receptors (PPARs). We hypothesize that PCBs contribute to an endothelial inflammatory response in part by down-regulating PPAR signaling. Our data so far support our hypothesis that antioxidant nutrients and related bio-active compounds common in fruits and vegetables protect against environmental toxic insult to the vascular endothelium by down-regulation of signaling pathways involved in inflammatory responses and atherosclerosis. The concept that nutrition may modify or ameliorate the toxicity of environmental chemicals is provocative but certainly warrants further study, since the implications for human health could be significant.

INTRODUCTION

Based on recent evidence, we hypothesize that nutrition can modulate the toxicity of environmental pollutants and thus modulate health and disease outcome associated with chemical insult. Atherosclerosis, a chronic inflammatory disease, is still the number one cause of death in the United States. Numerous risk factors for the development of atherosclerosis have been identified, including obesity and hypertriglyceridemia. Interestingly, Superfund chemicals such as PCBs also have been shown to increase the risk and incidence of cardiovascular diseases. Activation and dysfunction of the vascular endothelium are critical events in the early pathology of atherosclerosis. Our research suggests that nutrition can modulate PCB toxicity to the vascular endothelium. Most of all, we have evidence that both selected PCBs and fatty acids can induce and accelerate endothelial cell dysfunction and inflammation, and that antioxidant nutrients and related plant-derived bioactive compounds can reduce such inflammatory events during PCB-induced endothelial cell activation and atherosclerosis. Our data suggest that certain dietary fats may increase the risk to environmental insult induced by PCBs and that fruits and vegetables, rich in antioxidant and anti-inflammatory nutrients or bioactive compounds, may provide protection. Nutritional awareness in environmental toxicology is critical, because of opportunities to develop dietary guidelines which specifically target the exposed populations. Nutrition may provide the most sensible means to develop primary prevention strategies for vascular diseases associated with many environmental toxic insults.

PCBS AND ATHEROSCLEROSIS

From epidemiological studies, there is substantial evidence that cardiovascular diseases are linked to environmental pollution. For example, there was a significant increase in mortality from cardiovascular diseases among Swedish capacitor manufacturing workers exposed to PCBs for at least five years (Gustavsson and Hogstedt 1997), and most excess deaths were due to cardiovascular disease in power workers exposed to phenoxy herbicides and PCBs in waste transformer oil (Hay and Tarrel 1997). Furthermore, an increase in cardiovascular disease was detected in studies on the population of Seveso, Italy, after the industrial accident in 1976 (Bertazzi et al. 1998). These studies suggest that populations near Superfund sites in the United States are at increased risk to develop cardiovascular diseases, in particular in the presence of additional risk factors such as hypertriglyceridemia. Animal models which mimic human atherosclerosis allow for the study of mechanisms of environmental risk factors for cardiovascular diseases (Dalton et al. 2001). There is evidence linking the aryl hydrocarbon receptor (AhR) with mechanisms associated with cardiovascular diseases (Savouret, Berdeaux, and Casper 2003) and that AhR ligands may be atherogenic by disrupting the functions of endothelial cells in blood vessels.

PCBS AND ENDOTHELIAL CELL ACTIVATION

The lining of blood vessels is protected by the endothelium, and endothelial cells play an active role in physiological processes such as regulation of vessel tone, blood coagulation and vascular permeability. Dysfunction of endothelial cells is a critical underlying cause of the initiation of cardiovascular diseases. Severe endothelial cell activation and injury can lead to necrotic and apoptotic cytotoxicity, and ultimately to disruption of endothelial integrity. In addition to endothelial barrier dysfunction, another functional change in atherosclerosis is the activation of the endothelium that is manifested as an increase in the expression of specific cytokines and adhesion molecules. These cytokines and adhesion molecules are proposed to mediate the inflammatory aspects of the disease by regulating the vascular entry of leukocytes. We have demonstrated previously that coplanar as well as noncoplanar PCBs can cause endothelial cell dysfunction as determined by markers such as expression of cytokines and adhesion molecules (reviewed in Hennig et al. 2005).

The mechanisms by which environmental chemicals induce endothelial cell activation are not fully understood, and little is known about how PCB-mediated cell dysfunction can be prevented. Oxidative stress-induced transcription factors, which regulate inflammatory cytokine and adhesion molecule production, play critical roles in the induction of inflammatory responses. It appears that NF-kB, AP-1 and STATs are most significant in these regulatory processes. The binding sites for these transcription factors were identified in the promoter regions of a variety of inflammatory genes (Kunsch and Medford 1999; Muller, Rupec, and Baeuerle 1997) such as IL-6, VCAM-1 or COX-2, all of which are upregulated during PCB toxicity (Choi et al. 2003; Dalton et al. 2001; Hennig et al. 2002; Kwon et al. 2002).

DIETARY FAT, PCBS, AND ENDOTHELIAL CELL DYSFUNCTION

There is considerable evidence that exposure to PCBs can lead to lipid changes in plasma and tissues and that this may be linked to lipophilic properties of PCBs (Kakela and Hyvarinen 1999; Kamei et al. 1996) and/or interference with glucogenic and lipogenic enzymes or pathways (Boll et al. 1998). There is also evidence that elevated levels of linoleic acid may facilitate the cellular availability of PCBs (Doi et al. 2000). Furthermore, coplanar PCBs can suppress delta 5 and 6 desaturase activities, thus disrupting the synthesis of fatty acid precursors for eicosanoid metabolism (Matsusue et al. 1999). Our own preliminary data from plasma and livers of LDL receptor-deficient mice support the hypothesis that treatment with PCBs can facilitate clearance of linoleic acid from plasma into vascular tissues (Henning et al. 2005). Such a change in lipid milieu could exacerbate fatty acid- and/or PCB-induced oxidative stress and a vascular inflammatory response.

There is clear evidence that hypertriglyceridemia is an independent risk factor of cardiovascular diseases such as atherosclerosis (Austin et al. 2000; Malloy and Kane 2001). Even though diets high in n-6 and n-3 fatty acids may lead to a decrease in serum cholesterol, replacing saturated with unsaturated lipids may not be desirable because of their ability to oxidize easily. The high intake of linoleic acid-rich fats will lead to an increase in cellular oxidative stress, which has been implicated in most age-related diseases. We hypothesize that similar signaling pathways critical in the pathology of atherosclerosis may be induced both by n-6 fatty acids (primarily linoleic acid) and PCBs, thus explaining in part their interactive and amplified effects.

CAVEOLAE, FATTY ACIDS, PCBS, AND ENDOTHELIAL CELL FUNCTION

There is increasing evidence that caveolae play a critical role in the pathology of atherosclerosis (Frank et al. 2003) and that the lack of the caveolin-1 gene may provide protection against the development of atherosclerosis (Frank et al. 2004). Caveolae are particularly abundant in endothelial cells, where they are believed to play a major role in the regulation of endothelial vesicular trafficking and are involved in the uptake of lipids, as well as related compounds/chemicals and possibly environmental pollutants (Matveev et al. 2001). The lipid receptors CD36 and scavenger receptor class B1 (SR-B1) have been reported to be associated with caveolae (Smart et al. 1999). Furthermore, caveolin-1 itself has been described as a fatty acid (Trigatti, Anderson, and Gerber 1999) and cholesterol (Schroeder et al. 2001) binding protein and is a structural component in lipid body formation and thus cellular lipid uptake (Pol et al. 2004). The role of caveolae in cellular PCB-trafficking is not known. In our preliminary experiments we found caveolin-1 and CD36 to be upregulated in endothelial cells exposed to PCBs (Majkova et al. 2004).

Besides their role in cellular uptake of lipophilic substances, caveolae contain an array of cell signaling molecules (Frank et al. 2003). In fact, numerous genes involved in endothelial cell dysfunction, inflammation and PCB toxicity are associated with caveolae (Frank et al. 2003). Noncoplanar PCBs have been reported to activate phospholipase C (Machala et al. 2003) and its downstream target protein kinase C (27), both of which are located in caveolae (Frank et al. 2003). Also, our own data indicate an activation of caveolae-associated signaling molecules. Both coplanar and noncoplanar PCBs increased expression of COX-2 (Majkova et al. 2004). We hypothesize that caveolae are critical cell-surface plasma membrane invaginations, which facilitate the lipid and PCB-mediated cellular uptake and subsequent endothelial inflammatory response and cytotoxicity.

Furthermore, the internalization of caveolae can be suppressed by kinase inhibitors (Nabi and Le 2003) that lead to a disruption of the budding-process of caveolae (Nabi and Le 2003). We have reported recently that the dietary flavonoids quercetin and catechin can protect against 3,3′,4,4′-tetrachlorobiphenyl (PCB 77)-induced CYP1A1 induction and subsequent generation of oxidative stress (Ramadass et al. 2003). Both quercetin and catechin have been described to inhibit kinases (Maekawa et al. 2000), which suggests a potential inhibition of caveolae-mediated uptake.

PCBS AND PPAR SIGNALING

PPARs appear to possess potent anti-inflammatory signaling properties (Tedgui and Mallat 2001; Tham, Wang, and Rutledge 2003). In addition to regulating gene transcription via PPAR responsive elements, PPARs have recently been shown to modulate gene transcription by interfering with other transcription factor pathways. For example, PPARs have been shown to down-regulate inflammatory response genes by negatively interfering with the NF-kB, AP-1, and STAT transcriptional pathways (Delerive et al. 1999; Jiang, Ting, and Seed 1998; Marx et al. 1999). Both PPARa and PPARg are expressed in the vasculature, including endothelial cells, and their agonists such as fibrates (for a) and glitazones (for g) have repeatedly been shown to reduce inflammation in response to a variety of stimuli. For example, rosiglitazone markedly reduced inflammation, which may be due in part to down-regulation of proinflammatory genes such as COX-2 (Cuzzocrea et al. 2004; Na and Surh 2003).

Very little is known about the effect of environmental contaminants on PPAR signaling. There is evidence that the classical AhR ligand 2,3,7,8-tetrachlorodibenzo-p-dioxin (TCDD) can suppress PPARg expression (Alexander et al. 1998; Hanlon et al. 2003). It has been suggested that AhR functions may antagonize PPAR functions (Remillard and Bunce 2002). Our preliminary data suggest that PCBs can down-regulate PPAR activation. We also have preliminary data that show PPAR agonist-mediated protection against COX-2 and IL-6 expression induced by PCB 77. We hypothesize that PCBs contribute to an endothelial inflammatory response in part by down-regulating PPAR signaling.

DIET AND PROTECTION AGAINST PCB-INDUCED CYTOTOXICITY

We and others have demonstrated that PCBs can induce oxidative stress (Hennig et al. 2005). PCBs also can alter the activity of critical antioxidant enzymes, such as catalase, superoxide dismutase (Rodriguez-Ariza et al. 2003), glutathione reductase, glutathione peroxidase, and glutathione transferase activities (Jin et al. 2001; Twaroski et al. 2001; Twaroski, O'Brien, and Robertson 2001). Our data, demonstrating that vitamin E protects against PCB-induced vascular endothelial cell dysfunction (Slim et al. 1999), have critical health implications and suggest unique protective properties of this nutrient against toxic effects of certain environmental contaminants.

Furthermore, plant-based compounds including phenolics, flavonoids, isoflavones, terpenes, and so on, are reported to have antioxidant and antiatherogenic properties (Kris-Etherton et al. 2002). Little is known about the effect of

168   Hennig et al.

antioxidant and anti-inflammatory plant-derived nutrients on environmental contaminant-induced cytotoxicity. Recent studies suggest that potential protective mechanisms of nutrients against cytotoxicity induced by environmental chemicals, and especially polycyclic aromatic hydrocarbons, involve modification of AhR and CYP1A1 expression and activation (Casper et al. 1999; Henry et al. 1999; Shertzer et al. 1999; Williams et al. 2000). Our own data suggest that protective effects of flavonoids, like catechin and quercetin, are initiated upstream from CYP1A1 and that these flavonoids may be of value for inhibiting the toxic effects of PCBs on vascular endothelial cells (Ramadass et al. 2003). We hypothesize that antioxidant nutrients and related bioactive compounds common in fruits and vegetables protect against environmental toxic insult to the vascular endothelium by down-regulation of signaling pathways involved in inflammatory responses and atherosclerosis.

SUMMARY

There is a great need to explore the nutritional paradigm in environmental toxicology to improve our understanding of the relationship between nutrition and exposure to environmental toxins and diseases such as atherosclerosis (Hennig et al. 2004a). Many environmental contaminants exhibit human toxicity and disease via oxidative stress-sensitive signaling pathways. Thus, it is conceivable that nutrients which can contribute to cellular oxidative stress also can exacerbate or amplify environmental toxicity. On the other hand, nutrients which have antioxidant and/or anti-inflammatory activity could reduce or prevent compromised health or disease induction as a result of exposure to certain environmental pollutants. Our data suggest that PCBs and fatty acids are taken up by lipid rafts and that they can induce oxidative stress and inflammatory responses (fig. 16.1). We also propose that specific PCBs can decrease PPAR activation and thus contribute to inflammatory responses within the vascular endothelium. Coexposure to antioxidants may protect against these effects. In addition to nutrition as it relates to health and disease, one of the emerging issues in modern toxicological sciences is the modification of environmental toxicity by nutrients. Conversely, alterations of the biological or metabolic activity of nutrients by environmental pollutants may be equally important. Humans are increasingly

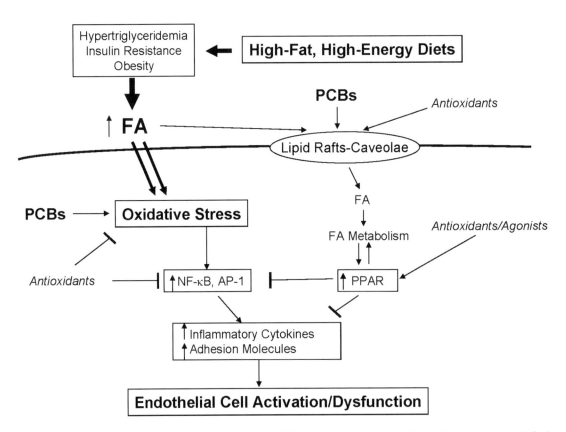

Figure 16.1.   We propose that PCBs and fatty acids (FA) cross membranes and contribute to an amplified induction of oxidative stress and inflammatory response. We also propose that specific PCBs can decrease PPAR activation and thus contribute to inflammatory responses within the vascular endothelium. Coexposure to antioxidants may protect against these effects.

exposed to environmental toxins, mostly as the result of modern industrial development. These emerging issues of nutritional biomarkers, coupled with new research technologies, appear to be of critical significance, because nutrition may be the most sensible means to develop primary prevention strategies of diseases associated with many environmental toxic insults.

ACKNOWLEDGMENTS

This work was supported in part by grants from NIH/NIEHS (ES 07380) and the Kentucky Agricultural Experiment Station.

REFERENCES

Alexander, D. L., L. G. Ganem, P. Fernandez-Salguero, F. Gonzalez, and C. R. Jefcoate. 1998. "Aryl-Hydrocarbon Receptor Is an Inhibitory Regulator of Lipid Synthesis and of Commitment to Adipogenesis." *J. Cell. Sci.* Pt. 22, 111:3311–22.

Austin, M. A., B. McKnight, K. L. Edwards, C. M. Bradley, M. J. McNeely, B. M. Psaty, J. D. Brunzell, and A. G. Motulsky. 2000. "Cardiovascular Disease Mortality in Familial Forms of Hypertriglyceridemia: A 20-Year Prospective Study." *Circulation* 101 (24): 2777–82.

Bertazzi, P. A., I. Bernucci, G. Brambilla, D. Consonni, and A. C. Pesatori. 1998. "The Seveso Studies on Early and Long-Term Effects of Dioxin Exposure: A Review." *Environ. Health Perspect.* 106 (Suppl. 2.): 625–33.

Boll, M., L. W. Weber, B. Messner, and A. Stampfl. 1998. "Polychlorinated Biphenyls Affect the Activities of Gluconeogenic and Lipogenic Enzymes in Rat Liver: Is There an Interference with Regulatory Hormone Actions?" *Xenobiotica* 28 (5): 479–92.

Casper, R. F., M. Quesne, I. M. Rogers, T. Shirota, A. Jolivet, E. Milgrom, and J. F. Savouret. 1999. "Resveratrol Has Antagonist Activity on the Aryl Hydrocarbon Receptor: Implications for Prevention of Dioxin Toxicity." *Mol. Pharmacol.* 56 (4): 784–90.

Choi, W., S. Y. Eum, Y. W. Lee, B. Hennig, L. W. Robertson, and M. Toborek. 2003. "PCB 104-Induced Proinflammatory Reactions in Human Vascular Endothelial Cells: Relationship to Cancer Metastasis and Atherogenesis." *Toxicol. Sci.* 75 (1): 47–56.

Cuzzocrea, S., B. Pisano, L. Dugo, A. Ianaro, P. Maffia, N. S. Patel, R. Di Paola, A. Ialenti, T. Genovese, P. K. Chatterjee, M. Di Rosa, A. P. Caputi, and C. Thiemermann. 2004. "Rosiglitazone, a Ligand of the Peroxisome Proliferator-Activated Receptor-Gamma, Reduces Acute Inflammation." *Eur. J. Pharmacol.* 483 (1): 79–93.

Dalton, T. P., J. K. Kerzee, B. Wang, M. Miller, M. Z. Dieter, J. N. Lorenz, H. G. Shertzer, D. W. Nerbert, and A. Puga. 2001. "Dioxin Exposure Is an Environmental Risk Factor for Ischemic Heart Disease." *Cardiovasc. Toxicol.* 1 (4): 285–98.

Delerive, P., K. De Bosscher, S. Besnard, W. Vanden Berghe, J. M. Peters, F. J. Gonzalez, J. C. Fruchart, A. Tedgui, G. Haegeman, and B. Staels. 1999. "Peroxisome Proliferator-Activated Receptor Alpha Negatively Regulates the Vascular Inflammatory Gene Response by Negative Cross-Talk with Transcription Factors NF-kappaB and AP-1." *J. Biol. Chem.* 274 (45): 32048–54.

Doi, A. M., Z. Lou, E. Holmes, C. Li, C. S. Venugopal, M. O. James, and K. M. Kleinow. 2000. "Effect of Micelle Fatty Acid Composition and 3,4,3′, 4′-Tetrachlorobiphenyl (TCB) Exposure on Intestinal [(14)C]-TCB Bioavailability and Biotransformation in Channel Catfish in Situ Preparations." *Toxicol. Sci.* 55 (1): 85–96.

Frank, P. G., S. E. Woodman, D. S. Park, and M. P. Lisanti. 2003. "Caveolin, Caveolae, and Endothelial Cell Function." *Arterioscler. Thromb. Vasc. Biol.* 23 (7): 1161–68.

Frank, P. G., H. Lee, D. S. Park, N. N. Tandon, P. E. Scherer, and M. P. Lisanti. 2004. "Genetic Ablation of Caveolin-1 Confers Protection Against Atherosclerosis." *Arterioscler. Thromb. Vasc. Biol.* 24 (1): 98–105.

Gustavsson, P., and C. Hogstedt. 1997. "A Cohort Study of Swedish Capacitor Manufacturing Workers Exposed to Polychlorinated Biphenyls (PCBs)." *Am. J. Ind. Med.* 32 (3): 234–39.

Hanlon, P. R., L. G. Ganem, Y. C. Cho, M. Yamamoto, and C. R. Jefcoate. 2003. "AhR- and ERK-Dependent Pathways Function Synergistically to Mediate 2,3,7,8-Tetrachlorodibenzo-*p*-Dioxin Suppression of Peroxisome Proliferator-Activated Receptor-Gamma1 Expression and Subsequent Adipocyte Differentiation." *Toxicol. Appl. Pharmacol.* 189 (1): 11–27.

Hay, A., and J. Tarrel. 1997. "Mortality of Power Workers Exposed to Phenoxy Herbicides and Polychlorinated Biphenyls in Waste Transformer Oil." *Ann. NY Acad. Sci.* 837:138–56.

Hennig, B., P. Meerarani, R. Slim, M. Toborek, A. Daugherty, A. E. Silverstone, and L. W. Robertson. 2002. "Proinflammatory Properties of Coplanar PCBs: In Vitro and In Vivo Evidence." *Toxicol. Appl. Pharmacol.* 181 (3): 174–83.

Hennig, B., G. Reiterer, M. Toborek, P. Meerarani, V. Saraswathi, A. Daugherty, E. Smart, S. Matveev, and L. W. Robertson. 2004a. "Dietary Lipids Modify PCB-Induced Changes in Plasma and Tissue Lipid Profiles in LDL Receptor Deficient Mice." Abstract. *FASEB J.* 18:A870.

Hennig, B., M. Toborek, L. G. Bachas, and W. A. Suk. 2004b. "Emerging Issues: Nutritional Awareness in Environmental Toxicology." *J. Nutr. Biochem.* 15 (4):194–95.

Hennig, B., M. Toborek, P. Ramadass, G. Ludewig, and L. W. Robertson. 2005. "Polychlorinated Biphenyls, Oxidative Stress and Diet." In *Reviews in Food and Nutrition Toxicity,* edited by V. R. Preedy and R. R. Watson, 93–128. Boca Raton, Fla.: CRC Press.

Hennig, B., G. Reiterer, M. Toborek, S. V. Matveev, A. Daugherty, E. Smart, and L. W. Robertson. 2005. "Dietary Fat Interacts with PCBs to Induce Changes in Lipid Metabolism in Mice Deficient in Low-Density Lipoprotein Receptor." *Environ. Health. Perspect.* 113(1):83–87.

Henry, E. C., A. S. Kende, G. Rucci, M. J. Totleben, J. J. Willey, S. D. Dertinger, R. S. Pollenz, J. P. Jones, and T. A. Gasiewicz. 1999. "Flavone Antagonists Bind Competitively with 2,3,7, 8-Tetrachlorodibenzo-*p*-Dioxin (TCDD) to the Aryl Hydrocarbon Receptor but Inhibit Nuclear Uptake and Transformation." *Mol. Pharmacol.* 55 (4): 716–25.

Jiang, C., A. T. Ting, and B. Seed. 1998. "PPAR-Gamma Agonists Inhibit Production of Monocyte Inflammatory Cytokines." *Nature* 391 (6662): 82–86.

Jin, X., S. W. Kennedy, T. Di Muccio, and T. W. Moon. 2001. "Role of Oxidative Stress and Antioxidant Defense in 3,3′,4,4′,5-Pentachlorobiphenyl-Induced Toxicity and Species-Differential Sensitivity in Chicken and Duck Embryos." *Toxicol. Appl. Pharmacol.* 172 (3): 241–48.

Kakela, R., and H. Hyvarinen. 1999. "Fatty Acid Alterations Caused by PCBs (Aroclor 1242) and Copper in Adipose Tissue around Lymph Nodes of Mink." *Comp. Biochem. Physiol. C. Pharmacol. Toxicol. Endocrinol.* 122 (1): 45–53.

Kamei, M., S. Ohgaki, T. Kanbe, M. Shimizu, S. Morita, I. Niiya, I. Matsui-Yuasa, and S. Otani. 1996. "Highly Hydrogenated Dietary Soybean Oil Modifies the Responses to Polychlorinated Biphenyls in Rats." *Lipids* 31 (11): 1151–56.

Kris-Etherton, P. M., K. D. Hecker, A. Bonanome, S. M. Coal, A. E. Binkoski, K. F. Hilpert, A. E. Griel, and T. D. Etherton. 2002. "Bioactive Compounds in Foods: Their Role in the Prevention of Cardiovascular Disease and Cancer." *Am. J. Med.* 113 (Suppl. 9B): 71S–88S.

Kunsch, C., and R. M. Medford. 1999. "Oxidative Stress as a Regulator of Gene Expression in the Vasculature." *Circ. Res.* 85 (8): 753–66.

Kwon, O., E. Lee, T. C. Moon, H. Jung, C. X. Lin, K. S. Nam, S. H. Baek, H. K. Min, and H. W. Chang. 2002. "Expression of Cyclooxygenase-2 and Pro-Inflammatory Cytokines Induced by 2,2′,4,4′,5,5′-Hexachlorobiphenyl (PCB 153) in Human Mast Cells Requires NF-Kappa B Activation." *Biol. Pharm. Bull.* 25 (9): 1165–68.

Machala, M., L. Blaha, J. Vondrácek, J. E. Trosko, J. Scott, and B. L. Upham. 2004. "Inhibition of Gap Junctional Intercellular Communication by Noncoplanar Polychlorinated Biphenyls: Inhibitory Potencies and Screening for Potential Mode(s) of Action." *Toxicol. Sci.* 76 (1): 102–11.

Maekawa, T., S. Kosuge, A. Karino, T. Okano, J. Ito, H. Munakata, and K. Ohtsuki. 2000. "Biochemical Characterization of 60S Acidic Ribosomal P Proteins from Porcine Liver and the Inhibition of Their Immunocomplex Formation with Sera from Systemic Lupus Erythematosus (SLE) Patients by Glycyrrhizin in Vitro." *Biol. Pharm. Bull.* 23 (1): 27–32.

Majkova, Z., H. Guo, G. Reiterer, W. Everson, E. Smart, M. Toborek, and B. Hennig. 2004. "Role of Caveolin in Proatherogenic Inflammation Caused by PCBs in Vascular Endothelial Cells." Abstract. *FASEB J.* 18:A870.

Malloy, M. J., and J. P. Kane. 2001. "A Risk Factor for Atherosclerosis: Triglyceride-Rich Lipoproteins." *Adv. Intern. Med.* 47:111–36.

Marx, N., G. K. Sukhova, T. Collins, P. Libby, and J. Plutzky. 1999. "PPARalpha Activators Inhibit Cytokine-Induced Vas-

cular Cell Adhesion Molecule-1 Expression in Human Endothelial Cells." *Circulation* 99 (24): 3125–31.

Matsusue, K., Y. Ishii, N. Ariyoshi, and K. Oguri. 1999. "A Highly Toxic Coplanar Polychlorinated Biphenyl Compound Suppresses Delta5 and Delta6 Desaturase Activities Which Play Key Roles in Arachidonic Acid Synthesis in Rat Liver." *Chem. Res. Toxicol.* 12 (12): 1158–65.

Matveev, S., X. Li, W. Everson, and E. J. Smart. 2001. "The Role of Caveolae and Caveolin in Vesicle-Dependent and Vesicle-Independent Trafficking." *Adv. Drug Deliv. Rev.* 49 (3):237–50.

Muller, J. M., R. A. Rupec, and P. A. Baeuerle. 1997. "Study of Gene Regulation by NF-Kappa B and AP-1 in Response to Reactive Oxygen Intermediates." *Methods* 11 (3): 301–12.

Na, H. K., and Y. J. Surh. 2003. "Peroxisome Proliferator-Activated Receptor Gamma (PPARgamma) Ligands as Bifunctional Regulators of Cell Proliferation." *Biochem. Pharmacol.* 66 (8): 1381–91.

Nabi, I. R., and P. U. Le. 2003. "Caveolae/Raft-Dependent Endocytosis." *J. Cell. Biol.* 161 (4): 673–77.

Pol, A., S. Martin, M. A. Fernandez, C. Ferguson, A. Carozzi, R. Luetterforst, C. Enrich, and R. G. Parton. 2004. "Dynamic and Regulated Association of Caveolin with Lipid Bodies: Modulation of Lipid Body Motility and Function by a Dominant Negative Mutant." *Mol. Biol. Cell.* 15 (1):99–110.

Quadri, S. A., A. N. Qadri, M. E. Hahn, K. K. Mann, and D. H. Sherr. 2000. "The Bioflavonoid Galangin Blocks Aryl Hydrocarbon Receptor Activation and Polycyclic Aromatic Hydrocarbon-Induced Pre-B Cell Apoptosis." *Mol. Pharmacol.* 58 (3): 515–25.

Ramadass, P., P. Meerarani, M. Toborek, L. W. Robertson, and B. Hennig. 2003. "Dietary Flavonoids Modulate PCB-Induced Oxidative Stress, CYP1A1 Induction, and AhR-DNA Binding Activity in Vascular Endothelial Cells." *Toxicol. Sci.* 76 (1): 212–19.

Remillard, R. B., and N. J. Bunce. 2002. "Linking Dioxins to Diabetes: Epidemiology and Biologic Plausibility." *Environ. Health Perspect.* 110 (9): 853–58.

Rodriguez-Ariza, A., M. J. Rodriguez-Ortega, J. L. Marenco, O. Amezcua, J. Alhama, and J. Lopez-Barea. 2003. "Uptake and Clearance of PCB Congeners in Chamaelea Gallina: Response of Oxidative Stress Biomarkers." *Comp. Biochem. Physiol. C. Toxicol. Pharmacol.* 134 (1): 57–67.

Savouret, J. F., A. Berdeaux, and R. F. Casper. 2003. "The Aryl Hydrocarbon Receptor and Its Xenobiotic Ligands: A Fundamental Trigger for Cardiovascular Diseases." *Nutr. Metab. Cardiovasc. Dis.* 13 (2): 104–13.

Schroeder, F., A. M. Gallegos, B. P. Atshaves, S. M. Storey, A. L. McIntosh, A. D. Petrescu, H. Huang, O. Starodub, H. Chao, H. Yang, A. Frolov, and A. B. Kier. 2001. "Recent Advances in Membrane Microdomains: Raft, caveolae, and Intracellular Cholesterol Trafficking." *Exp. Biol. Med.* 226:873–90.

Shertzer, H. G., A. Puga, C. Chang, P. Smith, D. W. Nebert, K. D. Setchell, and T. P. Dalton. 1999. "Inhibition of CYP1A1 Enzyme Activity in Mouse Hepatoma Cell Culture by Soybean Isoflavones." *Chem. Biol. Interact.* 123 (1): 31–49.

Slim, R., M. Toborek, L. W. Robertson, and B. Hennig. 1999. "Antioxidant Protection against PCB-Mediated Endothelial Cell Activation." *Toxicol. Sci.* 52 (2): 232–39.

Smart, E. J., G. A. Graf, M. A. McNiven, W. C. Sessa, J. A. Engelman, P. E. Scherer, T. Okamoto, and M. P. Lisanti. 1999. "Caveolins, Liquid-Ordered Domains, and Signal Transduction." *Mol. Cell. Biol.* 19 (11): 7289–7304.

Tedgui, A., and Z. Mallat. 2001. "Anti-Inflammatory Mechanisms in the Vascular Wall." *Circ. Res.* 88 (9): 877–87.

Tham, D. M., Y. X. Wang, and J. C. Rutledge. "Modulation of Vascular Inflammation by PPARs." *Drug News Perspect.* 16 (2): 109–16.

Tilson, H. A., P. R. Kodavanti, W. R. Mundy, and P. J. Bushnell. 1998. "Neurotoxicity of Environmental Chemicals and Their Mechanism of Action." *Toxicol. Lett.* 102–3:631–35.

Trigatti, B. L., R. G. Anderson, and G. E. Gerber. 1999. "Identification of Caveolin-1 as a Fatty Acid Binding Protein." *Biochem. Biophys. Res. Commun.* 255:34–39.

Twaroski, T. P., M. L. O'Brien, and L. W. Robertson. 2001. "Effects of Selected Polychlorinated Biphenyl (PCB) Congeners on Hepatic Glutathione, Glutathione-Related Enzymes, and Selenium Status: Implications for Oxidative Stress." *Biochem. Pharmacol.* 62 (3): 273–81.

Twaroski, T. P., M. L. O'Brien, N. Larmonier, H. P. Glauert, and L. W. Robertson. 2001. "Polychlorinated Biphenyl-Induced Effects on Metabolic Enzymes, AP-1 Binding, Vitamin E, and Oxidative Stress in the Rat Liver." *Toxicol. Appl. Pharmacol.* 171 (2): 85–93.

Williams, S. N., H. Shih, D. K. Guenette, W. Brackney, M. S. Denison, G. V. Pickwell, and L. C. Quattrochi. 2000. "Comparative Studies on the Effects of Green Tea Extracts and Individual Tea Catechins on Human CYP1A Gene Expression." *Chem. Biol. Interact.* 128 (3): 211–29.

17

Developmental Neurotoxicity of PCBs in Mice:
Critical Period of Brain Development and Effects of Interaction

Per Eriksson, *Uppsala University*

In the field of developmental toxicology it is essential to identify critical stages when chemical agents can be harmful. Several epidemiological studies indicate that exposure to environmental pollutants during early human development can have deleterious effects on cognitive development in childhood. Such exposure may also be involved in the slow, implacable induction of neurodegenerative disorders and/or interfere with the normal aging process.

By using the mouse as an animal model we can study the effect of a single toxicant administered directly to animals during different stages of the brain growth spurt (BGS). Interacting effects between different toxicants and the interaction between neonatal and adult exposure can also be studied in a controlled manner. Our investigations have shown that low-dose exposure to certain polychlorinated biphenyls (PCBs) (both ortho- and coplanar) during a defined period of rapid development of the neonatal brain and cholinergic system, in the neonatal mouse, can give rise to irreversible changes in adult brain function. This early exposure to PCB can also potentiate and/or modify reactions to adult exposure to xenobiotics.

The disturbed spontaneous behavior and impaired learning and memory were shown to develop over time, indicating a time-response/time-dependent effect. This suggests that certain PCBs might be involved in the slow, implacable induction of neurodegenerative disorders and/or interfere with normal aging processes.

A recent study also indicates that developmental exposure to both PCB and PBDE (a brominated flame retardant, or BFR) can cause enhanced developmental neurotoxic effects suggesting possible interactive effects between PCBs and the other environmental agents like PBDEs.

INTRODUCTION

Several epidemiological studies indicate that exposure to environmental pollutants during early human development can have deleterious effects on cognitive development in childhood. Such exposure may also be involved in the slow, implacable induction of neurodegenerative disorders and/or interfere with the normal aging process.

Developmental toxicology is a relatively new science, but its roots are firmly embedded in teratology. The choice of a test system and selection of endpoints that produce information about integrated reactions in an organism that are of importance for its functioning is vital for the prediction of developmental toxicity effects. In the field of developmental toxicology it is essential to identify critical stages when chemical agents can be harmful. Much of our current knowledge concerning the adverse effects of chemicals on mammalian development and human brain development relates to events during the early stages of development, during the embryonic part of gestation. The gestation period is divided into two major periods, the embryonic period and the fetal period. In humans the embryonic period constitutes 20 percent of the whole gestation period and the fetal period 80 percent. In research animals, such as mouse and rat, the opposite is seen: the embryonic period constitutes 80 percent of the gestation period and the fetal period 20 percent.

As the (central nervous system) CNS develops, every region of and structure in the brain follows an intricately planned and precisely timed developmental sequence, for example neurogenesis, differentiation and synaptogenesis. Vulnerable periods during ontogenesis of the CNS can nevertheless be divided into two major courses of events. The first includes early brain development, a period during which the brain acquires its general adult shape and when the spongioblasts and neuroblasts, precursors of glia cells and neurons, respectively, proliferate (Rodier 1980; Rogers and Kavlock 1996). Interference by xenobiotics during this period can cause malformation of the brain (Rodier et al. 1996; Rogers and Kavlock 1996). The second period coincides with the BGS (fig. 17.1). It is then that the brain undergoes a series of rapid fundamental developmental changes, including the maturation of axonal and dendritic outgrowth; establishment of neural connections; synaptogenesis; cell, axon and dendrite death; and the proliferation of glia cells with accompanying myelinization (Davison and Dobbing 1968; Kolb and Whishaw 1989). These cytoarchitectural changes are accompanied by a vast number of biochemical changes that transform the feto-neonatal brain into that of the mature adult. This is also the stage of development when animals acquire many new motor and sensory faculties (Bolles and Woods 1964), including advances in spontaneous motor behavior (Campbell, Lytle, and Fibiger 1969). The BGS does not take place at the same time point in all mammalian species. In the human, this period begins during the third trimester of pregnancy and continues throughout the first two years of life. In mice and rats the BGS is neonatal, spanning the first three to four weeks of life. One of the major neurotransmitters in the CNS is acetylcholine (ACh), which acts as the transmitter in the cholinergic pathways. In rodents, this transmitter system in the CNS undergoes rapid development during the first three to four weeks after birth (Coyle and Yamamura 1976; Fiedler, Marks, and Collins 1987), when gradually increasing numbers of muscarinic and nicotinic receptors appear in the cerebral cortex and hippocampus (Kuhar et al. 1980; Falkeborn et al. 1983; Slotkin, Orband-Miller, and Queen 1987; Fiedler, Marks, and Collins 1987). The cholinergic transmitter system is involved in many behavioral phenomena (Karczmar 1975) and is closely related to cognitive functions (Drachman 1977; Bartus et al. 1982).

It is known that exposure during the fetal period can cause functional anomalies of CNS that results in behavioral, cognitive, and motor defects. Furthermore, it is also known from developmental neuroscience that many potentially sensitive processes occur during the early postnatal period of brain maturation.

Therefore in the evaluation of developmental effects in mammals it is important to consider these differences between animals used in research and humans. In order to cover critical developmental phases occurring during the fetal and newborn periods, we have selected the mouse as the animal model, where we can follow this BGS.

By using the mouse as an animal model we can study the effect of a single toxicant administered directly to animals

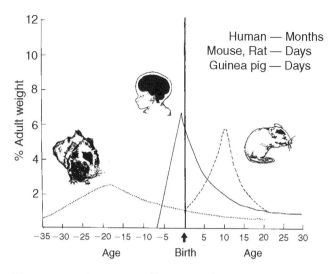

Figure 17.1.   Rate curves of brain growth in relation to birth in different species. Values are calculated at different time intervals for each species. (Data from Davison and Dobbing 1968 and Eriksson, unpublished). Illustrations made by Ylva Stenlund.

during different stages of the BGS. Interacting effects between different toxicants and the interaction between neonatal and adult exposure can also be studied in a controlled manner. This animal model allows us to isolate the effects of certain toxicants and to specify certain issues that can be difficult to solve in traditional developmental toxicity tests and also in epidemiological studies.

EXPOSURE TO PCBS DURING A CRITICAL PHASE OF BRAIN DEVELOPMENT IN THE NEONATAL MOUSE

In earlier studies we have shown that there is a critical and limited phase in the neonatal development of the mouse when the brain is vulnerable to insults from toxic agents, such as DDT (dichlorodiphenyltrichloroethane), organophosphorous compounds, and nicotine, leading to permanent disturbances of adult brain function and behavior (Ahlbom, Fredriksson, and Eriksson 1995; Eriksson, Ahlbom, and Fredriksson 1992; Eriksson 1997; Ankarberg, Fredriksson, and Eriksson 2001). The induction of behavioral disturbances by neonatal exposure to PCB in the mouse also seems to be limited to a short period of time during neonatal development. Two different types of PCBs congeners have been studied, one coplanar PCB, 3,3′,4,4′-tetrachlorobiphenyl (PCB 77), the other di-ortho-substituted PCB, 2,2′,5,5′-tetrachlorobiphenyl (PCB 52). In our earlier studies we have shown that neonatal exposure to PCB 77 can cause deranged spontaneous behavior and affect muscarinic cholinergic receptors in hippocampus in adult mice (Eriksson 1988; Eriksson, Lundkvist, and Fredriksson 1991), while PCB 52 affects spontaneous behavior and nicotinic cholinergic receptors in the cerebral cortex (Eriksson and Fredriksson 1996a). In order to study a critical phase during the BGS, the compounds were given as one single oral dose to mice at the age of either 3, 10, or 19 days. At the adult age of four months the mice were observed for spontaneous motor behavior and habituation ability. Normal habituation is defined here as a decrease in the spontaneous behavioral variables of locomotion, rearing, and total activity in response to the diminished novelty of the test chamber over a 60-minute test period, divided into three 20-minute periods, and is seen for control mice. However, in neonatal mice exposed to the coplanar PCB 77 (4.1 mg/kg b.wt.) a significant change was observed in spontaneous motor behavior and habituation at the adult age of four months, but only in mice given PCB 77 at the age of 10 days (fig. 17.2A). Similarly, in mice exposed to the di-ortho-substituted PCB 52 (4.1 mg/kg b.wt.), a significant behavioral aberration was observed in adult mice given PCB 52 at an age of either 3 or 10 days (fig. 17.2B). The adult mice (four months) exposed on postnatal day 10 displayed a nonhabituating behavioral profile, as earlier seen for both of these congeners (Eriksson and Fredriksson 1996a; Eriksson, Lundkvist, and Fredriksson 1991), namely, a hypoactive condition during the first part of the 60-minute test period, while toward the end of the period they became demonstrably hyperactive. Mice receiving PCB 52 on postnatal day 3 showed also a significant change in spontaneous behavior, but were able to habituate.

The amount of a toxic agent that is present in the brain at different neonatal ages might vary. Previous studies have shown a pronounced retention of lipophilic chlorinated hydrocarbons or their metabolites, such as DDT, PCB 153, and chlorinated paraffins, in the brain when administered on postnatal day 10 (Eriksson 1984; Eriksson and Darnerud 1985). The amounts of radioactivity found in the brain 24 hours and 7 days after administration of ^{14}C-labeled PCB 77 and PCB 52 at the three different ages are given in figure 17.3. The amount of PCB 77 in neonatal brain appears not to be higher in mice exposed on postnatal day 10 than on day 3 or 19. The absence of behavioral disturbances in mice exposed on postnatal day 3 or 19 therefore appears not to be due to differences in the amount of PCB 77 in brain. The retention of PCB 52 in neonatal brain at the different neonatal ages differs to some extent from that seen after exposure to PCB 77. Retention of PCB 52 is more pronounced compared to that seen after exposure to PCB 77. Retention of PCB 52 is similar to that found after neonatal exposure to PCB 153, showing a pronounced retention in animals receiving the substance on postnatal day 10 (Eriksson and Darnerud 1985). The behavioral disturbances seen after neonatal exposure on day 3 might be attributable to the amount of PCB 52 present on day 10 being enough to induce behav-

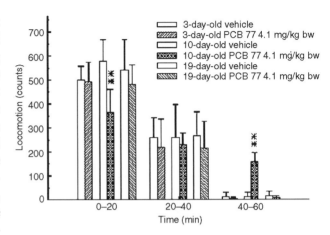

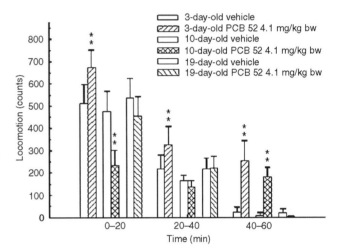

Figure 17.2.   Spontaneous behavior (locomotion) in four-month-old male NMRI mice exposed to a single oral dose of 3,3′,4,4′-tetrachlorobiphenyl (PCB 77, 4.1 mg [14 μmol]) (A), 2,2′,5,5′-tetrachlorobiphenyl (PCB 52, 4.1 mg [14 μmol]) (B), or the 20 percent fat-emulsion vehicle (10 ml) per kg body weight at an age of either 3, 10, or 19 days. Plain bars denote controls, hatched bars denote PCB 77 (A) or PCB 52 (B) treated mice (mean ± SD). The statistical evaluation was made by ANOVA with a split-plot design and pairwise testing with the Tukey HSD test. Statistical difference versus control is indicated by ** $P < 0.01$. From Eriksson (1998).

ioral disturbances, as the calculated amount present on day 10 was about 175 pmol/g brain. These studies indicate that certain PCBs can affect the developing brain, leading to irreversible effects in the adult mammal, when present during a critical stage of brain development.

This critical window for the induction of irreversible effects of PCBs is similar to that in our earlier studies. In various investigations we have observed the developing cholinergic system to be sensitive to environmental agents (Eriksson 1997). In a recent study we noticed a critical stage in the development of cholinergic nicotinic receptor subtypes, with

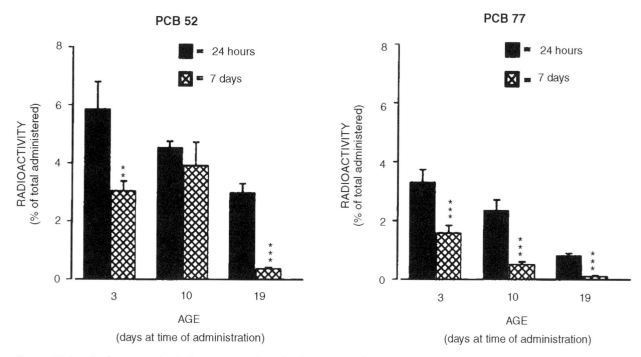

Figure 17.3.   Radioactivity levels (percentage of total radioactivity administered) in the neonatal mouse brain 24 hours and 7 days after a single oral dose of 1.5 MBq/kg b.wt. of either 3,3′,4,4′-tetrachloro[U-^{14}C]biphenyl (PCB 77) or 2,2′5,5′-tetrachloro[U-^{14}C]biphenyl (PCB 52) at an age of either 3, 10, or 19 days. The height of the bars represents the mean ± SD from four or five animals. The statistical difference between 24 hours and 7 days is indicated by * $P < 0.05$, ** $P < 0.01$, *** $P < 0.001$. From Eriksson (1998).

consequences for behavioral response at adult age. After neonatal mice had been exposed to nicotine at three different ages during the neonatal period, low-affinity nicotinic binding sites could not be found at any time, though the persistence of this effect was evident only in adult mice exposed on days 10–14. A spontaneous behavior test performed on adult mice (four months) did not reveal any difference between vehicle-treated and nicotine-treated mice, whereas when challenged with nicotine and observed for spontaneous behavior, their response to nicotine was hypoactive, though only in mice treated with nicotine on days 10–14. The response to nicotine by control mice and the other age categories was increased activity (Eriksson 1997; Eriksson, Ankarberg, and Fredriksson 2000). This observed change in both nicotinic receptors and nicotine-induced behavior also has been seen in adult mice neonatally exposed to PCB 52 (Eriksson and Fredriksson 1996a, 1996b).

NEUROTOXIC EFFECTS OF NEONATAL EXPOSURE TO ORTHO-SUBSTITUTED AND COPLANAR PCB CONGENERS

In several studies we found that neonatal exposure to different single PCB congeners during this defined critical phase of neonatal brain development can induce neurotoxic effects

that become functionally evident in the adult animal. Induction of permanent aberration in spontaneous behavior has been observed following neonatal exposure to ortho-substituted PCBs such as 2,4,4′-tri- (PCB 28), 2,2′,5,5′-tetra- (PCB 52) (fig. 17.4) and 2,2′,4,4′,5,5′-hexachlorobiphenyls (PCB 153) (Eriksson and Fredriksson 1996a, 1996b, 2004), and after neonatal exposure to coplanar PCBs such as 3,3′,4,4′-tetra- (PCB 77), 3,3′,4,4′,5-penta- (PCB 126), and 3,3′,4, 4′,5,5′-hexachlorobiphenyl (PCB 169) (Eriksson, Lundkvist, and Fredriksson 1991; Eriksson and Fredriksson 1998, 2004). Moreover, this effect seems to worsen with age, as evident after neonatal exposure to PCB 153, PCB 126, and PCB 169. Furthermore, neonatal exposure to the di-ortho substituted PCBs, PCB 52, and PCB 153, and the coplanar PCBs, PCB 126, and PCB 169, also affected learning and memory functions in the adult animal. In animals with deficits in learning and memory function following neonatal exposure to PCB 52, the cholinergic nicotinic receptors in the cerebral cortex were affected, while after neonatal exposure to PCB 126, the cholinergic nicotinic receptors in the hippocampus were affected. By contrast, exposure to 2,2′,4,4′-tetra (PCB 47), 2,3,3′,4,4′-penta (PCB 105), 2,3′,4,4′,5-penta- (PCB 118), and 2,3,3′,4,4′,5-hexachlorobiphenyl (PCB 156), in the same dose-range, did not cause any significant change in the behavioral variables investigated (Eriksson 1998). PCB

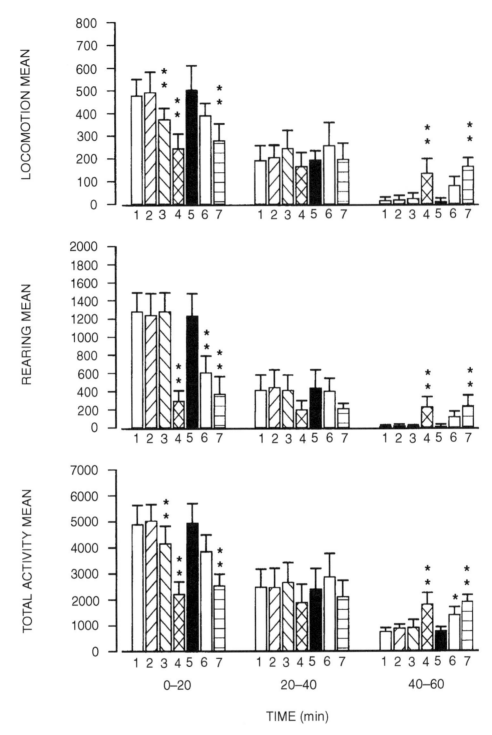

Figure 17.4. Spontaneous behavior of four-month-old NMRI male mice exposed to a single oral dose of either PCB 28 (2,4,4′-trichlorobiphenyl), PCB 52 (2,2′,5,5′-tetrachlorobiphenyl), or the 20 percent fat emulsion vehicle at a neonatal age of 10 days. The statistical evaluation was by ANOVA with a split-plot design and pairwise testing with the Tukey HSD test. The treatment groups are indicated by (1) control, 10 ml fat emulsion vehicle/kg body weight, (2) PCB 28, 0.18 mg (0.7 μmol)/kg body weight, (3) PCB 28, 0.36 mg (1.4 μmol)/kg body weight, (4) PCB 28, 3.6 mg (14 μmol)/kg body weight, (5) PCB 52, 0.20 mg (0.7 μmol)/kg body weight, (6) PCB 52, 0.41 mg (1.4 μmol)/kg body weight, (7) PCB 52, 4.1 mg (14 μmol)/kg body weight. The height of each bar represents the mean + SD of eight animals. The statistical difference versus control is indicated by **$P < 0.01$. From Eriksson and Fredriksson (1996a).

Table 17.1. Effects of Neonatal Exposure to PCB 28 or PCB 52 on Nicotinic Binding Sites and Affinity Constants in the Cerebral Cortex of Adult Mice.

Treatment (mg/kg b.wt.)	(N)	High-Affinity Site		Low-Affinity Site	
		(%)	k (nM)	(%)	k (μM)
Vehicle	(4)	78.5 ± 9.5	9.2	21.5 ± 9.5	18.2
PCB 28 (3.6)	(3)	78.3 ± 2.8	7.9	22.6 ± 4.3	7.4
PCB 52 (4.1)	(4)	98.4 ± 2.3	13.3	n.d.	n.d.

Note: Male NMRI mice received a single oral dose of either PCB 28, PCB 52, or the 20% fat emulsion vehicle on day 10. The mice were killed at an adult age of about six months. (^{3}H)nicotine/(-)nicotine competition curves performed on P2 fraction. Statistical evaluation using goodness of fit to one- and two-site models. The percentage values are means ± S.D., affinity constants (k) are geometric means, and n.d. = not detected. From Eriksson and Fredriksson 1996a.

105, PCB 118, and PCB 156 are examples of mono-ortho congeners that are "coplanar-like."

In mice exposed neonatally to PCB 52 (4.1 mg (14μmol)/kg b.wt.) and showing deficits in the learning and memory in Morris water maze and 8-ram maze tests (Eriksson and Fredriksson 1996a), the nicotinic cholinergic receptors were found to be affected (table 17.1). In these animals only the high affinity (HA) binding sites of nicotinic receptors were present in the cerebral cortex, whereas in controls and mice exposed to PCB 28 (3.6 mg(14μmol)/kg b.wt.), both HA and low affinity (LA) binding sites were present in proportions of about 80 and 20 percent, respectively. These proportions of HA- and LA binding sites are in agreement with previously reported proportions of nicotinic binding sites in adult mice (Nordberg et al. 1991; Eriksson, Ankarberg, and Fredriksson 2000). No significant effect was observed on the density of muscarinic acetylcholine receptors in hippocampus, which indicates that nicotinic acetylcholine receptors may be affected in animals with deficits in memory and learning. The effect on the cholinergic system was further supported from nicotine-induced behavior where PCB 52 treated mice responded with decreased activity whereas controls responded with increased activity. Moreover, the behavioral profile of new/reversal learning in a Morris water maze test in mice given PCB 126 (0.46 mg/kg b.wt.) showed similarities to those of hippocampally lesioned animals (Morris et al. 1982). This study showed that in mice exposed neonatally to PCB 126 (0.46 mg(1.4μmol)/kg b.wt.) and showing deficits in the learning and memory test, the density of nicotinic cholinergic receptors in hippocampus was significantly reduced (table 17.2) (Eriksson and Fredriksson 1998).

NEONATAL EXPOSURE AND BRAIN TISSUE LEVELS OF PCBS

The amount of PCB 77, PCB 52, and PCB 153 found in the brain 24 hours after a single oral administration to 10-

Table 17.2. Effects of Neonatal Exposure to PCB 126 on the Densities of Nicotinic Receptors in the Hippocampus of the Adult Mouse.

Treatment (mg/kg b.wt.)	(N)	(^{3}H)(-)Nicotine (pmol/g Protein)
Vehicle	(5)	19.6 ± 1.0
PCB 126, 0.046	(5)	19.9 ± 3.4
PCB 126, 0.46	(5)	15.5 ± 1.4[a]

[a]$p \leq 0.05$.

Note: Male NMRI mice received a single oral dose of PCB 126 or the 20% fat emulsion vehicle on day 10. The mice were killed at an adult age of six months. (^{3}H)-(-)nicotine binding (pmol/g protein, mean ± S.D.) was assessed in P2 fraction. Statistical evaluation was made by one-way ANOVA, and pairwise testing between treated groups and control by Duncan's test. From Eriksson and Fredriksson 1998a.

day-old mice is about 3–5 per mil of the administered dose (see fig. 17.2A and 17.2B; Eriksson and Darnerud 1985). Data on actual tissue levels of PCBs in infants are few. The amounts of the different PCBs given in the studies resulted in a brain tissue concentration (ppb levels) that can be of the same order of magnitude that has been observed in infants less than one year old (for ref see Gallenberg and Vodicnik 1989). In two postmortem studies from Japan and the United Kingdom an average value of 7 ppb was found in the cerebrum. In our studies the amount present in the brain during the critical phase and to induce behavioral defects is about 20 ppb. In human studies it is difficult to distinguish between exposure of offspring by transplacental and by breast milk transfer. However, both human and animal data from a variety of species suggest that accumulation of highly persistent chemicals via milk far exceeds the contribution made by maternal-fetal transfer (for ref see Gallenberg and Vodicnik 1989). In animal studies, using PCB 153, it was found that about 60 percent of the body burden was eliminated via milk during the first 5 days of lactation and virtually all by day 20 (Vodicnik and Lech 1980; Gallenberg and Vodicnik 1989). Various epidemiological studies on the developmental neurotoxic effects on gestational and/or lactational exposure to PCBs show the difficulty of predicting when developmental disorders can be induced. When considering the critical window for induction of permanent neurotoxic derangement during the BGS in mice, a corresponding period in humans starts during the third trimester of gestation and continues for several months after birth, and according to the "window" in our mice studies the critical phase would roughly be perinatal.

Many of the ortho-substituted PCB congeners studied, present in the environment, can accumulate in mammalian tissue. According to a report by Norén (1993) the concentration of ortho-substituted PCBs in human milk fat, measured in native Swedish mothers, was about 100-fold greater than coplanar PCBs such as PCB 77, PCB 126, and PCB 169. PCB 52 is of particular interest by virtue of its concentration in human milk fat, which has remained constant from 1972 to 1989. Thus the results of recent studies and observations in epidemiological studies (see Tilson and Harry 1994; Seegal 1996; Schantz 1999) indicate that the ortho-substituted PCBs could well be of importance when evaluating the risk of different PCB congeners to induce cognitive disorders.

LINKS BETWEEN BEHAVIOR AND FINDINGS REGARDING CHOLINERGIC RECEPTORS

The cholinergic system plays an important role in many behavioral phenomena, for example, learning and memory, neurological syndromes, audition, vision, and aggression (see Karczmar 1975). Pharmacological manipulations of the cholinergic system have been correlated in several studies with altered cognitive behavior (Decker and McGaugh 1991; Decker et al. 1995; Murray and Fibiger 1985).

The spontaneous behavior test showed that mice exposed neonatally to certain ortho-substituted and coplaner PCBs displayed a changed habituating behavior profile with regard to the three variables: locomotion, rearing, and total activity. Spontaneous behavior reflects a function dependent on the integration of a sensoric input into a motoric output and thus reveals the ability of animals to habituate to an environment and integrate new information with information previously attained and thereby be a measure of cognitive function. It is known that lesions of cholinergic nuclei, or cholinergic neurons projecting to hippocampus or cortex, can cause learning and memory deficits (see Berger-Sweeney 1994; Nabeshima 1993). The swim maze of Morris water-maze type, with its submerged platform, is designed to measure spatial learning, which is suggested to be corre-

lated with cholinergic function (Lindner and Schallert 1988; Whishaw 1985). Behavioral performance of tasks requiring attention and rapid processing of information in Man, and new/reversal learning and working memory in animals, have both been suggested to involve cholinergic transmission (see Hodges et al. 1991) and the cholinergic system is one of the major transmitter systems that correlate closely with cognitive function (Drachman 1977; Fibiger, Damsma, and Day 1991).

Regionally specific effects were observed following neonatal exposure to PCBs; PCB 126 affected cholinergic nicotinic receptors in the hippocampus of adult mice, but the proportions of HA and LA nicotinic receptors in the cerebral cortex were unaltered (Eriksson and Fredriksson 1998). Neonatal exposure to the ortho-substituted PCB 52 affected cholinergic nicotinic receptors in the cerebral cortex of adult mice, where only HA nicotinic receptors were found (Eriksson and Fredriksson 1996a). In the latter study, no significant effects on cholinergic muscarinic receptors were observed in hippocampus. Regionally specific changes in cholinergic receptors have also been found in adult mice exposed neonatally to the coplanar PCB 77 (3,3',4,4'-tetrachlorobiphenyl), where the density of muscarinic receptors was affected in hippocampus, but not in cerebral cortex (Eriksson, Lundkvist, and Fredriksson 1991). Although it has been shown that developmental exposure to PCB causes an even distribution of different PCB congeners in the brain, regardless of planarity (Eriksson and Darnerud 1985; Ness, Schantz, and Hansen 1994), our results, together with our earlier findings, suggest that there may be regionally specific effects of different PCB congeners in the brain. Whether or not these effects can be linked to varying effects of different PCBs during brain development in the medial septal area (MSA), with cholinergic projections to hippocampus, and the nucleus basalis magnocellularis (nBM), with cholinergic projections to cerebral cortex, remains to be established. It is known that the cholinergic basal forebrain system is important in the neural circuitry of mnemonic processing, and lesions of MSA and nBM can lead to defects in spatial memory (see Hodges et al. 1991; Berger-Sweeney et al. 1994).

Human epidemiological studies suggest that perinatal exposure to PCBs can have developmental neurotoxic effects (Fein et al. 1984; Jacobson, Jacobson, Humphrey 1990; Jacobson and Jacobson 1996; Rogan et al. 1988). Experimental studies in animals have shown that commercial mixtures of PCBs can cause behavioral aberrations and changes in brain neurotransmitter metabolism (see Seegal and Shain 1992; Seegal and Schantz 1994; Seegal 1996). Exposure of mice, rats, and monkeys to commercial mixtures of PCBs during development has been shown to produce long-term neurobehavioral changes (see Tilson, Jacobson, and Rogan 1990; Tilson and Harry 1994). Whether exposure to PCB can contribute to neuronal disorders is of special interest. The development of neurodegenerative disorders appears to be a complex interaction between genetic and environmental factors, e.g. the development of Alzheimer´s disease (AD) and Parkinson´s disease (PD) (James and Nordberg 1995). Neurodegenerative disorders such as AD are characterized by the impairments of memory and cognitive functions. The cholinergic system in particular has been implicated as an important role in aging and memory defect disorders. Dysfunctions of the cholinergic system have been shown to cause learning and memory impairment (Bartus et al. 1982; Overstreet and Russell 1991). Although AD is a disorder involving several neurotransmitter systems, the cholinergic system is severely and consistently affected (for review, see Whitehouse and Au 1986). In normal aging there is usually a change and/or decrease in cholinergic receptors (Nordberg and Winblad 1986; Nordberg, Alafuzoff, and Winblad 1992; Narang 1995).

POSSIBLE INTERACTIVE EFFECTS BETWEEN PCB AND OTHER ENVIRONMENTAL TOXICANTS

We have in different investigations seen that several environmental toxicants such as PCBs, nicotine, and certain BFRs and polybrominated diphenyl ethers (PBDEs) can affect the cholinergic receptor in the cerebral cortex and/or hippocampus, namely, the nicotinic receptors (Eriksson 1998; Ankarberg 2003; Viberg 2004).

Nicotine is an agent known to affect the cholinergic transmitter function (Nordberg et al. 1989; Nordberg 1993). Nicotine, which can be found in certain pesticides, makes its impact on human health as a component in tobacco products. Nicotine is known to be one of the most commonly used dependence engendering substances (Henningfield and Woodson 1988). Nicotine is an agonist for nicotinic receptors, but can also mediate the release of neurotransmitters such as ACh, dopamine, norepinephrine and serotonin (Summers and Giacobini 1995; Wonnacott 1997). The effects of neonatal exposure to low doses of nicotine on spontaneous behavior and nicotine-induced behavior in four-month-old mice and on the development of nicotinic receptors in the brain showed that nicotine can prevent the development of LA nicotinic binding sites in the cerebral cortex and that this exposure induces a different behavioral response to nicotine in adult animals (Nordberg et al. 1991; Eriksson, Ankarberg, and Fredriksson 2000; Ankarberg, Fredriksson, and Eriksson 2001) and can lead to learning and memory defects in adult mice (Ankarberg, Fredriksson, and Eriksson 2001). These effects are similar to those observed after neonatal exposure to PCB 52 (Eriksson and Fredriksson 1996a, 1996b). In a recent study we observed that coexposure to

both nicotine and PCB 52 affects the development of nicotinic receptors, measured by binding of alfa-bungarotoxin. Alfa-bungarotoxin, an antagonist to ACh, is suggested to bind to nicotinic receptor subtype consisting of alfa$_7$ subunits (Seguela et al. 1993), showing similarities to the LA nicotine binding sites. Alfa-bungarotoxin bound significantly less in mice neonatally exposed to both nicotine and PCB 52. This interactive effect was also evident in the spontaneous behavior response to nicotine at adult age, where the neonatally coexposed animals showed an increased response to nicotine (Ankarberg, Fredriksson, and Eriksson 1998).

BFRs are a novel group of global environmental contaminants (de Boer et al. 1998; de Wit 2002; Andersson et al. 1981). Within this group, PBDEs constitute a class that are found in electrical appliances, building materials, and textiles. PBDEs are persistent compounds that appear to have an environmental dispersion similar to that of polychlorinated biphenyls and DDT (de Wit 2002). While we observe a decrease for PCBs and DDT, the PBDEs have been found to increase in the environment and in human mother's milk (Manchester-Neesvig, Valters, and Sonzogni 2001; Meironyté, Norén, and Bergman 1999; Norén and Meironyté 2000).

In recent studies we have seen that certain PBDEs, such as 2,2′,4,4′-tetrabromodiphenyl ether (PBDE 47), 2,2′,4, 4′,5-pentabromodiphenyl ether (PBDE 99), 2,2′,4,4′,5,5′-hexabromodiphenyl ether (PBDE 153), and 2,2′,3,3′,4,4′, 5,5′,6,6′-decabromodiphenyl ether (PBDE 209) can cause developmental neurotoxic effects, like those observed for certain PCBs. Neonatal exposure to those PBDEs have been shown to cause deranged spontaneous behavior, for example, hyperactivity, reduced habituation capability and learning and memory deficits. Neonatal exposure to PBDE 99 or PBDE 153 has also been shown to affect the cholinergic system, manifested as altered behavioral response to the cholinergic agent nicotine and also reduced amount of nicotinic receptors in hippocampus (Eriksson, Jakobsson, and Fredriksson 2001; Eriksson et al. 2002; Viberg, Fredriksson, and Eriksson 2002; Viberg, Fredriksson, and Eriksson 2003).

In a recent study we have seen that animals neonatally exposed to the combined low dose of PCB 52 (1.4 µmol/kg bw) + PBDE 99 (1.4 µmol/kg bw) showed significantly impaired spontaneous motor behavior at the age of four months and six months (fig 17.5) (Eriksson, Fischer, and Fredriksson 2003). This deranged spontaneous behavior was also seen in mice exposed to the high dose of PCB 52 (14 µmol/kg bw) and the high dose of PBDE 99 (14 µmol/kg bw), but not at the low dose (1.4 µmol/kg bw). The effect on spontaneous behavior was even more pronounced in mice receiving the combined dose of PCB 52 (1.4 µmol/kg bw) + PBDE 99 (1.4 µmol/kg bw) compared to mice neonatally exposed to just the high dose of PCB 52 (14 µmol/kg

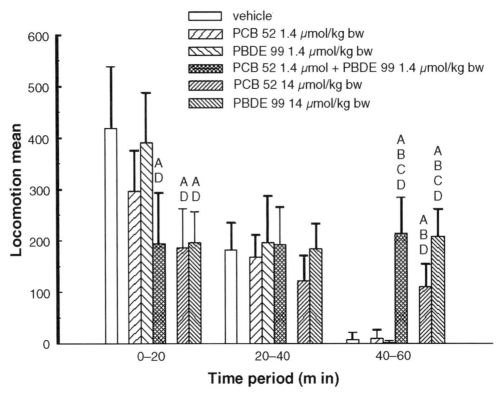

Figure 17.5.   Spontaneous behavior in four-month-old NMRI male mice exposed to a single oral dose of PCB 52 (2,2′,5,5′-tetrachlorobiphenyl), PBDE 99 (2,2′,4,4′,5-pentabromodiphenyl ether), coexposure to PCB 52 and PBDE 99, or the 20 percent fat emulsion vehicle at a neonatal age of 10 days. The statistical evaluation was by ANOVA with a split-plot design and pairwise testing with the Tukey HSD test. The heights of the bars represents mean ± SD. A = significant difference from vehicle ($P < 0.01$); B = significant difference from PCB 52 1.4 μmol/kg bw. ($P < 0.01$); C = significant difference from PCB 52 14 μmol/kg bw; D = significant difference from PBDE 991.4 μmol/kg bw. ($P < 0.01$). From Eriksson et al. (2004).

bw). Furthermore, in animals exposed to the combined dose of PCB 52 (1.4 μmol/kg bw) + PBDE 99 (1.4 μmol/kg bw), and the high dose of PCB 52, the defects worsen with age as the habituation capability was significantly worse in six-month-old mice compared to four-month-old mice.

That developmental effects on behavioral and cholinergic variables are similar between PCB and PBDE suggests similar mechanisms of action between PBDEs and PCBs. However, the observed interaction between PBDE 99 and PCB 52, with an effect significantly more pronounced than the 5 times higher dose of just PCB 52, indicates that additional mechanisms can be involved and/or that different brain regions are affected.

INCREASED SUSCEPTIBILITY IN ADULTS NEONATALLY EXPOSED TO ORTHO-SUBSTITUTED PCBS

A naturally occurring exposure circumstance is a combination of neonatal (perinatal) exposure and later adult exposure to

various environmental toxicants. In earlier studies we have noticed that neonatal exposure to an environmental neurotoxic agent such as DDT can lead to increased susceptibility in adults to short-acting insecticides such as bioallethrin (Eriksson et al. 1993; Johansson, Fredriksson, and Eriksson 1995) and paraoxon (Johansson, Fredriksson, and Eriksson 1996). Among the effects observed were behavioral disturbances, including learning and memory deficits, and changes in cholinergic receptors. A time-response effect was also observed; thus the aberrant behavior in mice exposed neonatally to DDT and to bioallethrin or paraoxon as adults was more pronounced two months after the adult exposure, compared with that 24 hours after adult exposure. Recently, we have observed that neonatal exposure to PCB 52 (both 0.8 and 4.1 mg/kg b.wt.) can potentiate the susceptibility of adult mice on renewed exposure to PCB 52 (4.1 mg/kg b. wt.). It was particularly interesting that, 24 hours after the adult exposure to PCB 52, there were no further disturbances in spontaneous behavior, yet such additional changes were observed two months later (fig. 17.6). These animals became significantly more active than mice exposed only to PCB 52 during the neonatal period. This aberration in spontaneous behavior was found to develop over time, indicating a time-response/time-dependent effect. The dose used for adult exposure had no significant effect on neonatally untreated animals. This indicates that in adult mice, susceptibility to PCB may be acquired, in varying degree, from PCB exposure during perinatal life when the maturational

processes of the developing brain and CNS are at a stage of critical vulnerability. These results indicate that PCB might be involved in the slow, implacable induction of neurodegenerative disorders and/or interfere with normal aging processes. This short period of low-dose exposure to PCB during the neonatal period seems sufficient to potentiate adult susceptibility to a new interventive exposure not having any persistent effect in untreated animals, and changes

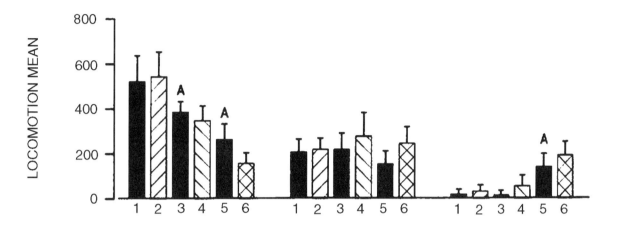

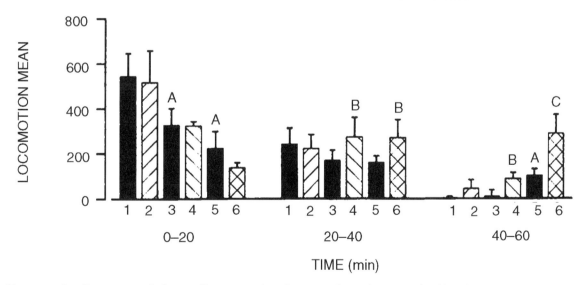

Figure 17.6.   Spontaneous behavior (locomotion) in four-month- and six-month-old male NMRI mice following (i) a single oral exposure to 2,2′,5,5′-tetrachlorobiphenyl (PCB 52, 0.8 or 4.1 mg/kg body weight) on day 10 after birth and (ii) a single oral dose of PCB 52 (4.1 mg/kg body weight) at age four months. Controls received the 20 percent fat-emulsion vehicle (10 ml/kg body weight). The treatment groups are denoted as (*1*) vehicle—vehicle, (*2*) vehicle—PCB 52 (4.1 mg/kg), (*3*) PCB 52 (0.8 mg/kg)—vehicle, (*4*) PCB 52 (0.8 mg/kg)—PCB 52 (4.1 mg/kg), (*5*) PCB 52 (4.1 mg/kg)—vehicle, (*6*) PCB 52 (4.1 mg/kg)—PCB 52 (4.1 mg/kg). The statistical evaluation was by ANOVA with a split-plot design and pairwise testing with the Tukey HSD test. A = significantly different from vehicle-vehicle, $P < 0.01$; B = significantly different from its respective control, $P < 0.05$; C = significant different from its respective control, $P < 0.01$. From Eriksson (1998).

developing over time are an indication of an accelerated dysfunctional process caused by the adult exposure. This indicates that susceptibility to different agents in adult life is not necessarily inherited. It might be acquired during a critical period of neonatal brain development through exposure to environmental toxicants.

CONCLUDING REMARKS

Our investigations have shown that low-dose exposure to certain PCBs during the period of rapid development of the neonatal brain (so-called brain growth spurt) and cholinergic system, in the mouse, can give rise to irreversible changes in adult brain function.

Susceptibility to developing such irreversible disturbances can be limited to a defined developmental phase of the brain growth spurt, and of the developing cholinergic system in the perinatal/neonatal brain. The increased susceptibility to PCB at adult age indicates that neonatal exposure to PCB can potentiate and/or modify reactions to adult exposure to xenobiotics. In addition, the disturbed spontaneous behavior, and impaired learning and memory, were shown to develop over time, indicating a time-response/time-dependent effect. This indicates that certain PCBs might be involved in the slow, implacable induction of neurodegenerative disorders and/or interfere with normal aging processes.

The recent findings that developmental exposure to both PCB and PBDE can cause enhanced developmental neurotoxic effects indicate possible interactive effects between PCBs and the newer environmental agents PBDEs.

ACKNOWLEDGMENTS

Financial support was provided by the Swedish Environmental Protection Board, the Foundation for Strategic Environmental Research, and the Swedish Research Council for Environmental, Agricultural Sciences and Spatial Planning.

REFERENCES

Ahlbom, J., A. Fredriksson, and P. Eriksson. 1995. "Exposure to an Organophosphate (DFP) during a Defined Period in Neonatal Life Induces Permanent Changes in Brain Muscarinic Receptors and Behavior in Adult Mice." *Brain Res.* 677:13–19.

Andersson, Ö., and G. Blomkvist. 1981. "Polybrominated Aromatic Pollutants Found in Fish in Sweden." *Chemosphere* 10: 1051–60.

Ankarberg, E. 2003. "Neurotoxic Effects of Nicotine During Neonatal Brain Development: Critical Period and Adult Susceptibility." In *Comprehensive Summaries of Uppsala Dissertation from the Faculty of Science and Technology,* 48. Uppsala: Uppsala University.

Ankarberg, E., A. Fredriksson, and P. Eriksson. 1998. "Interactive Effects of PCB and Nicotine Administered during the Neonatal Brain Development." *Organohalogen Compounds* 37: 93–96.

Ankarberg, E., A. Fredriksson, and P. Eriksson. 2001. "Neurobehavioral Defects in Adult Mice Neonatally Exposed to Nicotine: Changes in Nicotine-Induced Behavior and Maze Learning Performance." *Behav. Brain Res.* 123:185–92.

Bartus, R. T., R. L. D. Dean, B. Beer, and A. S. Lippa. 1982. "The Cholinergic Hypothesis of Geriatric Memory Dysfunction." *Science* 217:408–14.

Berger-Sweeney, J., S. Heckers, M. M. Mesulam, R. G. Wiley, D. A. Lappi, and M. Sharma. 1994. "Differential Effects on Spatial Navigation of Immunotoxin-Induced Cholinergic Lesions of the Medial Septal Area and Nucleus Basalis Magnocellularis." *J. Neurosci.* 14:4507–19.

Bolles, R. G., and P. J. Woods. 1964. "The Ontogeny of Behavior in the Albino Rat." *Animal Behavior* 12:427–41.

Campbell, B. A., L. D. Lytle, and H. C. Fibiger. 1969. "Ontogeny of Adrenergic Arousal and Cholinergic Inhibitory Mechanisms in the Rat." *Science* 166:635–37.

Coyle, J. T., and H. I. Yamamura. 1976. "Neurochemical Aspects of the Ontogenesis of Cholinergic Neurons in the Rat Brain." *Brain Res.* 118:429–40.

Davison, A. N., and J. Dobbing. 1968. *Applied Neurochemistry.* Oxford: Blackwell.

de Boer, J., P. G. Wester, H. J. Klamer, W. E. Lewis, and J. P. Boon. 1998. Do Flame Retardants Threaten Ocean Life? Letter. *Nature* 394:28–29.

Decker, M. W., J. D. Brioni, A. W. Bannon, and S. P. Arneric. 1995. "Diversity of Neuronal Nicotinic Acetylcholine Receptors: Lessons from Behavior and Implications for CNS Therapeutics." *Life Sci.* 56:545–70.

Decker, M. W., and J. L. McGaugh. 1991. "The Role of Interactions between the Cholinergic System and Other Neuromodulatory Systems in Learning and Memory." *Synapse* 7: 151–68.

de Wit, C. A. 2002. "An Overview of Brominated Flame Retardants in the Environment." *Chemosphere* 46:583–624.

Drachman, D. A. 1977. "Cognitive Function in Man: Does Cholinergic System Have a Special Role?" *Neurology* 27:783–90.

Eriksson, P. 1984. "Age-Dependent Retention of [14C]DDT in the Brain of the Postnatal Mouse." *Toxicol. Lett.* 22:323–28.

Eriksson, P. 1988. "Effects of 3,3′,4,4′-tetrachlorobiphenyl in the Brain of the Neonatal Mouse." *Toxicology* 49:43–48.

Eriksson, P. 1997. "Developmental Neurotoxicity of Environmental Agents in the Neonate." *Neurotoxicology* 18:719–26.

Eriksson, P. 1998. *Perinatal Developmental Neurotoxicity of PCBs.* Stockholm: Swedish Environmental Protection Agency.

Eriksson, P., J. Ahlbom, and A. Fredriksson. 1992. "Exposure to DDT during a Defined Period in Neonatal Life Induces Permanent Changes in Brain Muscarinic Receptors and Behavior in Adult Mice." *Brain Res.* 582:277–81.

Eriksson, P., A. Ankarberg, and A. Fredriksson. 2000. "Exposure to Nicotine during a Defined Period in Neonatal Life Induces Permanent Changes in Brain Nicotinic Receptors and in Behavior of Adult Mice." *Brain Res.* 853:41–48.

Eriksson, P., and P. O. Darnerud. 1985. "Distribution and Retention of Some Chlorinated Hydrocarbons and a Phthalate in the Mouse Brain during the Pre-Weaning Period." *Toxicology* 37:189–203.

Eriksson, P., and A. Fredriksson. 1996a. "Developmental Neurotoxicity of Four Ortho-Substituted Polychlorinated Biphenyls in the Neonatal Mouse." *Environ. Toxicol. Pharmacol.* 1: 155–65.

Eriksson, P., and A. Fredriksson. 1996b. "Neonatal Exposure to 2,2′,5,5′-Tetrachlorobiphenyl Causes Increased Susceptibility in the Cholinergic Transmitter System at Adult Age." *Environ. Toxicol. Pharmacol.* 1:217–220.

Eriksson, P., and A. Fredriksson. 1998. "Neurotoxic Effects in Adult Mice Neonatally Exposed to 3,3′,4,4′,5-Pentachlorobiphenyl or 2,3,3′,4,4′-Pentachlorobiphenyl: Changes in Brain Nicotinic Receptors and Behavior." *Environmental Toxicology and Pharmacology* 5:17–27.

Eriksson, P., and A. Fredriksson. 2004. "Neonatal Exposure to 2,2′,4,4′,5,5′-Hexachlorobiphenyl or 3,3′,4,4′,5,5′-Hexachlorobiphenyl Causes Behavioral Derangements in Mouse That Deteriorate with Age." Unpublished results.

Eriksson, P., C. Fischer, and A. Fredriksson 2003. "Co-exposure to a Polybrominated Diphenyl Ether (PBDE 99) and an Ortho-Substituted PCB (PCB 52) Enhances Developmental Neurotoxic Effects." *Organohalogen Compounds* 61:81–83.

Eriksson, P., E. Jakobsson, and A. Fredriksson. 2001. "Brominated Flame Retardants: A Novel Class of Developmental Neurotoxicants in Our Environment?" *Environ. Health Perspect.* 109: 903–8.

Eriksson, P., U. Johansson, J. Ahlbom, and A. Fredriksson. 1993. "Neonatal Exposure to DDT Induces Increased Susceptibility to Pyrethroid (Bioallethrin) Exposure at Adult Age—Changes in Cholinergic Muscarinic Receptor and Behavioral Variables." *Toxicology* 77:21–30.

Eriksson, P., U. Lundkvist, and A. Fredriksson. 1991. "Neonatal Exposure to 3,3′,4,4′-Tetrachlorobiphenyl: Changes in Spontaneous Behavior and Cholinergic Muscarinic Receptors in the Adult Mouse." *Toxicology* 69:27–34.

Eriksson, P., H. Viberg, E. Jakobsson, U. Orn, and A. Fredriksson. 2002. "A Brominated Flame Retardant, 2,2′,4,4′,5-Pentabromodiphenyl Ether: Uptake, Retention, and Induction of Neurobehavioral Alterations in Mice during a Critical Phase of Neonatal Brain Development." *Toxicol. Sci.* 67:98–103.

Falkeborn, Y., C. Larsson, A. Nordberg, and P. Slanina. 1983. "A Comparison of the Regional Ontogenesis of Nicotine- and Muscarine-Like Binding Sites in Mouse Brain." *J. Devl. Neuroscience* 1:187–90.

Fein, G. G., J. L. Jacobson, S. W. Jacobson, P. M. Schwartz, and J. K. Dowler. 1984. "Prenatal Exposure to Polychlorinated Biphenyls: Effects on Birth Size and Gestational Age." *J. Pediatr.* 105:315–20.

Fibiger, H. C., G. Damsma, and J. C. Day. 1991. "Behavioral Pharmacology and Biochemistry of Central Cholinergic Neurotransmission." *Adv. Exp. Med. Biol.* 295:399–414.

Fiedler, E. P., M. J. Marks, and A. C. Collins. 1987. "Postnatal Development of Cholinergic Enzymes and Receptors in Mouse Brain." *J. Neurochem.* 49:983–90.

Gallenberg, L. A., and M. J. Vodicnik. 1989. "Transfer of Persistent Chemicals in Milk." *Drug Metab. Rev.* 21:277–317.

Henningfield, J. E., and P. P. Woodson. 1988. "Dose Related Action of Nicotine on Behavior and Physiology: Review and Implication for Replacement Therapy for Nicotine Dependence." *J. Subst. Abuse* 1:301–17.

Hodges, H., Y. Allen, T. Kershaw, P. L. Lantos, J. A. Gray, and J. Sinden. 1991. "Effects of Cholinergic-Rich Neural Grafts on Radial Maze Performance of Rats after Excitotoxic Lesions of the Forebrain Cholinergic Projection System—I. Amelioration of Cognitive Deficits by Transplants into Cortex and Hippocampus but Not into Basal Forebrain." (Published erratum appears in *Neuroscience* 47 (1): 249.) *Neuroscience* 45:587–607.

Jacobson, J. L., and S. W. Jacobson. 1996. "Intellectual Impairment in Children Exposed to Polychlorinated Biphenyls in Utero." *N. Engl. J. Med.* 335:783–89.

Jacobson, J. L., S. W. Jacobson, and H. E. Humphrey. 1990. "Effects of Exposure to PCBs and Related Compounds on Growth and Activity in Children." *Neurotoxicol. Teratol.* 12: 319–26.

James, J. R., and A. Nordberg. 1995. "Genetic and Environmental Aspects of the Role of Nicotinic Receptors in Neurodegenerative Disorders: Emphasis on Alzheimer's Disease and Parkinson's Disease." *Behav. Gen.* 25:149–59.

Johansson, U., A. Fredriksson, and P. Eriksson. 1995. "Bioallethrin Causes Permanent Changes in Behavioral and Muscarinic Acetylcholine Receptor Variables in Adult Mice Exposed Neonatally to DDT." *Eur. J. Pharmacol.* 293:159–66.

Johansson, U., A. Fredriksson, and P. Eriksson. 1996. "Low-Dose Effects of Paraoxon in Adult Mice Exposed Neonatally to DDT: Changes in Behavioral and Cholinergic Receptor Variables." *Environmental Toxicology and Pharmacology* 2: 307–14.

Karczmar, A. G. 1975. *Cholinergic Influences on Behavior.* New York: Raven Press.

Kolb, B., and I. Q. Whishaw. 1989. "Plasticity in the Neocortex: Mechanisms Underlying Recovery from Early Brain Damage." *Prog. Neurobiol.* 32:235–76.

Kuhar, M. J., N. J. Birdsall, A. S. Burgen, and E. C. Hulme. 1980. "Ontogeny of Muscarinic Receptors in Rat Brain." *Brain Res.* 184:375–83.

Lindner, M. D., and T. Schallert. 1988. "Aging and Atropine Effects on Spatial Navigation in the Morris Water Task." *Behav. Neurosci.* 102:621–34.

Manchester-Neesvig, J. B., K. Valters, and W. C. Sonzogni. 2001. "Comparison of Polybrominated Diphenyl Ethers (PBDEs) and Polychlorinated Biphenyls (PCBs) in Lake Michigan Salmonids." *Environ. Sci. Technol.* 35:1072–77.

Meironyté, D., K. Norén, and A. Bergman. 1999. "Analysis of Polybrominated Diphenyl Ethers in Swedish Human Milk: A Time Dependent Trend Study, 1972–1997." *Journal of Toxicology and Environmental Health,* Pt. A, 58:329–41.

Morris, R. G., P. Garrud, J. N. Rawlins, and J. O'Keefe. 1982. "Place Navigation Impaired in Rats with Hippocampal Lesions." *Nature* 297:681–83.

Murray, C. L., and H. C. Fibiger. 1985. "Learning and Memory Deficits after Lesions of the Nucleus Basalis Magnocellularis: Reversal by Physostigmine." *Neuroscience* 14:1025–32.

Nabeshima, T. 1993. "Behavioral Aspects of Cholinergic Transmission: Role of Basal Forebrain Cholinergic System in Learning and Memory." *Progress in Brain Research* 98:405–11.

Narang, N. 1995. "In Situ Determination of M1 and M2 Muscarinic Receptor Binding Sites and mRNAs in Young and Old Rat Brains." *Mech. Ageing Dev.* 78:221–39.

Ness, D. K., S. L. Schantz, and L. G. Hansen. 1994. "PCB Congeners in the Rat Brain: Selective Accumulation and Lack of Regionalization." *J. Toxicol. Environ. Health* 43:453–68.

Nordberg, A. 1993. "Neuronal Nicotinic Receptors and Their Implication in Aging and Neurodegenerative Disorders in Mammals." *J. Reprod. Fert. Suppl.* 46:145–54.

Nordberg, A., I. Alafuzoff, and B. Winblad. 1992. "Nicotinic and Muscarinic Subtypes in the Human Brain: Changes with Aging and Dementia." *J. Neurosci. Res.* 31:103–11.

Nordberg, A., K. Fuxe, B. Holmstedt, and A. Sundwall. 1989. "Nicotinic Receptors in the CNC—Their Role in Synaptic Transmission." In *Progress in Brain Research* 79:366. Amsterdam: Elsevier.

Nordberg, A., and B. Winblad. 1986. "Brain Nicotinic and Muscarinic Receptors in Normal Aging and Dementia." In *Alzheimer's and Parkinson's: Advances in Behavioral Biology,* edited by A. Fisher, I. Hanin, and C. Lachman, 95–108. New York: Plenum Press.

Nordberg, A., X. A. Zhang, A. Fredriksson, and P. Eriksson. 1991. "Neonatal Nicotine Exposure Induces Permanent Changes in Brain Nicotinic Receptors and Behavior in Adult Mice." *Brain Res. Dev. Brain Res.* 63:201–7.

Norén, K. 1993. "Contemporary and Retrospective Investigations of Human Milk in the Trend Studies of Organochlorine Contaminants in Sweden." *Sci. Tot. Environ.* 139/140:347–55.

Norén, K., and D. Meironyté. 2000. "Certain Organochlorine and Organobromine Contaminants in Swedish Human Milk in Perspective of Past 20–30 Years." *Chemosphere* 40:1111–23.

Overstreet, D. H., and R. W. Russel. 1991. "Animal Models of Memory Disorders." In *Animal Models in Psychiatry II,* edited by A. A. Boulton, G. B. Baker, and M. T. Martin-Iversen, 315–68. Atlantic Highlands, N.J.: Humana Press.

Rodier, P. M. 1980. "Chronology of Neuron Development: Animal Studies and Their Clinical Implications." *Dev. Med. Child. Neurol.* 22:525–45.

Rodier, P. M., J. L. Ingram, B. Tisdale, S. Nelson, and J. Romano. 1996. "Embryological Origin for Autism: Developmental Anomalies of the Cranial Nerve Motor Nuclei." *J. Comp. Neurol.* 370:247–61.

Rogan, W. J., B. C. Gladen, K. L. Hung, S. L. Koong, L. Y. Shih, J. S. Taylor, Y. C. Wu, D. Yang, N. B. Ragan, and C. C. Hsu. 1988. "Congenital Poisoning by Polychlorinated Biphenyls and Their Contaminants in Taiwan." *Science* 241:334–36.

Rogers, J. M., and R. L. Kavlock. 1996. "Developmental Toxicology." In *Casarett and Doull's Toxicology: The Basic Science of Poisons,* edited by C. D. Klassen, 301–31. New York: McGraw-Hill.

Schantz, S. L. 1999. "Neurotoxic Food Contaminants: Polychlorinated Biphenyls (PCBs) and Related Compounds." In *Introduction to Neurobehavioral Toxicology: Food and Environment,* edited by R. J. M. Niesink, R. M. A. Jaspers, L. M. W. Kornet, J. M. van Hee, and H. A. Tilson, 252–82. Boca Raton, Fla.: CRC Press.

Seegal, R., and S. L. Schantz. 1994. *Neurochemical and Behavioral Sequele of Exposure to Dioxins and PCBs.* New York: Plenum Press.

Seegal, R., and W. Shain. 1992. "Neurotoxicity of Polychlorinated Biphenyls: The Role of Ortho-Substituted Congeners in Altering Neurochemical Function." *Vulnerable Brain Environ. Risks.* 2:169–95.

Seegal, R. F. 1996. "Epidemiological and Laboratory Evidence of PCB-Induced Neurotoxicity." *Crit. Rev. Toxicol.* 26:709–37.

Seguela, P., J. Wadiche, K. Dineley-Miller, J. A. Dani, and J. W. Patrick. 1993. "Molecular Cloning, Functional Properties, and Distribution of Rat Brain Alpha 7: A Nicotinic Cation Channel Highly Permeable to Calcium." *J. Neurosci.* 13(2): 596–604.

Slotkin, T. A., L. Orband-Miller, and K. L. Queen. 1987. "Development of [3H] Nicotine Binding Sites in Brain Regions of Rats Exposed to Nicotine Prenatally via Maternal Injections or Infusions." *J. Pharmacol. Exp. Ther.* 242:232–37.

Summers, K. L., and E. Giacobini. 1995. "Effects of Local and Repeated Systemic Administration of (-)Nicotine on Extracellular Levels of Acetylcholine, Norepinephrine, Dopamine, and Serotonin in Rat Cortex." *Neurochem. Res.* 20:753–59.

Tilson, H. A., and G. J. Harry. 1994. "Developmental Neurotoxicology of Polychlorinated Biphenyls and Related Compounds." In *The Vulnerable Brain and Environmental Risks,* vol. 2, edited by R. L. Isaacson and K. F. Jensen, 267–79. New York: Plenum Press.

Tilson, H. A., J. L. Jacobson, and W. J. Rogan. 1990. "Polychlorinated Biphenyls and the Developing Nervous System: Cross-Species Comparisons." *Neurotoxicol. Teratol.* 12:239–48.

Viberg, H. 2004. *Developmental Neurotoxicity of Polybrominated Diphenyl Ethers.* Acta Univ. Ups., Comprehensive Summaries of Uppsala Dissertations from the Faculty of Science and Technology. Uppsala: Uppsala University.

Viberg, H., A. Fredriksson, and P. Eriksson. 2002. "Neonatal Exposure to the Brominated Flame Retardant 2,2′,4,4′,5-Pentabromodiphenyl Ether Causes Altered Susceptibility in the Cholinergic Transmitter System in the Adult Mouse." *Toxicol. Sci.* 67:104–7.

Viberg, H., A. Fredriksson, and P. Eriksson. 2003. "Neonatal Exposure to Polybrominated Diphenyl Ether (PBDE 153) Disrupts Spontaneous Behavior, Impairs Learning and Memory, and Decreases Hippocampal Cholinergic Receptors in Adult Mice." *Toxicol. Appl. Pharmacol.* 192:95–106.

Vodicnik, M. J., and J. J. Lech. 1980. "The Transfer of 2,4,5,2′,4′,5′-Hexachlorobiphenyl to Fetuses and Nursing Offspring." Pt. 1, "Disposition in Pregnant and Lactating Mice and Accumulation in Young." *Toxicol. Appl. Pharmacol.* 54:293–300.

Whishaw, I. Q. 1985. "Cholinergic Receptor Blockade in the Rat Impairs Locale but Not Taxon Strategies for Place Navigation in a Swimming Pool." *Behav. Neurosci.* 99:979–1005.

Whitehouse, P. J., and K. S. Au. 1986. "Cholinergic Receptors in Aging and Alzheimer's Disease." *Prog. Neuropsychopharmacol. Biol. Psychiatry* 10:665–76.

Wonnacott, S. 1997. "Presynaptic Nicotinic ACh Receptors." *Trends Neurosci.* 20:92–98.

18

Comparison of Potencies of Individual PCB Congeners on Behavioral Endpoints in Animal Studies

Deborah C. Rice, *Maine Center for Disease Control and Prevention*

Polychlorinated biphenyls (PCBs) produce toxicity to multiple organ systems; however, for environmental exposures, the effects of primary importance are developmental neurotoxicity, developmental endpoints, immunotoxicity, and effects on thyroid function. Early studies in animal models used commercial mixtures, documenting toxicity in all these systems. The congener profile, to which humans are exposed following environmental exposure, and the profile in human tissues, is considerably different from those in commercial mixtures. It would be advantageous to understand the relative potencies of individual congeners to inform the risk assessment process. Unfortunately, relatively few congeners have been studied on any endpoint, and there is often little consistency between studies with regard to dosing protocol or outcome measures, making comparisons of potency between congeners difficult. Several tentative conclusions may be drawn, however:

1. Most relevant studies of neurotoxicity have identified LOAELs but not NOAELs, making determination of relative potency impossible. But it is clear that both dioxin-like and non-dioxin-like congeners are neurotoxic at doses that do not produce overt toxicity in dams or pups.
2. Both non-dioxin-like and dioxin-like congeners produce adverse effects on thyroid function.
3. Dioxin-like congeners produce effects on development, including decreased litter size and weight gain in the pups, and effects in pups related to perturbation of sex hormones. Non-dioxin-like congeners have been little studied, and sex-hormone mediated overt effects in pups have not been observed.

4. For immune effects, most congeners studied have been dioxin-like, and effects on immune function are documented. Non-dioxin-like congeners may also have effects on immune function, however.
5. The relative toxicity of the various congeners is somewhat dependent upon the system/endpoint. Effects have been observed at the lowest doses for congener 126; however, unbounded LOAELs have been documented for other congeners for some endpoints. There are not enough data to reach conclusions regarding relative toxicity of the congeners that have been studied.

PCBs are known to be toxic to multiple organ systems. The poisoning episodes in Taiwan and Japan identified developmental effects, including developmental neurotoxicity, as sensitive endpoints (Chen et al. 1992; Harada 1976). Epidemiological studies in a number of cohorts have identified adverse neuropsychological outcomes associated with developmental exposure to PCBs (Schantz, Widholm, and Rice 2003), including effects on IQ, executive function, and sexually dimorphic play behavior. Animal studies using commercial mixtures have also documented behavioral (Lilienthal et al. 1990; Lilienthal and Winneke 1991; Newland and Paletz 2000; Hany et al. 1999a; Taylor, Crofton, and MacPhail 2002) and sensory (Goldey et al. 1995; Herr, Goldey, and Crofton 1996; Herr et al. 2001) deficits following developmental exposure.

Hazard assessment for PCBs presents significant challenges. PCB mixtures consist of potentially as many as 209 congeners, with differing physiochemical and pharmacokinetic properties, different mechanisms of action, and differing effects and potencies in various organ systems. A dozen of the 209 congeners activate the aryl hydrocarbon (Ah) receptor, and so have dioxin-like properties and effects. These congeners have been assigned dioxin toxic equivalency factors (TEFs) by the World Health Organization (Van den Berg et al. 1998), which indicates potencies relative to each other and to tetrachlorinated dibenzo-p-dioxin (TCDD). This allows determination of the theoretical toxicity of any mixtures of dioxins and dioxin-like PCBs and other compounds. In contrast, there is not just a single mechanism of action responsible for the effects of the remaining non-dioxin-like PCB congeners, or, for that matter, for the effects of dioxin-like congeners that are not manifested through Ah receptor activation.

This presents a significant dilemma with respect to hazard assessment, since PCBs always occur together (and with dioxins), and the relative proportion will vary depending on source and medium (e.g., sediment, fish, human breast milk). Moreover, relative toxicity of various congeners will undoubtedly vary depending on endpoint. Unfortunately, there is no TEF-like approach for rank ordering these effects. This includes neurotoxicity, which is not believed to be Ah-receptor mediated. The mechanisms of neurotoxicity produced by PCBs are not completely elucidated. Effects on ryanodine receptor binding, calcium transport, and second messenger systems have been documented, including exploration of the relative potency of individual congeners (Wong and Pessah 1997; Wong et al. 1997, 2000; Kodavanti et al. 1995, 1996, 1998). However, the relationship of those relative potencies to *in vivo* behavioral or other neurotoxic effects is unknown. At present, the most straightforward method of determining relative potency is to assess effects of individual congeners on behavior. Unfortunately, few studies have been performed, both in terms of studies on one congener and the number of congeners that have been studied.

It is established that the fetus is more sensitive to the neurotoxic effects of PCBs than the mother (Schantz, Widholm, and Rice 2003), and therefore most experimental research has focused on developmental rather than adult exposure. This review compares the doses at which effects were observed for individual congeners in developmental behavioral studies. It is not a detailed review of the neurotoxic effects of PCBs in general or of PCB congeners in particular. Dioxin-like and non-dioxin-like congeners are discussed separately, since dioxin-like PCBs can currently be included in a dioxin risk assessment.

Doses chosen for neurotoxicity studies were obviously constrained by effects in other organ systems, that is, overt toxicity. The goal for neurotoxicity testing is to use doses that have no or minimal toxicity to the dam and no or minimal overt effects (e.g., reduced birth weight, retardation of weight gain, changes in attainment of developmental milestones) in the offspring.

DIOXIN-LIKE CONGENERS

Susan Schantz's laboratory has studied several dioxin- and non-dioxin-like congeners using the same protocol and behavioral endpoints. Long-Evans dams were gavaged from gestation day (GD) 10–16 in all studies. Litter size, pup mortality, gestation length, maternal and pup weight gain, thyroid hormone levels in the pups at weaning, and thymus weights in the pups were also recorded. One male and female per litter were tested on radial arm maze and delayed spatial alternation tasks using a maze. Radial arm maze performance is considered a test of working (short-term) memory, whereas the delayed spatial alternation task tests spatial memory and ability to change response strategy by requiring the subject to change response position following a "successful" (reinforced) response. The consistency of dosing regimen and behavioral tests allows direct comparison of individual congeners.

Effects were identified at the lowest dose for congener 126 (table 18.1 and fig. 18.1); the lowest doses studied were also for 126 because of its potent toxicity. Schantz et al. (1996) dosed rats with 1.0 or 0.25 ug/kg/day of congener 126. They reported decreased errors on radial arm maze performance at an unbounded LOAEL of 0.25 ug/kg/day, an effect which is considered to be adverse because it may result from failure of normal exploratory behavior. No effect was observed on delayed spatial alternation. A LOAEL of 1.0 ug/kg/day and a NOAEL of 0.25 ug/kg/day were found for changes in saccharine preference, a sexually dimorphic behavior (Amin et al. 2000). Females exhibited a decreased preference for saccharine, which represents a masculinization of response. A nonsignificant decrease in T_4 was observed in the high-dose group (Seo et al. 1995).

Rice and colleagues began dosing Long-Evans rats orally with 1.0 or 0.25 ug/kg/day of congener 126 several weeks before breeding and dosed the dams until weaning at postnatal day (PND) 21. One male and one female per litter were tested on a number of behavioral and sensory endpoints (Rice 1999; Rice and Hayward 1998, 1999a; Crofton and Rice 1999; Bushnell and Rice 1999; Geller, Bushnell, and Rice 2000). No effects were observed on sustained attention, visuospatial ability, spatial memory and adaptability, temporal organization, ability to inhibit responding, and willingness to expend increasing effort for a reward (i.e., performance on fixed interval, DRL, progressive ratio, spatial delayed alternation, cued target detection, and a signal detection task). A LOAEL of 1.0 ug/kg/day and a NOAEL

Table 18.1. Summary of Behavioral Effects of PCB Congeners in Developmental Behavioral Studies.

Congener	Dosing Protocol	LOAEL/NOAEL for Behavioral Endpoints	Effect	References
Dioxin-like congeners				
77	SD rat 0, 2, 8 mg/kg/day GD 10-16	LOAEL 2 mg/kg/day unbounded	↓ errors on radial arm maze, ↓ T$_4$ in females at weaning in high dose, n.e. reproductive/ developmental outcomes	Seo et al. 1995; Schantz et al. 1996
	SD rat 0, 2, 8 ug/kg/day GD 10-16	LOAEL 8 mg/kg/day; NOAEL 2 mg/kg/day	↓ saccharine preference females (masculinization)	Amin et al. 2000
	CD-1 mouse 0, 32 mg/kg/day GD 10-16	LOAEL 32 mg/kg/day unbounded	↑ activity, change in grip strength, impaired active avoidance and neuromuscular function	Tilson et al. 1979
	NMRI mouse 0, 0.41, 41 mg/kg PND 10	LOAEL 0.41 mg/kg; invalid experimental design/statistical analysis**	↓ habituation of activity	Eriksson, Lundkvist, and Fredriksson 1991
	CD-1 mouse 0, 32 mg/kg/day GD 10-16	LOAEL 32 mg/kg/day; invalid experimental design/statistical analysis**	↑ overall activity	Agrawal, Tilson, and Bondy 1981
118	SD rat 1, 4, 16 mg/kg/day GD 10-16	LOAEL 16 mg/kg/day; NOAEL 4 mg/kg/day	↑ errors delayed spatial alternation, ↓ body weight in pups at high dose, ↓ T$_4$ at weaning at both doses, ↓ body weight throughout nursing in the high dose	Schantz et al. 1995; Ness et al. 1993
	Lewis rat 0, 1, 5 mg/kg/2 days GD 10-20	LOAEL 1 mg/kg/day; invalid experimental design/statistical analysis for behavioral and developmental effects**	↑ reinforcers earned on random ratio visual discrimination task	Holene et al. 1995
	NMR1 mouse 0, 0.23, 0.46, 4.6 mg/kg PND 10	NOAEL 4.6 mg/kg; invalid experimental design/statistical analysis**	no effect activity, Morris water maze	Eriksson and Fredriksson 1996a
	Lewis rat 0, 2 ug/kg/2 days GD 10-20	LOAEL 2 ug/kg/2 days; unbounded; invalid experimental design/statistical analysis**	↑ FI response rate; ↑ rate and ↑ reinforcers earned, random ratio schedule	Holene et al. 1995, 1998

126	SD rat 0, 0.25, 1.0 ug/kg/day GD 10-16	LOAEL 0.25 ug/kg/day unbounded	↓ errors on radial arm maze, n.e. reproductive/developmental outcomes, n.e. on T_4	Seo et al. 1995; Schantz et al. 1996
	SD rat 0, 0.25, 1.0 ug/kg/day GD 10-16	LOAEL 1.0 ug/kg/day; NOAEL 0.25 ug/kg/day	↓ saccharine preference females (masculinization)	Amin et al. 2000
	LE rats 0, 0.25, 1.0 ug/kg/day prebreeding to weaning	LOAEL 1.0 ug/kg/day; NOAEL 0.25 ug/kg/day	Transient impairment on concurrent random interval-random interval, no impairment on several other tasks; ↑ auditory thresholds at 1.0 ug/kg/day, no effect on visual thresholds; ↓ anogenital distance high dose males ↓T_4 and TU, ↓ hemoglobin, ↓ RBC, ↓ hematocrit, ↑ cholesterol, ↑ GGT, ↑ BUN; ↓ platelets in dams, ↑ cholesterol in dams at higher dose; no effect on reproduction, developmental milestones, body weights; ↓T_4 in offspring at weaning	Rice 1999; Rice and Hayward 1998, 1999; Crofton and Rice 1999; Geller et al. 2000; Bushnell and Rice 1999
	Lewis rat 0, 10, 20 ug/kg/2 days GD 9-19	NOAEL 20 ug/kg/2 days unbounded (10 ug/kg/day);	No effect on visual discrimination	Bernhoft et al. 1994
	Lewis rat 0, 2 ug/kg/2 days GD 10-20	LOAEL 2 ug/kg/2 days unbounded; invalid experimental design/statistical analysis for behavioral tasks**	↑ FI response rate, ↑ and ↑ reinforcers earned, random ratio schedule; n.e. body weight, developmental milestones	Holene et al. 1995, 1998, 1999
156	NMRI mouse 0, 0.25, 0.51, or 5.1 mg/kg PND 10	NOAEL 5.1 mg/kg; invalid experimental design/statistical analysis**	No effect on activity or Morris water maze	Eriksson and Fredriksson 1996a
Non-dioxin-like congeners				
28	SD rat 0, 8, 32 mg/kg/day GD 10-16	LOAEL 32 mg/kg/day females; NOAEL 8 mg/kg/day	↑ errors delayed spatial alternation; ↓ weight gain in female pups, ↓ birth weight at high dose	Schantz et al. 1995
	NMRI mouse 0, 0.18, 0.36, 3.6 mg/kg PND 10	LOAEL 3.6 mg/kg; NOAEL 0.36 mg/kg; invalid experimental design/statistical analysis**	↓ habituation of activity	Eriksson and Fredriksson 1996a
52	NMRI mouse 0, 0.2, 0.41, or 4.1 mg/kg PND 10	LOAEL 4.1 mg/kg; invalid experimental design/statistical analysis**	↓ habituation of activity	Eriksson and Fredriksson 1996a, 1996b

(continued)

Table 18.1. Continued

Congener	Dosing Protocol	LOAEL/NOAEL for Behavioral Endpoints	Effect	References
95	SD rat 0, 8, 32 mg/kg/day GD 10-16	LOAEL 8 mg/kg/day unbounded	↓ errors radial arm maze; n.e. on reproductive/developmental measures; T_4 not measured	Schantz et al. 1997
153	SD rat 0, 16, 64 mg/kg/day GD 10-16	LOAEL 64 mg/kg/day; NOAEL 16 mg/kg/day	↑ errors on delayed spatial alternation in females; ↓ body weight during nursing; ↓T_4 at weaning in both doses	Schantz et al. 1995; Ness et al. 1993
	Lewis rat 5 mg/kg/2 days PND 3-13	LOAEL 5 mg/kg/2 days; invalid experimental design/statistical analysis for behavioral**	↑ fixed interval responding, ↓ acquisition of another intermittent schedule of reinforcement; ↓ body weight at weaning in males, n.e. on developmental milestones	Holene et al. 1998, 1999
Mixtures				
Mixture representative of breast milk	LE rats PCB 28 – 5.9%; PCB 77 – 0.00179%; PCB 101 – 1.4%; PCB 105 – 2.5%; PCB 118 – 7.3%; PCB 126 – 0.00828%; PCB 138 – 22.1%; PCB 146 – 3.13%; PCB 153 – 27.6%; PCB 156 – 3.82%; PCB 169 – 0.00373%; PCB 170 – 7.4%; PCB 180 – 14.03%; PCB 187 – 4.77% 40 ppm 50 days before mating-birth (approx. 4 mg/kg/day)	LOAEL 40 ppm unbounded (4 mg/kg/day)	↑ sweet preference in males (feminization)	Hany et al. 1999a
Mixture representative of breast milk	LE rats PCB 28 – 5.9%; PCB 77 – 0.00179%; PCB 101 – 1.4%; PCB 105 – 2.5%; PCB 118 – 7.3%; PCB 126 – 0.00828%; PCB 138 – 22.1%; PCB 14 – 3.13%;	LOAEL 5 ppm unbounded (0.5 mg/kg/day)	↑ sweet preference in males (feminization)	Kaya et al. 2002

Compound	Dose/Species	Effects	LOAEL	Reference
Mixture representative of breast milk	PCB 153 – 27.6%; PCB 156 – 3.82%; PCB 169 – 0.00373%; PCB 170 – 7.4%; PCB 180 – 14.03%; PCB 187 – 4.77% 0, 5, 20, or 40 ppm 50 days before mating–birth (0, 0.5, 2, or 4 mg/kg/day) Cynomolgus monkey 7.5 ug/kg/day birth – 20 weeks (52 – 1.5%; 66 – 2.9%; 74 – 10.4%; 105 – 4.4%; 118 – 2.8%; 138 – 17.5%; 153 – 18.6%; 156 – 4.7%; 157 – 1.5%; 180 – 2. 8%; 183 – 2.3%; 187 – 4.7%; 189 – 0.5%; 194 – 2.9%; 203 – 2.3%)		LOAEL 7.5 ug/kg/day unbounded	Rice 1997; Rice and Hayward 1999, 1997
Metabolites 4-OH-CB107	Wistar rat 0.5 or 5 mg/kg/day or 25 mg/kg/day Aroclor 1254 GD 10-16	↓ habituation open field, males; impaired passive avoidance, low dose males; ↑ auditory thresholds, females; ↓ T_4 PND 4, all groups; n.e., TSH	LOAEL 0.5 mg/kg/day unbounded	Meerts et al. 2004
Systemic administration 77, 47 or 77 + 47	LE rat 1.5 mg/kg/day s.c. 77 or 47 or 1.0 mg/kg/day 47 + 0.5 mg/kg/day 77 GD 7-18 by s.c. injection	↓ scotopic vision in females with 77 only, assessed electrophysiologically, ↓ birth weight and weight at weaning	LOAEL 1.5 mg/kg/day unbounded for 77 only	Kremer et al. 1999
77, 47 or 77 + 47	LE rat 0.5, 1.5 mg/kg/day 77 1.5 mg/kg/day 47 or mixture GD 7-18 by s.c. injection	differences in spatial pattern of activity PND 70 with 77, decreased latency in passive avoidance with 77 and combined treatment, ↑ activity all groups PND 340	LOAEL 0.5 mg/kg/day 77; LOAEL 1.5 mg/kg/day 47 or combination unbounded	Hany et al. 1999b

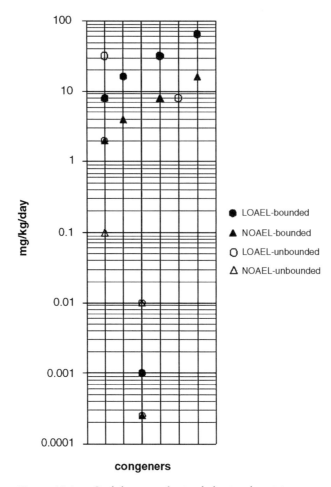

Figure 18.1.   Oral doses producing behavioral toxicity.

of 0.25 ug/kg/day were identified on a concurrent random-interval task, which requires changing response strategy in response to changes in environmental (in this case reinforcement) contingencies. Using littermates from the Rice study, increased auditory thresholds were identified at 1.0 but not 0.25 ug/kg/day (Crofton and Rice 1999), whereas no effects were observed on visual function as measured by flash-evoked potentials or a behavioral task (Geller, Bushnell, and Rice 2000). In the Rice cohort of rats, effects were found on other, nonbehavioral measures. Effects included decreased anogenital distance in the high-dose males, as well as decreased T_4, hemoglobin, red blood cell count, and hematocrit, and increased cholesterol, GGT, and BUN at both doses in the pups.

Schantz et al. (1996) dosed dams with 8 or 2 mg/kg/day of congener 77. An unbounded LOAEL of 2 mg/kg/day was identified on radial arm maze performance, with no effects on the delayed spatial alternation task. As with congener 126, rats exhibited decreased errors on the radial arm maze. A NOAEL of 2 mg/kg/day was identified by this group for changes in saccharine preference (Amin et al. 2000). LOAEL/NOAELs were 8 and 2 mg/kg/day for decreased thymus weight and decreased T_4 in females at weaning (Seo et al. 1995), with no effect on maternal or pup weight gain or other measures of reproductive success.

For congener 118, Schantz, Moshtaghian, and Ness (1995) identified a LOAEL of 16 mg/kg/day and a NOAEL of 4 mg/kg/day for increased errors on a delayed spatial alternation task. No effects were observed on radial arm maze. This is the opposite pattern to that observed with the other dioxin-like congeners. A decrease in T_4 at weaning was observed in the pups at both doses (Ness et al. 1993). Body weight was decreased in the high-dose pups throughout nursing (Ness et al. 1993).

Other studies identified effects at lower doses for 77 and 118, as well as providing data for congener 156 (table 18.1). However, the results are uninterpretable because littermates were treated as independent observations statistically, which may inflate α (Holson and Pearce 1992), such that the accuracy of p values cannot be known. Even though some studies used inbred strains of mice which presumably shared an identical genotype, other variables that may influence behavior such as intrauterine environment, maternal behavior, interactions with litter mates, and physical environment were shared within litters.

Based on the limited data available from these studies, it appears that the order of potency is 126 > 77 > 118. Based on the TEFs, 126 should have been 1,000 times more potent than 77 or 118. The lowest unbounded LOAEL of 2 mg/kg/day for 77 is 8,000 times higher than the unbounded LOAEL of 0.25 ug/kg/day for 126. For 118, the NOAEL of 4 mg/kg/day is 16,000 times greater than the 126 LOAEL of 0.25 ug/kg/day. These ratios are not strikingly dissimilar to results expected on a TEF basis. Nonetheless, it appears that congeners 77 and 118 are not equipotent on these behavioral tasks, equivalent TEFs notwithstanding.

NON-DIOXIN-LIKE CONGENERS

Schantz and colleagues studied the effects of prenatal exposure to congeners 28, 95 or 153 on radial arm maze and delayed spatial alternation performance. They administered congener 95 at 8 or 32 mg/kg/day and identified an unbounded LOAEL of 8 mg/kg/day for decreased errors on the radial arm maze (Schantz et al. 1997). No effect was observed on delayed spatial alternation performance. No effects were observed at either dose on any measure of reproduction or development. Thyroid hormone was not measured. For congener 28, a LOAEL of 32 and a NOAEL of 8 mg/kg/day were identified for a decrease in correct responses in female but not male offspring on the delayed spatial alternation task (Schantz, Moshtaghian, and Ness 1995). No effect was observed on radial arm maze performance. Decreased birth weight was found in high-dose females and decreased weight gain in female pups was observed at both doses. No effect on T_4 was observed at either dose. For 153, a LOAEL of 64 mg/kg/day and a NOAEL of 16 mg/kg/day were identified for increased errors on the delayed alternation task in females (Schantz, Moshtaghian, and Ness 1995), with no effect on radial arm maze performance. Decreased T_4 at weaning and decreased weight gain in the pups during nursing was observed in both groups (Ness et al. 1993).

From these limited data from one laboratory, it appears that the order of potency may be 95 > 28 > 153.

MIXTURES

The focus of the Hany et al. (1999a) study using a congener mixture formulated to represent human breast milk was effects on reproduction and measures of development in the offspring. However, sweet (saccharine) preference was also measured as a way of studying sexually dimorphic behavior. An approximate dose of 4 mg/kg/day resulted in increased preference for saccharine in males (feminization), whereas the same dose of Arochlor 1254 had no effect. This effect was replicated at an unbounded LOAEL of 0.5 mg/kg/day for the same mixture (Kaya et al. 2002). [Both 126 and 77 resulted in masculinization of female behavior with respect to saccharine preference (Amin et al. 2000).] The high dose produced decreased body weight, whereas both doses produced effects on sex steroids (Hany et al. 1999a; Kaya et al. 2002).

The Rice monkey study (Rice 1998, 1997; Rice and Hayward 1999b, 1997) dosed male monkeys from birth to 20 weeks of age with 7.5 ug/kg/day of a mixture designed to represent human breast milk. Blood PCB concentrations peaked at from 1.8–2.8 ppb for individual animals, and declined rapidly following cessation of dosing. All the congeners in the PCB mixture were ortho-substituted, and included dioxin-like and non-dioxin-like congeners. Behavioral testing began at about three years of age, more than two years after blood concentrations of the treated monkeys were at background. Deficits were observed on most of the behavioral tasks assessed. PCB-exposed monkeys exhibited learning/performance decrements on a delayed spatial alternation task, and tended to exhibit less accurate performance on a nonspatial discrimination reversal task, which measures adaptability (Rice and Hayward 1997). They made more responses on a fixed-interval schedule of reinforcement (Rice 1997), and failed to inhibit inappropriate responses on a DRL task (Rice 1998). Treated monkeys also performed differently than controls on a progressive ratio task, which requires progressively more responses for reinforcement (Rice and Hayward 1999b). In contrast, PCB-treated monkeys were not different from controls on a concurrent random interval-random interval task. The deficits observed in monkeys contrast with the findings in rats with congener 126 on the same tasks. In the rat studies, transient effects were observed on the concurrent random interval-random interval, with no effects on the other tasks. This is the opposite of the pattern observed in the monkey study.

No differences in body weight gain were observed in the monkey study; thyroid hormone levels were not measured during the period of dosing.

METABOLITES

The metabolite 4-OH-2,3,3′,4′,5-pentachlorobiphenyl (4-OH-CB107) and Aroclor 1254 were compared on several endpoints in Wistar rats (Meerts et al. 2004). Females in both dose groups failed to habituate in an open field (i.e., failed to decrease their activity over time as much as controls) at PND 130. Low-dose males had shorter response latencies in a passive avoidance task, indicating poorer learning of this simple task. Brain stem auditory evoked potentials (BAEPs) revealed increased thresholds at low frequencies in the Aroclor 1254 and high-dose 4-OH-CB107 groups. All groups exhibited a decrease in serum T_4 at four days of age.

SYSTEMIC ADMINISTRATION

The ability of systemic administration to predict effects produced via environmentally relevant exposure pathways is unknown, since normal absorption is bypassed, and metabolism may be different. Two studies were identified in which PCBs were administered systemically (Hany et al. 1999b, 2004). Congener 77 produced effects on vision and activity, with an unbounded LOAEL of 0.5 mg/kg/day administered on GD 7-18. Congener 47 produced an unbounded LOAEL of 1.5 mg/kg/day. The combination of 77 + 47 also produced unbounded LOAELs. Therefore no conclusions may be drawn concerning interaction of these congeners, since dose-response functions were not generated.

CONCLUSIONS

Effects on behavior were found for each of the individual congeners tested, as well as for the one metabolite studied and the congener mixtures. The pattern of deficits, on the two tasks used in Susan Schantz's laboratory, differs between congeners. Congeners 77, 126, and 95 produced the same pattern of effects: decreased errors on radial arm maze performance with no effect on delayed spatial alternation. In contrast, congeners 28, 118, and 153 produced increased errors on delayed spatial alternation, with no effect on radial arm maze learning. It is impossible with current knowledge to hypothesize potential mechanisms for this differential effect, since clearly it does not follow the coplanar/ortho-substituted or dioxin-like/non-dioxin-like dichotomies.

In a less straightforward comparison, the pattern of effects of 126 in rats and an environmentally relevant congener mixture (not including 126) in monkeys produced an opposite pattern of effects across several tasks. Clearly our understanding of the effects of these congeners, and the underlying mechanisms of action, is incomplete.

The dioxin-like congener 126 produced an unbounded LOAEL of 10^{-4} mg/kg/day. The two other dioxin-like congeners (77 and 118) produced effects at 10^0–10^1 mg/kg/day. The non-dioxin-like congeners 28 and 153 produced effects at 10^1–10^2 mg/kg/day, and 95 had an unbounded LOAEL of about 10^1 mg/kg/day. It appears, therefore, based on the very limited information available, that there is little difference between dioxin-like and non-dioxin-like congeners, with the exception of 126.

An important ultimate outcome would be to determine whether determination of the TEQ for any particular exposure would protect against neurotoxicity and other health effects, both for typical congener patterns in the human body and site-specific exposure.

REFERENCES

Agrawal, A. K., H. A. Tilson, and S. C. Bondy. 1981. "3,4,3',4'-Tetrachlorobiphenyl Given to Mice Prenatally Produces Long-Term Decreases in Striatal Dopamine and Receptor Binding Sites in the Caudate Nucleus." *Toxicol. Lett.* 7:417–24.

Amin, S., R. W. Moore, R. E. Peterson, and S. L. Schantz. 2000. "Gestational and Lactational Exposure to TCDD or Coplanar PCBs Alters Adult Expression of Saccharin Preference Behavior in Female Rats." *Neurotoxicol. Teratol.* 22:675–82.

Bernhoft, A., I. Nafstad, P. Engen, and J. U. Skaare. 1994. "Effects of Pre- and Postnatal Exposure to 3,3',4,4',5-Pentachlorobiphenyl on Physical Development, Neurobehavior and Xenobiotic Metabolizing Enzymes in Rats." *Environ. Toxicol. Chem.* 13:1589–97.

Bushnell, P. J., and D. C. Rice. 1999. "Behavioral Assessments of Learning and Attention in Rats Exposed Perinatally to 3,3',4,4',5-Pentachlorobiphenyl (PCB 126)." *Neurotoxicol. Teratol.* 21:381–92.

Chen, Y. C. J., M. L. Yu, C. C. Hsu, and W. J. Rogan. 1992. "Cognitive Development of Yu-Cheng (Oil Disease) Children Prenatally Exposed to Heat-Degraded PCBs." *J. Am. Med. Assoc.* 268:3213–18.

Crofton, K. M., and D. C. Rice. 1999. "Low-Frequency Hearing Loss Following Perinatal Exposure to 3,3',4,4',5-Pentachlorobiphenyl (PCB 126) in Rats." *Neurotoxicol. Teratol.* 21:299–301.

Eriksson, P., and A. Fredriksson. 1996a. "Developmental Neurotoxicity of Four Ortho-Substituted Polychlorinated Biphenyls in the Neonatal Mouse." *Environ. Toxicol. Pharmacol.* 1:155–65.

Eriksson, P., and A. Fredriksson. 1996b. "Neonatal Exposure to 2,2',5,5'-Tetrachlorobiphenyl Causes Increased Susceptibility in the Cholinergic Transmitter System in Adult Age." *Environ. Toxicol. Pharmacol.* 1:217–20.

Eriksson, P., U. Lundkvist, and A. Fredriksson. 1991. "Neonatal Exposure to 3,3',4,4'-Tetrachlorobiphenyl: Changes in Spontaneous Behavior and Cholinergic Muscarinic Receptors in the Adult Mouse." *Toxicol.* 69:27–34.

Geller, A. M., P. J. Bushnell, and D. C. Rice. 2000. "Behavioral and Electrophysiological Estimates of Visual Thresholds in Awake Rats Treated with 3,3',4,4',5-Pentachlorobiphenyl (PCB 126)." *Neurotoxicol. Teratol.* 22:521–31.

Goldey, E. S., L. S. Kehn, C. Lau, G. L. Rehnberg, and K. M. Crofton. 1995. "Developmental Exposure to Polychlorinated Biphenyls (Aroclor 1254) Reduces Circulating Thyroid Hormone Concentrations and Causes Hearing Deficits in Rats." *Toxicol. Appl. Pharmacol.* 135:77–88.

Hany, J., H. Lilienthal, A. Sarasin, A. Roth-Harer, A. Fastabend, L. Dunemann, W. Lichtensteiger, and G. Winneke. 1999a. "Developmental Exposure of Rats to a Reconstituted PCB Mixture or Aroclor 1254: Effects on Organ Weights, Aromatase Activity, Sex Hormone Levels, and Sweet Preference Behavior." *Toxicol. Appl. Pharmacol.* 158 (3): 231–43.

Hany, J., H. Lilienthal, A. Roth-Harer, G. Ostendor, B. Heinzow, and G. Winneke. 1999b. "Behavioral Effects Following Single and Combined Maternal Exposure to PCB 77 (3,4,3′,4′-Tetrachlorobiphenyl) and PCB 47 (2,4,2′,4′-Tetrachlorobiphenyl) in Rats." *Neurotoxicol. Teratol.* 21:147–56.

Hany, J., H. Lilienthal, A. Roth-Harer, G. Ostendor, B. Heinzow, and G. Winneke. 2004. Erratum to "Behavioral Effects Following Single and Combined Maternal Exposure to PCB 77 (3,4,3′,4′-Tetrachlorobiphenyl) and PCB 47 (2,4,2′,4′-Tetrachlorobiphenyl) in Rats." *Neurotoxicol. Teratol.* 26:615.

Harada, M. 1976. "Intrauterine Poisoning." *Bull. Inst. Constit. Med.* 25:38–61.

Herr, D. W., E. S. Goldey, and K. M. Crofton. 1996. "Developmental Exposure to Aroclor 1254 Produces Low-Frequency Alterations in Adult Rat Brainstem Auditory Evoked Responses." *Fundam. Appl. Toxicol.* 33:120–28.

Herr, D. W., J. E. Graff, E. C. Derr-Yellin, K. M. Crofton, and P. R. S. Kodavanti. 2001. "Flash-, Somatosensory-, and Peripheral Nerve-Evoked Potentials in Rats Perinatally Exposed to Aroclor 1254." *Neurotoxicol. Teratol.* 23:591–601.

Holene, E., I. Nafstad, J. U. Skaare, A. Bernhoft, P. Engen, and T. Sagvolden. 1995. "Behavioral Effects of Pre- and Postnatal Exposure to Individual Polychlorinated Biphenyl Congeners in Rats." *Environ. Toxicol. Chem.* 14:967–76.

Holene, E., I. Nafstad, J. U. Skaare, and T. Sagvolden. 1998. "Behavioural Hyperactivity in Rats Following Postnatal Exposure to Sub-Toxic Doses of Polychlorinated Biphenyl Congeners 153 and 126." *Behav. Brain Res.* 94:213–24.

Holene, E., I. Nafstad, J. U. Skaare, H. Krogh, and T. Sagvolden. 1999. "Behavioural Effects in Female Rats of Postnatal Exposure to Sub-Toxic Doses of Polychlorinated Biphenyl Congener 153." *Acta Paediatrica,* Supplement, 88:55–63.

Holson, R. R., and B. Pearce. 1992. "Principles and Pitfalls in the Analysis of Prenatal Treatment Effects in Multiparous Species." *Neurotoxicol. Teratol.* 14:221–28.

Kaya, H., J. Hany, A. Fustabend, A. Roth-Harer, G. Winneke, and H. Lilienthal. 2002. "Effects of Maternal Exposure to a Reconstituted Mixture of Polychlorinated Biphenyls on Sex-Dependent Behaviors and Steroid Hormone Concentrations in Rats: Dose-Response Relationship." *Toxicol. Applied Pharmacol.* 178:71–81.

Kodavanti, P. R. S., T. R. Ward, J. D. McKinney, and H. A. Tilson. 1995. "Increased [^{3}H]phorbol Ester Binding in Rat Cerebellar Granule Cells by Polychlorinated Biphenyl Mixtures and Congeners: Structure-Activity Relationships." *Toxicol. Appl. Pharmacol.* 130:140–48.

Kodavanti, P. R. S., T. R. Ward, J. D. McKinney, C. L. Waller, and H. A. Tilson. 1996. "Increased [^{3}H]phorbol Ester Binding in Rat Cerebellar Granule Cells and Inhibition of ^{45}Ca^{2+} Sequestration in Rat Cerebellum by Polychlorinated Diphenyl Ether Congeners and Analogs: Structure-Activity Relationships." *Toxicol. Appl. Pharmacol.* 138:251–61.

Kodavanti, P. R. S., E. C. Derr-Yellin, W. R. Mundy, T. J. Shafer, D. W. Herr, S. Barone Jr., N. Y. Choski, R. C. MacPhail, and H. A. Tilson. 1998. "Repeated Exposure of Adult Rats to Aroclor 1254 Causes Brain Region-Specific Changes in Intracellular Ca^{2+} Buffering and Protein Kinase C Activity in the Absence of Changes in Tyrosine Hydroxylase." *Toxicol. Appl. Pharmacol.* 153:186–98.

Kremer, H., H. Lilienthal, J. Hany, A. Roth-Härer, and G. Winneke. 1999. "Sex-Dependent Effects of Maternal PCB Exposure on the Electroretinogram in Adult Rats." *Neurotoxicol. Teratol.* 21:13–19.

Lilienthal, H., and G. Winneke. 1991. "Sensitive Periods for Behavioral Toxicity of Polychlorinated Biphenyls: Determination by Cross-Fostering in Rats." *Fundam. Appl. Toxicol.* 17:368–75.

Lilienthal, H., M. Neuf, C. Munoz, and G. Winneke. 1990. "Behavioral Effects of Pre- and Postnatal Exposure to a Mixture of Low Chlorinated PCBs in Rats." *Fundam. Appl. Toxicol.* 15:457–67.

Meerts, I. A., H. Lilienthal, S. Hoving, J. H. Van den Berg, B. M. Weijers, A. Bergman, J. H. Koeman, and A. Brouwer. 2004. "Developmental Exposure to 4-Hydroxy-2,3,3′,4′,5-Pentachlorobiphenyl (4-OH-CB107): Long-Term Effects on Brain Development, Behavior and Brain Stem Auditory Evoked Potentials in Rats." *Toxicol. Sci.* 82:207–18.

Ness, D. K., S. L. Schantz, J. Moshtaghian, and L. G. Hansen. 1993. "Effects of Perinatal Exposure to Specific PCB Congeners on Thyroid Hormone Concentrations and Thyroid Histology in the Rat." *Toxicol. Lett.* 68:311–23.

Newland, M. C., and E. M. Paletz. 2000. "Animal Studies of Methylmercury and PCBs: What Do They Tell Us About Expected Effects in Humans?" *Neurotoxicol.* 21:1003–28.

Rice, D. C. 1997. "Effect of Postnatal Exposure to a PCB Mixture in Monkeys on Multiple Fixed Interval-Fixed Ratio Performance." *Neurotoxicology and Teratology* 19:429–34.

———. 1998. "Effects of Postnatal Exposure of Monkeys to a PCB Mixture on Spatial Discrimination Reversal and DRL Performance." *Neurotoxicology and Teratology* 20:391–400.

———. 1999. "Effect of Exposure to 3,3′,4,4′,5-Pentachlorobiphenyl (PCB 126) throughout Gestation and Lactation on Development and Spatial Delayed Alternation Performance in Rats." *Neurotoxicol.. Teratol.* 21:59–69.

Rice, D. C., and S. Hayward. 1997. "Effects of Postnatal Exposure to a PCB Mixture in Monkeys on Nonspatial Discrimination Reversal and Delayed Alternation Performance." *Neurotoxicology* 18:479–94.

———. 1998. "Lack of Effect of 3,3′,4,4′,5-Pentachlorobiphenyl (PCB 126) throughout Gestation and Lactation on Multiple Fixed Interval-Fixed Ratio and DRL Performance in Rats." *Neurotoxicol. Teratol.* 20:645–50.

———. 1999a. "Effects of Exposure to 3,3′,4,4′,5-Pentachlorobiphenyl (PCB 126) throughout Gestation and Lactation on

Behavior (Concurrent Random Interval-Random Interval and Progressive Ratio Performance) in Rats." *Neurotoxicol. Teratol.* 21:679–87.

———. 1999b. "Effects of Postnatal Exposure of Monkeys to a PCB Mixture on Concurrent Random Interval-Random Interval and Progressive Ratio Performance." *Neurotoxicology and Teratology* 21:47–58.

Schantz, S. L., J. Moshtaghian, and D. K. Ness. 1995. "Spatial Learning Deficits in Adult Rats Exposed to Ortho-Substituted PCB Congeners during Gestation and Lactation." *Fundam. Appl. Toxicol.* 26:117–26.

Schantz, S. L., J. J. Widholm, and D. C. Rice. 2003. "Effects of PCB Exposure on Neuropsychological Function in Children." *Environmental Health Perspectives* 111:357–76.

Schantz, S. L., B-W. Seo, J. Moshtaghian, R. E. Peterson, and R. W. Moore. 1996. "Effects of Gestational and Lactational Exposure to TCDD or Coplanar PCBs on Spatial Learning." *Neurotoxicol. Teratol.* 18:305–13.

Schantz, S. L., B-W. Seo, P. W. Wong, and I. N. Pessah. 1997. "Long-Term Effects of Developmental Exposure to 2,2′,3,5′,6-Pentachlorobiphenyl (PCB 95) on Locomotor Activity, Spatial Learning and Memory and Brain Ryanodine Binding." *Neurotoxicol.* 18:457–67.

Seo, B-W., M-H Li, L. G. Hansen, R. W. Moore, R. E. Peterson, and S. L. Schantz. 1995. "Effects of Gestational and Lactational Exposure to Coplanar Polychlorinated Biphenyl (PCB) Congeners or 2,3,7,8-Tetrachlorodibenzo-*p*-Dioxin (TCDD) on Thyroid Hormone Concentrations in Weanling Rats." *Toxicol. Lett.* 78:253–62.

Taylor, M. M., K. M. Crofton, and R. C. MacPhail. 2002. "Schedule-Controlled Behavior in Rats Exposed Perinatally to the PCB Mixture Aroclor 1254." *Neurotoxicol. Teratol.* 24:511–18.

Tilson, H. A., G. J. Davis, J. A. McLachlan, and G. W. Lucier. 1979. "The Effects of Polychlorinated Biphenyls Given Prenatally on the Neurobehavioral Development of Mice." *Environ. Res.* 18:466–74.

Van den Berg, M., L. Birnbaum, A. T. Bosveld, B. Brunstrom, P. Cook, M. Feeley, J. P. Giesy, A. Hanberg, R. Hasegawa, S. W. Kennedy, T. Kubiak, J. C. Larsen, F. X. van Leeuwen, A. K. Liem, C. Nolt, R. E. Peterson, L. Poellinger, S. H. Safe, D. Schrenk, D. Tillitt, M. Tysklind, M. Younes, F. Waern, and T. Zacharewski. 1998. "Toxic Equivalency Factors (TEFs) for PCBs, PCDDs, PCDFs for Humans and Wildlife." *Environ. Health Perspect.* 106:775–92.

Wong, P. W., and I. N. Pessah. 1997. "Noncoplanar PCB 95 Alters Microsomal Calcium Transport by an Immunophilin FKBP12-Dependent Mechanism." *Molec. Pharmacol.* 51:693–702.

Wong, P. W., R. M. Joy, T. E. Albertson, S. L. Schantz, and I. N. Pessah. 1997. "Ortho-Substituted 2,2′,3,5′,6-Pentachlorobiphenyl (PCB 95) Alters Rat Hippocampal Ryanodine Receptors and Neuroplasticity in Vitro: Evidence for Altered Hippocampal Function." *Neurotoxicology* 18:443–56.

Wong, P. W., E. Mai, L. G. Hansen, C. E. Garner, and I. Pessah. 2000. "Structure-Activity Relationship between Selected Polychlorinated Biphenyl Congeners and Metabolites Towards Activation of Ryanodine Receptor Type 1." *Toxicol. Sci.* 54:77.

Appendix

12:00–1:00 PM Lunch
1:00 PM Reconvene

II.   **Human Exposures—Health Effects and Characteristic Congener Profiles**
       Chairs, LARRY HANSEN, DONALD PATTERSON
       (Focus on Anniston, AL, and Slovakian former Production Facilities)

1.   **Anniston Alabama PCB Community Consortium**
      Charles Sherrer

2.   **Human and Environmental Contamination with Polychlorinated Biphenyls in Anniston, Alabama**
      Kenneth G. Orloff,[1] Steve Dearwent,[1] Susan Metcalf,[1] and Wayman Turner[2]
      [1]Agency for Toxic Substances and Disease Registry, Atlanta, GA
      [2]Centers for Disease Control and Prevention National Center for Environmental Health, Atlanta, GA

3.   **Toxicokinetic Extrapolation to Former Residues in Residents of Anniston, AL**
      Tomas Martin-Jimenez and Larry G. Hansen
      Department of Veterinary Biosciences, University of Illinois, Urbana, IL

4.   **PCBs in East Slovakia and the Structure and Function of the PCBRISK Project**
      Tomas Trnovec, Anton Kocan, Jan Petrik, Jana Chovacova, Beata Drobna, Stanislav Jursa and Kamil Conka
      Slovak Medical University, Bratislava, Slovak Republic

5.   **PCB Metabolites in Blood from Humans Living in a PCB Contaminated Area and
      Two Background Districts in the Slovak Republic**
      Åke Bergman, Lotta Hovander, Linda Linderholm, Maria Athanasiadou and Ioannis Athanassiadis
      Department of Environmental Chemistry, Stockholm University, Stockholm, Sweden

2:45 PM Break
3:15 PM Reconvene

6.   **A Birth Cohort in Eastern Slovakia and Exposure to PCBs**
      Irva Hertz-Picciotto,[1] Z. Yu,[1] A. Kocan,[2] J. Petrik,[2] and T. Trnovec[2]
      [1]University of California–Davis
      [2]Institute of Preventive and Clinical Medicine, Bratislava

7.   **Neurobehavioral Output of 8–9-Year-Old Children from the Michalovce Region**
      Eva Sovcikova, Tomas Trnovec, Anton Kocan, Ladislava Wsolova
      Slovak Medical University

8.   **Increased Prevalence of Diabetes Mellitus and Other Dysglycemias in a Population Chronically
      Exposed to Polychlorinated Biphenyls and Other Persistent Organochlorine Pollutants**
      *Z. Radikova,[1] J. Koska,[1] L. Ksinantova,[1] R. Imrich,[1] A. Kocan,[2] J. Petrik,[2] M. Huckova,[1] J. Chovancova,[2]
      B. Drobna,[2] S. Jursa,[2] S. Wimmerova,[2] L. Wsolova,[2] P. Langer,[1] T. Trnovec,[2] E. Sebokova,[1] I. Klimes[1]*
      [1]Institute of Experimental Endocrinology, Slovak Academy of Sciences and [2]Institute of Preventive and
      Clinical Medicine, Slovak Health University, Bratislava, Slovakia

9.   **Fundamental Views on the Association of High PCB Levels with Changes of Thyroid
      Volume and Thyroid Hormones and Antibody Levels in an Exposed Population**
      *P. Langer,[1] M. Tajtáková,[3] A. Kočan,[2] J. Petrík,[2] J. Koška,[1] L. Kšinantová,[1] Ž. Rádiková,[1]
      R. Imrich,[1] M. Hučková,[1] J. Chovancová,[2] B. Drobná,[2] S. Jursa,[2] S. Wimmerová,[2] Y. Shishiba,[4]
      T. Trnovec,[2] E. Šeböková,[1] I. Klimeš[1]*
      [1]Slovak Academy of Sciences and [2]Slovak Health University, Bratislava, Slovakia;
      [3]P. J. Šafárik University, Košice, Slovakia;
      [4]Mishuku Hospital, Tokyo, Japan

4:45 PM Dinner (on your own)
6 to 8 PM POSTER SESSION

Monday, June 14th

8:30 AM Convene

III. Actions of PCBs; Emphasis on Endocrine Effects
Chairs, REX HESS, PAUL COOKE

1. **Cross-talk between the Aryl Hydrocarbon Receptor and Estrogen Receptor Signaling Pathways**
    Paul S. Cooke, Motoko Mukai and David L. Buchanan
    Department of Veterinary Biosciences, University of Illinois, Urbana, IL

2. **Estrogen Receptor Ligand Binding, Promiscuous or Eclectic: Opportunities for Pharmaceutical Design, but Perils for Endocrine Disruption**
    John A. Katzenellenbogen
    Department of Chemistry, University of Illinois, Urbana, IL

3. **Genetic and Genomic Approaches to Evaluate Models of Ah Receptor Mediated Toxicity**
    Chris Bradfield
    McArdle Laboratory, University of Wisconsin, Madison, WI

10:00 AM Break
10:30 AM Reconvene

4. **Detection of Dioxin-like, Estrogenic and Antiestrogenic Activities in Human Serum Samples from Eastern Slovakia**
    Miroslav Machala,[1] Martina Plísková,[1] Rocio Fernandez Canton,[2] Jirí Neca,[1] Jan Vondrácek,[1,3] Anton Kocan,[4] Ján Petrík,[4] Tomás Trnovec,[4] Thomas Sanderson,[2] Martin van den Berg[2]
    [1]Veterinary Research Institute, Brno, Czech Republic;
    [2]IRAS, University of Utrecht, Utrecht, The Netherlands;
    [3]Institute of Biophysics, CAS, Brno, Czech Republic;
    [4]Institute of Preventive and Experimental Medicine, State Health University, Bratislava, Slovakia

5. **Interaction of Environmental Chemicals with the Nuclear Receptor PXR**
    Christopher H. Hurst and David J. Waxman
    Cell and Molecular Biology, Department of Biology, Boston University, Boston MA

6. **Polychlorinated Biphenyls (PCBs) Exert Thyroid Hormone-Like Effects in the Fetal Rat Brain but Do not Bind to Thyroid Hormone Receptors**
    Kelly J. Gauger,[1] Yoshihisa Kato,[2] Koichi Haraguchi,[3] Hans-Joachim Lehmler,[4] Larry W. Robertson,[4] Ruby Bansal,[1] and R. Thomas Zoeller[1]
    [1]Biology Department, Molecular and Cellular Biology Program, Morrill Science Center, University of Massachusetts, Amherst, MA
    [2]School of Pharmaceutical Sciences, University of Shizuoka, Shizuoka, Japan
    [3]Daiichi College of Pharmaceutical Sciences, Fukuoka, Japan
    [4]Department of Occupational and Environmental Health, University of Iowa, College of Public Health, Iowa City, Iowa

12:00 PM Lunch
1:30 PM Reconvene

IV. **Cardiovascular Targets of PCBs**
Chairs, NIGEL WALKER, DICK PETERSON

1. **AHR Agonist-Induced Cardiovascular Toxicity in Developing Zebrafish**
Warren Heideman, Dagmara S. Antkiewicz, Amy L. Prasch, Sara A. Carney, and Richard E. Peterson
University of Wisconsin, Madison, WI

2. **Nutrition Modulates PCB Toxicity: Implications in Atherosclerosis**
Bernhard Hennig,[1] Michal Toborek[1] and Larry W. Robertson[2]
[1]University of Kentucky, Lexington, KY; [2]University of Iowa, Iowa City, IA

2:30 PM Break
3:00 PM Reconvene

3. **Inhibition of Coronary Angiogenesis by AhR Agonists**
Mary K. Walker and Irena Ivnitski-Steele
College of Pharmacy, University of New Mexico, Albuquerque, NM

4. **Cardiotoxicity in Rats Following Chronic Exposure to Dioxins and PCBs**
Nigel J. Walker,[1] Micheal P. Jokinen,[2] Amy Brix,[3] Donald M. Sells,[4] and Abraham Nyska[1]
[1]National Institute of Environmental Health Sciences, Research Triangle Park, North Carolina, USA
[2]Pathology Associates—A Charles River Company, Durham, North Carolina, USA
[3]Experimental Pathology Laboratories, Research Triangle Park, North Carolina, USA
[4]Battelle Columbus Laboratories, Columbus, Ohio, USA

4:00 PM	**Adjourn**
5:30–6:30 PM	**Open Bar and Midwest Barbecue Dinner**
6:30–9:00 PM	**Cash Bar**
7:00–9:00 PM	**Music by FRAGMENT**
	The Blue Grass Band from Brno, CR via Nashville, TN

Tuesday, June 15th

8:30 AM Convene

V. **Combined Exposure to PCBs and Other Contaminants**
Chairs, SUE SCHANTZ, RICH SEEGAL, ISAAC PESSAH

1. **Developmental Exposure to PCBs in Humans: Questions and Controversies**
S. L. Schantz
Department of Veterinary Biosciences and Neuroscience Program, UIUC

2. **Auditory and Motor Impairments in Rats Exposed to PCBs and Methylmercury during Early Development**
J. J. Widholm,[1] B. E. Powers,[2] C. S. Roegge,[2] R. E. Lasky,[3] D. M. Gooler[4] and S. L. Schantz[5]
[1]Department of Psychology, College of Charleston, Charleston, South Carolina
[2]Neuroscience Program, University of Illinois at Urbana-Champaign, Urbana, Illinois
[3]University of Texas Health Science Center at Houston Medical School, Houston, TX
[4]Department of Speech and Hearing Science and Neuroscience Program, UIUC
[5]Department of Veterinary Biosciences and Neuroscience Program, UIUC

3. **Block of GABAergic Receptors in Rat Hippocampus Significantly Potentiates Non-coplanar PCB Excitotoxicity *in vitro* and *in vivo***
Isaac N. Pessah and Kyung Ho Kim
Department of Veterinary Molecular Biosciences, and Center for Children's Environmental Health, University of California, Davis CA

10:00 AM Break
10:30 AM Reconvene

4. **Age Matters: Developmental Effects of PCBs and Methylmercury on Striatal Synaptosomal Dopamine**
R. F. Seegal,[1,2] K. O. Brosch,[1] R. J. Okoniewski,[1] and M. Shan[2]
[1]Wadsworth Center, NYS Dept. of Health and [2]School of Public Health, University at Albany, Albany, NY

5. **Neurotoxicity of Developmental Exposure to PCBs and PBDEs: Evidence for Interaction**
Per Eriksson
Department of Environmental Toxicology, Uppsala University, Uppsala, Sweden

12:00 PM Lunch
1:00 PM Reconvene

VI. Risk Issues
Chairs, DEBORAH RICE, RIE MASHO (Japan)

1. **Relative Potency of Individual PCB Congeners: What Can the Experimental Literature Tell Us?**
Deborah C. Rice
Environmental Health Unit, Maine Bureau of Health, Augusta, ME

2. **Trends of PCB Isomeric Patterns in Japanese Environmental Media, Food, Breast Milk and Blood in View of Risk Assessment**
Takeshi Nakano,[1] Yoshimasa Konishi,[2] Rie Masho,[3] and Chiharu Tohyama[4]
[1]Hyogo Pref. Inst. of Public Health & Env. Sci., Kobe 654-0037, Japan
[2]Osaka Pref. Inst. of Public Health
[3]Center for Environmental Information Science
[4]National Institute for Environmental Studies

3. **Evaluating Dioxin Toxin Equivalency Factors (TEFs) for the Carcinogenicity of Dioxins and PCBs in Rodents**
Nigel J. Walker,[1] Patrick Crockett,[2] Ming Yin,[2] Abraham Nyska,[1] Amy Brix,[3] Micheal P. Jokinen,[4] Donald M. Sells,[5] James R. Hailey,[1] Joseph K. Haseman,[1] Michael E. Wyde,[1] John R. Bucher,[1] and Christopher J. Portier[1]
[1]National Institute of Environmental Health Sciences, Research Triangle Park, North Carolina
[2]Constella Group, Research Triangle Park, North Carolina
[3]Pathology Associates – A Charles River Company, Durham, North Carolina
[4]Experimental Pathology Laboratories, Research Triangle Park, North Carolina
[5]Battelle Columbus Laboratories, Columbus, Ohio

2:30 PM Break
2:45 PM Reconvene

SUM-UP: LINDA BIRNBAUM, USEPA, RTP, NC.

3:15 PM Announcements; Adjourn

Index

acetylcholine (Ache), 173

Achman, D. R., 55

Adelrahim, M., 146

Agency for Toxic Substances and Disease Registry (ATSDR), 2, 60. *See also* Anniston, Alabama, Agency for Toxic Substances and Disease Registry study

Agent Orange, 108

air (ambient), PCBs in, 15, 19, 20; in Michalovce and Stropkov districts, Slovak Republic, 83–84

Alabama Department of Public Health, 64

Altshul, L., 121

Alzheimer's disease, 4, 179

American Diabetes Association (ADA), 110

Anniston, Alabama, Agency for Toxic Substances and Disease Registry study, 1–2, 60, 61–62, 64; biological sampling and analyses, 61; congener-specific analyses, 62–64; environmental sampling and analyses, 61; target population, 60–61, 64

Anniston, Alabama, University of Illinois at Urbana-Champaign homologue serum profile study, 66–68, 70–71, 78–79; blood sampling and chemical assay, 72; results, 67–68, 77–78; study group, 71–72

anti-estrogenicity assays, 127, 128; ER-CALUX assay, 127, 129–30, 130, 132–35; processes associated with modulation of estrogen signaling detected by, 128

anti-thyroperoxidase anti-bodies (TPOab), 119–20

aquatic invertebrates, PCBs in, 16, 20, 38–39; amphipod (*Diporeia hoyi*), 40; blue mussel (*Mytilus edulis*), 39; crayfish (*Procambaraus* sp.), 39; freshwater bivalve (*Corbicula* sp.), 39; opossum shrimp (*Mysis relicta*), 40; zooplankton (*Calanus* sp.), 29–30

arene oxide intermediates, 31–32

Aroclor 1254, 98, 153, 154–55, 156, 157, 193, 193–94; and hearing loss, 97–98

Aroclor 1260, 15, 20, 62

Aroclor 1268, 62, 77, 78

Aroclors, 20, 31, 62, 63, 71, 78. *See also specific Aroclors (by number)*

aryl hydrocarbon receptor (AhR), 2–3, 4, 103, 139–40, 168, 187; AhR agonists, 128–29, 135, 146; AhR signaling, 142; and cardiovascular disease, 166; mechanism of anti-estrogenic effects by AhR agonists, 140–42; mediation of the anti-estrogenic effects of TCDD, 140; other

interactions between AhR and steroid hormone signaling pathways, 142–45; possible estrogenic effects of liganded AhR, 145–46. *See also* aryl hydrocarbon receptor (AhR) activity bioassays

aryl hydrocarbon receptor (AhR) activity bioassays, 127, 128; DR-CALUX assay, 127, 129, 130, 132

atherosclerosis, and PCB exposure, 165–66; pathogenetic mechanisms of action, 166–67; role of caveolae in, 167

atmosphere, interactions of with PCB cycling, 51: absorptive gas flux ($F_{gas,ab}$), 53–54; correlation of with urbanization, 52; direct atmospheric deposition (dry particle deposition/wet deposition/gaseous deposition), 53–55, 58; dry deposition flux (F_{dry}), 53; indirect atmospheric deposition (stormwater runoff and tributary inputs), 56, 58; volatilization (air-water exchange), 55–56; wet deposition flux (F_{wet}), 53

atropisomers/atropisomerism, 31–32; analytical separation of, 33–35

autoimmune thyroid disease (AITD), 117–18

Axys Laboratories (Sidney, British Columbia), 67, 72

Baccarelli, A., 119

Bahn, A. K., 115, 119

Bamford, H. A., 53, 54

Benická, E., 36

Benton Recognition test, 95, 97

Bernhoft, A., 119

bioallethrin, 180

biotransformation, 36, 38, 39, 72, 128; *in situ*, 37, 41; *in vivo*, 39, 41, 43

birds, PCBs in, 16, 19, 42; albatross, 42; American kestrel, 42; barn swallow (*Hirundo rustica*), 42; chick embryos, 44; ivory gull (*Pagophila eburnea*), 42; thick-billed murre (*Uria lomvia*), 42

blood (human), PCBs in, 15–16, 16, 19; decay of PCBs in blood, 73. *See also* homologue serum profiles

Bogazzi, F., 155

brain development (brain growth spurt [BGS]) in the neonatal mouse, and PCB exposure, 172–75; behavior and cholinergic receptors, 178–79; neonatal exposure and brain tissue level of PCBs, 177–78; neurotoxic effects of neonatal exposure, 175, 177

breast milk, PCBs in, 15–16, 16, 102, 119, 121, 178, 193

About the Editors

Larry G. Hansen is Professor Emeritus in the College of Veterinary Medicine at the University of Illinois-Urbana/Champaign. He earned his PhD (Entomology/Toxicology) with Ernest Hodgson at N.C. State University and came to the University of Illinois in 1970 to work with Robert Metcalf in Environmental Toxicology. In 1972, he transferred to the College of Veterinary Medicine where he taught in the newly created Environmental Toxicology Program and directed a research project on PCBs in Food Animals. In 1980, he worked 6 months in Wageningen with Jan Koeman, Anjo Strik, Louis Tuinstra and Kees Kan and was promoted to Professor. In 1982, he hosted an International Workshop on Organophosphorus-Induced Delayed Neurotoxicity. His research/teaching/consulting interests have been in the environmental disposition, toxicokinetics and toxicology of PCBs, organophosphorus pesticides and sewage sludge/cadmium in a number of species from flies and worms to swine and catfish. Since the mid-1980s he has focused primarily on PCBs, having written several reviews and a book. He has directed over 20 MS and PhD students and assisted with scores more.

Larry W. Robertson is a Professor in the Department of Occupational and Environmental Health, College of Public Health at The University of Iowa. His other responsibilities include Director of the Superfund Basic Research Program, and Director of the Interdisciplinary Graduate Program in Human Toxicology. He earned his Ph.D. (Environmental Health Sciences) and M.P.H. degrees from The University of Michigan. Postdoctoral research was carried out with Prof. Stephen Safe, first at Guelph University, and then at Texas A&M University. Dr. Robertson then joined the laboratory of Prof. Franz Oesch at the University of Mainz. In 1986 he was named Associate Professor in the Graduate Center for Toxicology at the University of Kentucky. Dr. Robertson led the effort to establish a Superfund Basic Research Program at the University. In 2003 an opportunity arose at the University of Iowa, and Dr. Robertson along with several colleagues, joined the Department of Occupational and Environmental Health in a newly created College of Public Health. A goal of bringing Dr. Robertson and his colleagues to Iowa was to resurrect a defunct toxicology training and research program that had known great success in the past (http://toxicology.grad.uiowa.edu). A secondary goal was to establish a new Superfund Basic Research Program at Iowa drawing on the broad faculty expertise in several disciplines (http://www.uiowa.edu/~isbrp/). Dr. Robertson's research on halogenated environmental pollutants include studies of their effects on gene regulation, their metabolism to electrophilic species and their binding to nucleophiles including amino acids, nucleotides and DNA, and the involvement of these processes in the initiation and promotion of carcinogenesis. Dr. Robertson has also an interest in the mechanistic relationships between PCB induced carcinogenesis and PCB-induced changes in vitamin, mineral and fatty acid patterns in living organisms. These research activities have resulted in the training of more than 35 pre- and postdoctoral students.

With common interests and long-standing friendship, Professors Hansen and Robertson began exchanging visits to the respective campuses in the 1990s, accompanied by their graduate students and postdocs. This led to the concept of the 1st PCB Workshop held in Lexington in 2000. Together, they planned and hosted additional PCB Workshops in Brno, CR (2002) and Urbana (2004). After Dr. Hansen retired in 2006, Dr. Robertson hosted a PCB Workshop in Zakopane, Poland in 2006 and is planning one for Iowa City in 2008.

The University of Illinois Press
is a founding member of the
Association of American University Presses.

———————————————————————

Composed in 10.5/13 Adobe Garamond Pro
by BookComp, Inc.
for the University of Illinois Press
Manufactured by Sheridan Books, Inc.

University of Illinois Press
1325 South Oak Street
Champaign, IL 61820-6903
www.press.uillinois.edu